Spatial Data on the Web

T0134817

Alberto Belussi · Barbara Catania
Eliseo Clementini · Elena Ferrari (Eds.)

Spatial Data on the Web

Modeling and Management

With 111 Figures

 Springer

Editors

Alberto Belussi

University of Verona
Department of Computer Science
Strada Le Grazie 15
37100 Verona
Italy
alberto.belussi@univr.it

Barbara Catania

University of Genova
Department of Computer
and Information Sciences
Via Dodecaneso 35
16146 Genova
Italy
barbara.catania@unige.it

Eliseo Clementini

University of L'Aquila
Department of Electrical Engineering
Monteluco di Roio
67040 L'Aquila
Italy
eliseo@ing.univaq.it

Elena Ferrari

University of Insubria
Via Mazzini 5
21100 Varese
Italy
elena.ferrari@uninsubria.it

ISBN 978-3-642-08930-5   e-ISBN 978-3-540-69878-4

ACM Computing Classification (1998): H.2, H.3, J.2

Springer is a part of Springer Science+Business Media
springer.com
© Springer-Verlag Berlin Heidelberg 2010

Cover design: KünkelLopka, Heidelberg

Preface

Data describing the Earth's surface and the phenomena that occur on it are frequently available on distributed environments, such as the Web, but the complexity of their spatial components raises many issues concerning data integration and many challenges regarding new applications and services that should be designed and implemented. This book addresses these topics by presenting key developments, directions, and challenges concerning representation, integration, protection, and innovative applications of geographical and spatial data in distributed environments.

The initiative that led to the publication of this book is an Italian research project, called "Management and Modelling of Geographical Data over the Web" (SPADA@WEB, http://homes.dico.unimi.it/dbandsec/spadaweb/), involving the following Italian universities: University of Genoa, University of Milan, University of Insubria, University of Udine, University "La Sapienza" of Rome, University of Trento, University of L'Aquila, and University of Verona. The book includes the results of this project as well as selected contributions from international experts.

The book could be used as a reference book for senior undergraduate or graduate courses on geographical information systems, which have a special focus on the Web as well as heterogeneous and distributed systems. It is also useful for technologists, managers, and developers who want to know more about emerging trends in geographical and spatial data management.

The book is organized into four parts. The first part consists of four chapters and deals with modeling and processing issues for geographical and spatial data in distributed environments, such as spatiotemporal data representation, multiresolution, and multiscale data representation and processing. The second part, composed of three chapters, deals with integration issues for spatial data sources, such as geographical information fusion, mediator systems, data quality in integration systems, and similarity-based processing over multiresolution maps. The third part consists of three chapters dealing with security issues for spatial data, such as access control, secure outsourcing, and information hiding. Finally, the fourth part contains two chapters dealing with innovative applications of spatial data for mobile devices.

The editors would like to thank all the people who made the successful completion of this project possible. First of all, we would like to thank the publishing team at Springer. In particular, we would like to thank Ralf Gerstner who gave us the opportunity to edit this book and constantly supported us throughout the whole process. We also want to express our gratitude to the authors of the chapters for their insights and excellent contributions to this book. Most of them also served as referees for chapters written by other authors. We wish to thank all of them, as well as all the other reviewers, for their constructive and comprehensive reviews.

Alberto Belussi, Barbara Catania,
May 2007   *Eliseo Clementini, and Elena Ferrari*

Contents

Using Qualitative Information in Query Processing over

Part III Spatial Data Protection

Part IV Innovative Applications for Mobile Devices

List of Contributors

Alberto Belussi
University of Verona
Strada le Grazie, 15
Verona - Italy, 37134
alberto.belussi@univr.it

Elisa Bertino
Purdue University
N. University Street, 305
West Lafayette - USA, IN 47907
bertino@cerias.purdue.edu

Michela Bertolotto
University College Dublin
Belfield
Dublin - Ireland, 4
michela.bertolotto@ucd.ie

Omar Boucelma
Université Aix-Marseille 3
Avenue Escadrille Normandie-Niemen
Marseille - France, 13397
omar.boucelma@lsis.org

Stefano Burigat
University of Udine
Via delle Scienze, 206
Udine - Italy, 33100
burigat@dimi.uniud.it

Maria Calagna
University of Rome "La Sapienza"
Via Salaria, 113
Rome - Italy, 00198
calagna@di.uniroma1.it

Barbara Carminati
University of Insubria
Via Mazzini, 5
Varese - Italy, 21100
barbara.carminati@uninsubria.it

Barbara Catania
University of Genoa
Via Dodecaneso, 35
Genoa - Italy, 16146
barbara.catania@unige.it

Luca Chittaro
University of Udine
Via delle Scienze, 206
Udine - Italy, 33100
chittaro@dimi.uniud.it

Alminas Čivilis
Vilnius University
Naugarduko Str., 24
Vilnius - Lithuania, 03225
alminas.civilis@maf.vu.lt

Eliseo Clementini
University of L'Aquila
Monteluco di Roio
L'Aquila - Italy, 67040
eliseo@ing.univaq.it

Carlo Combi
University of Verona
Strada le Grazie, 15
Verona - Italy, 37134
carlo.combi@univr.it

Maria Luisa Damiani
University of Milan
Via Comelico, 39
Milan - Italy, 20135
damiani@dico.unimi.it

Emanuele Danovaro
University of Genoa
Via Dodecaneso, 35
Genoa - Italy, 16146
danovaro@disi.unige.it

Leila De Floriani
University of Genoa
Via Dodecaneso, 35
Genoa - Italy, 16146
deflo@disi.unige.it

Marian de Vries
Delft University of Technology
Jaffalaan, 9
Delft - The Netherlands, 2628 BX
m.d.vries@otb.tudelft.nl

Matt Duckham
University of Melbourne
Australia, 3010
mduckham@unimelb.edu.au

Mehdi Essid
Université Aix-Marseille 3
Avenue Escadrille Normandie-Niemen
Marseille - France, 13397
mehdi.essid@lsis.org

Elena Ferrari
University of Insubria
Via Mazzini, 5
Varese - Italy, 21100
elena.ferrari@uninsubria.it

Christian S. Jensen
Aalborg University
Fredrik Bajers Vej, 7E
Aalborg Ost - Denmark, 9220
csj@cs.aau.dk

Yassine Lassoued
Université Aix-Marseille 3
Avenue Escadrille Normandie-Niemen
Marseille - France, 13397
yassine.lassoued@lsis.org

Luigi V. Mancini
University of Rome "La Sapienza"
Via Salaria, 113
Rome - Italy, 00198
lv.mancini@di.uniroma1.it

Sara Migliorini
University of Verona
Strada le Grazie, 15
Verona - Italy, 37100
sara.migliorini@tiscali.it

Barbara Oliboni
University of Verona
Strada le Grazie, 15
Verona - Italy, 37100
barbara.oliboni@univr.it

Stardas Pakalnis
Aalborg University
Fredrik Bajers Vej, 7E
Aalborg Ost - Denmark, 9220
stardas@cs.aau.dk

Paola Podestà
University of Genoa
Via Dodecaneso, 35
Genoa - Italy, 16146
paola.podesta@unige.it

Enrico Puppo
University of Genoa
Via Dodecaneso, 35
Genoa - Italy, 16146
puppo@disi.unige.it

Hanan Samet
University of Maryland
College Park - USA, MD 20742
hjs@umiacs.umd.edu

Peter van Oosterom
Delft University of Technology
Jaffalaan, 9
Delft - The Netherlands, 2628 BX
p.v.oosterom@otb.tudelft.nl

Mike Worboys
University of Maine
Orono - USA, ME 04469
worboys@spatial.maine.edu

1

Spatial Data on the Web: Issues and Challenges

Alberto Belussi[1], Barbara Catania[2], Eliseo Clementini[3], and Elena Ferrari[4]

[1] University of Verona, Verona (Italy)
[2] University of Genoa, Genoa (Italy)
[3] University of L'Aquila, L'Aquila (Italy)
[4] University of Insubria, Varese (Italy)

Spatial data are today needed in a wide range of application domains. Indeed, spatial properties are included in several application contexts requiring the management of very large data sets, such as, for instance, computer-aided design (CAD), very large scale integration (VLSI), robotics, and image processing. However, the primary target of systems dealing with spatial data remains *geographical applications*, since they served as the first motivation for the development of such technology and still represent the most challenging application environment [19]. *Spatial data* can be defined as pieces of information describing quantitative and/or qualitative properties that refer to space. Such properties can be represented as attributes of a set of objects (like the path of a given highway or the technical drawing of the new version of a car engine) or as functions of the space locations (like the temperature measured at a given location on the European continent or the measured infrared emissions in a remote sensing image).

This book considers spatial data in geographical applications; thus it is focused on *geographical data*. This means that spatial data are used to describe objects or, more generally, natural phenomena and human activities that occur on the Earth's surface. Often geographical data are described as composed of two parts: the spatial component, describing shape, extension, location, and orientation of an object (or a phenomenon) existing on the Earth's surface, and the non-spatial component (also called "thematic," "descriptive," or "alphanumeric component"), describing other properties of the considered object (or phenomenon), like traditional attributes (e.g. the name of the object).[5] Therefore, in the chapters of this book, the term "spatial data" refers to the spatial component of some geographical information.

Even if we focus on a specific category of spatial data, heterogeneity is still very high. Indeed, spatial data for geographical applications are often managed independently by various parties and specialized systems, such as systems for managing images from remote sensing, raster data (grids) coming from environmental monitoring devices, or vector data collected by cartographers in geographical maps. The

---

[5] Geographical objects are also called features in standards of International Organization for Standardization (ISO) and Open Geospatial Consortium (OGC) [9] [14].

recent and fast development of the Internet and the Web has made accessing the various information sources easier, since it has increased the interconnectivity among computers and information systems.

Despite such advances, however, data available on various Web sites are still highly heterogeneous. The problem is even worse for geographical data with respect to traditional data because they are quite simple to visualize for the user (they look like nice images or colored drawings describing portions of our planet). However, despite this apparent simplicity, they usually have a complex structure (space and often time dimensions are involved) and it is very hard to analyze and combine them using existing representation and processing standards. Additionally, often, entities of interest to specific applications have quite different representations at different sites. Such representations may differ, for example, with respect to the semantic definitions, the data structures, the resolution level, the time of validity, or the spatial reference system.

Another relevant problem to be addressed in a distributed environment is the development of techniques for efficient data transmission and visualization of spatial data at the client end. In addition, it is important to note that many innovative applications, such as location-based management systems, require access to spatial and geographical data from mobile clients. It is thus crucial to support efficient access to data from both fixed and mobile clients and to guarantee easy understanding and processing of those data by resorting to suitable information visualization and human–computer interaction techniques.

Finally, a relevant issue is the need to ensure data integrity and security to safeguard both users and owners of data. The goal is to ensure that data, even when made available on the Web and transmitted across sites, is accessed and modified only by authorized parties. There is thus a strong need for tools supporting secure access and use of spatial and geographical data from a large variety of client systems.

This book addresses these topics by presenting contributions related to modeling, processing, integration, and protection of geographical data, as well as innovative applications for mobile devices, under distributed architectures. The book covers those issues, devoting a separate part of the book to each of them, according to the following organization:

- **PART I**: Models for representing spatial semistructured, multiresolution, and multiscale data.
- **PART II**: Integration of spatial data sources, including issues concerning data quality in integration systems, mediation systems, geographical information fusion, and similarity-based processing techniques.
- **PART III**: Spatial data protection, including access control, secure outsourcing, and information hiding.
- **PART IV**: Innovative applications for mobile devices.

In the following, we dedicate a section to each of the book parts, providing a brief description of the background concepts and of the contributions presented.

1.1 Models for Representing Spatial Semistructured, Multiresolution, and Multiscale Data

Data models and data structures for representing and managing spatial data, especially in geographical applications, have been well studied in the past two decades in several research areas and with different aims.

In the area of *spatial databases*, much effort has been devoted to both the definition of a data model (and query language), to be adopted by database systems in order to handle spatial data and integrate them with traditional information, and the study of spatial data structures, access methods, and algorithms for increasing the performance of query processing when spatial properties are considered. In this area, spatial data means mainly *vector data*, i.e. the representation of the position, shape, and extension of an object made by means of a geometry, which is described by a finite set of points embedded in a reference space. Usually, for geographical data, only the linear geometry is adopted and the object shape is built up by drawing open or closed polylines. Recently, ISO has also devoted a complete series of standards to geographical data. In particular, the specification of a vector-based geometry data model is given in the standard 19107 (called "Spatial Schema") [9] and an object-oriented conceptual model for the design of spatial databases using the Spatial Schema as geometry model is given in the standard 19109 (called "Rules for Application Schemas") [10]. Moreover, the integration of temporal and spatial dimensions has enriched the data models of constructs for the representation and management of "moving objects." The reader can find many papers and books about data models, data structures, and access methods for spatial data; here we cite only some of them [13, 19, 20, 22].

Besides vector data, other types of spatial information are quite relevant for geographical applications: (1) remote sensing images and any other image of the Earth's surface; (2) grids of cells containing the measures of some physical parameters surveyed on the territory (they are often used in many environmental studies); (3) digital terrain models, describing, for example, terrain and geological information, like the well-known digital elevation models (DEMs) and triangulated irregular networks (TINs).

The first two types of the above listed spatial information are usually denoted using the general term *raster data* and have been well studied also in the research areas of *image databases* and *image processing*. Terrain representations based on DEMs and TINs have been investigated by researchers in the area of geometric data structures and algorithms.

The fast development of Web applications is providing to Internet users a huge amount of spatial information represented in one of the above listed types. However, such a distributed environment poses new problems with respect to spatial data modeling, in particular:

- *Multiresolution.* Each available data set can have a different resolution level, and this level may be very far from the resolution level the user is interested in. Therefore, it could be very useful for spatial data providers to be able to generate spatial data at the required resolution level on the fly. This could be very expensive in terms of computation time and thus "ad hoc" techniques have to be studied in order to improve the performance of such resolution level transformations. Moreover, each type of spatial data requires specific approaches for dealing with multiresolution representations and supporting resolution level transformation.

- *Progressive Data Transmission and Visualization.* Another problem that the publication of spatial data on the Web has to face is the band availability for data transmission. Spatial data set size is often measured in gigabytes or terabytes; thus the transmission of the whole data set at a finer resolution level could take a very long time. Progressive data transmission can therefore be very useful for spatial data; this technique is based on the following strategy: first, the most relevant spatial details are sent at a coarse resolution level, then the user can perform some preliminary operations and decide whether it is convenient to wait for the detailed representation at a finer resolution level or to interrupt the downloading. This allows one to save both time and disk space. A similar approach can be adopted for supporting the visualization of spatial data at different resolution levels. In both cases, the key issue is the definition of data structures and algorithms for increasing the performance of the resolution level transformations.

- *Semistructured Data Representation.* The Web has also promoted a new approach for representing information that derives from the use of hyper texts for presenting (using HTML) and for representing data (using XML). This idea has produced the development of a new research area in the context of databases and information systems, called "semistructured data management." As a consequence, new formal data models able to describe semistructured data without depending on a specific tag language have been defined. The integration of spatial data in a semistructured data model is an interesting issue that has been addressed in particular by the open geospatial consortium (OGC) through the definition of the geography markup language (GML). However, new effort is needed to integrate spatial and temporal properties in abstract semistructured data models.

Part I consists of four chapters presenting complementary issues in the context of spatial data modeling.

Chapter 2 presents a semistructured data model where spatial data are integrated in the multimedia temporal graphical model (MTGM). This is an example of an abstract data model for semistructured data with the ability to represent temporal and spatial properties of objects. Moreover, a possible mapping to XML is proposed that adopts the XML elements proposed by GML for spatial and temporal properties.

In Chap. 3, issues related to multiresolution representations for very large digital terrain data sets are discussed. In particular, the authors describe how to deal with very large terrain data sets through out-of-core techniques that explicitly manage I/O operations between levels of memory. After a brief discussion about digital terrain models, focusing on TINs, the authors review out-of-core techniques for simplification of triangle meshes and specific out-of-core multiresolution models for regularly distributed data and irregularly distributed data.

Progressive data transmission is presented in Chap. 4. First, the author illustrates well-established implementations for progressive transmissions of data over the Internet, including raster images and triangular meshes. Then the problem of transmitting geographical map data in vector format is analyzed and an overview of the prototype systems developed and discussed in the literature is presented. Moreover, the author presents a critical discussion of the main research and implementation challenges still associated with progressive vector data transmission for Web-mapping.

Finally, in Chap. 5, a specific new approach for the generation and display of spatial information at a given resolution level is presented. It is based on a variable scale data structure, the topological generalized area partitioning (GAP) structure called tGAP. The purpose of this structure is to store the geometry only once, at the most detailed resolution level, and to represent in addition the result of a generalization algorithm applied to the stored geometry. Then, the tGAP is used when data are requested by a Web client for deriving on-the-fly the geometry at the requested resolution level, based on the generalization preprocessing.

1.2 Integration of Spatial Data Sources

A geographical data set is an abstraction of the real world, according to a specific point of view and purpose. From this consideration, it follows that different processes (e.g. social, ecological, economical), based on different motivations, at different times (e.g. every 5 years), and possibly using different devices, may produce and may make available online different geographical data sets, concerning the same or overlapping areas to users and applications. In a distributed environment, several application contexts may require the ability to use together and compare such distinct data sets. The process of making distinct data sets usable in a homogeneous way by a given application is called "data integration process".

Integration processes should be able to cope with heterogeneity concerning how source data sets are represented and how they can be accessed. Heterogeneity in data representation arises when the same concept is represented differently in each data set, thus generating some semantic conflict. Conflicts can be related to the meaning assigned to concept names (same meaning but different concept names or different meaning and same concept name), descriptive attributes used to describe a concept (a road can be associated with an attribute length in a data set and with attributes length and type in another), data types assigned to spatial attributes associated with concepts (a road can be a polygon in a data set and a line in another), and units used to represent geometric attributes (road lengths can be expressed in kilometers in one data set and in meters in another). Some of the conflicts cited above, for example those related to concept meaning and descriptive attributes, are not specific to the spatial context, and can therefore arise whatever the application domain. On the other hand, those related to geometric attributes and spatial concepts are typical of the geographical and, more generally, of the spatial context. Therefore, specific solutions have to be provided for them.

Data integration aims at overcoming problems concerning data conflicts to provide homogeneous access to local sources. Data integration solutions rely on the identification of different objects, represented in distinct sources and related by some semantic link, and on the resolution of conflicts existing between such objects.

In order to describe the semantics of each data set, a schema or, more generally, an ontology can be used. An ontology can be defined as "an explicit specification of a conceptualization" [7]. An ontology, besides describing the structural characteristics of a data set, i.e. its schema, also provides logical systems to be used for defining and reasoning about relationships and constraints existing between the data set concepts. The process of identifying the relationships between corresponding elements in two heterogeneous ontologies is often called *ontology alignment* [18]; on the other hand, the process of constructing a single combined ontology based on the identified relationships, and therefore based on a given ontology alignment, is called "ontology integration" and, more specifically, *geographical information fusion* [5]. Ontology alignment solutions can be intensional, if they are based on concepts definitions (e.g. properties of roads), or extensional, if they consider concepts instances (e.g. the single roads in the data set).

Among the existing approaches for ontology integration, *mediation* has been investigated in depth in the past few years [24]. A mediation system can be defined as a system providing to the user a uniform interface by which different data sources can be accessed through a common global model. The mediation system relies on rules for mapping concepts in the global schema into concepts contained in each local source based on a given and predefined ontology alignment. The architecture of a mediation system is based on two main components: the mediator and the wrappers. The mediator allows *semantic translations* by rewriting, using the predefined mapping rules, the user's query into queries over data sources expressed in a common query language, which is specific to the mediator. Each data source is accessed through a wrapper. When a query is posed against a data source, the corresponding wrapper translates it according to the data source query language.

In order to increase interoperability, OGC has defined several standards for geographical data representation, transfer, and access to be used in mediator systems and in distributed architectures in general. OGC has adopted GML for the XML representation and transport of geographical data [16] and Web Feature Services (WFS) for describing or getting features from a spatial data source on the Web [17]. Although such standards can help the development of interoperable applications, they do not provide solutions to ontology alignment and integration problems, which must be addressed in order to provide adequate data integration approaches.

Another relevant integration issue concerns the quality of the accessed data. Indeed, high quality at each local source may correspond to low quality at the integration level. For geographical data, several models have been proposed by various organizations, converging to an ISO/TC 211 model [11, 12]. Each model proposes a set of parameters applicable to a certain class of concepts of a given schema (e.g. geometric properties, attributes). Among quality parameters, those concerning imperfections like accuracy, precision, and consistency play an important role. Accuracy refers to the correspondence between the geographical information and the real world they represent. Precision deals with the granularity level by which the real world is represented inside a database, i.e. its resolution. Consistency refers to the presence of contradictory concepts inside the same data set. Values for quality parameters can be used during ontology alignment, to improve the quality of the detected spatial relationships, and integration, in order to come up with a global schema having a certain quality level.

The quality problem can be tackled from the bottom, by extending data integration solutions to cope with quality parameters, or from the top, by extending query specification and processing to deal with quality issues. One of the key observations is that in the presence of data with different granularities, as in the case of multiresolution and multiscale data, the specification of equality-based queries, by which the user specifies in an exact way the constraints that data to be retrieved must satisfy, may not be the right choice, since multiresolution is not effectively used during such processing. In order to exploit multiresolution, a possible approach would be that of introducing some mechanism of query relaxation, by which the specific characteristics of multiresolution maps are taken into account during query execution, based on the actual resolution and scale of the data source; as a consequence, approximated answers are returned to the user, possibly introducing some false hits, but at the same time making query answers more satisfactory from the user point of view.

Part II of this book consists of three chapters that cover complementary issues in the context of spatial data integration, according to what was discussed above.

Ontology alignment is the topic of Chap. 6. In particular, after reviewing ontology alignment, ontology integration, and geographical information fusion problems, the ROSETTA system is presented, for extensional and automated geographical information fusion based on inductive inference. Discussions concerning how the system can cope with various types of uncertainty (inaccuracy, imprecision, vagueness) are also provided.

Chapter 7 deals with quality-enabled spatial mediation systems. Besides presenting the basic problems and the possible solutions, a quality-enabled spatial integration system called VirGIS/Q is presented. VirGIS/Q relies on the existing standards for geographical data representation and access, as well as for data quality parameters specification.

Chapter 8 presents some qualitative techniques for query relaxation over multiresolution spatial data sets, where different types could be assigned to the same geographical feature in distinct data sets. The considered information is qualitative in the sense that it corresponds to qualitative (topological and directional) relationships between spatial objects. The proposed techniques rely on some distance functions, one for each class of relationships, that can be used to relax user queries.

Examples and discussions concerning how such functions can be applied to mediation systems and consistency checking are also provided.

1.3 Spatial Data Protection

Recent developments in information system technologies have resulted in computerizing many applications in various business areas. Data have become a critical asset in many organizations, and, therefore, protecting data has become an urgent need. Data and information have to be protected from unauthorized access as well as from malicious corruption. The advent of the Internet has made the access to data and information much easier [6]. For example, users can now access large quantities of information in a short space of time. This has further exacerbated the need for protection.

Over the past three decades, various developments have been made on securing data, most of them in the field of Database Management Systems [3]. Much of the early work was on statistical database security. Then, in the 1970s, as research in relational databases began, attention was directed toward access control issues. In particular, work on discretionary access control models began. While some work on mandatory security started in the late 1970s, it was not until the air force summer study in 1982 that many of the efforts in multilevel secure database management systems were initiated. This resulted in the development of various secure database system prototypes and products [3]. In the 1990s, with the advent of new technologies such as digital libraries, the Web, and collaborative computing systems, there was much interest in security not only from government organizations but also from commerce and industry. Database systems are no longer stand-alone systems. They are being integrated into various applications such as multimedia, electronic commerce, mobile computing systems, digital libraries, and collaboration systems, and therefore more complex security mechanisms should be developed [6].

Traditionally [3], securing data requires dealing with three main issues: *authenticity*, *confidentiality*, and *integrity*. Satisfying data authenticity means that the subject receiving data is guaranteed that they actually come from the expected source. Related to this issue is that of *ownership protection*, that is, ensuring that the protected data cannot be misused and maliciously transferred to unauthorized users. Ensuring data confidentiality means that data contents can be disclosed only to authorized users. By contrast, the term integrity implies two different security properties. The first refers to data protection from unauthorized modification operations, whereas the second means that data contents are not altered during their transmission over the net. In past decades, several security mechanisms have been proposed to ensure such security properties in different environments and application domains. Usually, such solutions enforce integrity through access control mechanisms and encryption-based techniques, and confidentiality through access control mechanisms, whereas data authenticity requires the use of watermarking and digital signature techniques.

However, these traditional security mechanisms are not always adequate to meet the requirements of new data and emerging applications, which call for more flexibility or require taking into account additional security properties [2]. Such a need has resulted in various extensions to traditional security mechanisms over the past decade, which have been driven by the following three main factors:

- *New Data Types*. Information systems were originally defined to efficiently manage alphanumeric data. However, today most applications need to manage multimedia data in addition to traditional data. For example, medical applications, digital libraries, or applications in the cultural heritage domain should manage in an integrated way data from different media (such as audio, images, text, and video) and with totally different characteristics and requirements. The main difference with respect to traditional data is that such data are unstructured, and as such, new data models should be devised to represent their semantics within a data management system, and new security solutions should be devised for their protection. Additionally, the Web has increased the types of data that need to be managed, being XML [26] the de facto standard for data representation and transmission over the Web. Also spatial data have further security requirements with respect to traditional data, which must be taken into account when developing a security mechanism. For instance, when dealing with spatial data, it is important to take into account the spatial dimension that access permissions usually have: a user can be authorized to display only maps referring to a particular region, or can be authorized to access different data depending on his/her position.

- *New Architectures*. The widespread use of the Internet and other network facilities has made possible the adoption of new architectures, besides the traditional client–server one. In particular, database management systems are today moving from a centralized architecture to more distributed ones. Examples in this direction are parallel or distributed databases, peer-to-peer systems, third-party architectures, or architectures based on Web services. Clearly, dealing with distributed architectures makes data protection more difficult. In a centralized environment, the core component of any security mechanism is the *access control module* that is hosted on a trusted site, i.e. the server intercepts each request to access the data and authorizes only those meeting the requirements of the specified security policies. Such a simple solution cannot easily be applied when it is not possible to rely on a single trusted server.

- *New Applications*. Many advanced application environments, such as distributed digital libraries, workflow applications, groupware, data warehousing, and data mining applications, have new security requirements. For instance, in a digital library, environment ownership protection is a fundamental requirement, besides confidentiality and authenticity. The requirements of these new applications cannot be adequately supported by standard security mechanisms that are tailored to few, specific security policies. In most cases, either the organization is forced to adopt the specific mechanism built into the data management system, or the new requirements must be implemented as application programs. Both situations are clearly unacceptable. Thus, there is the need for developing security mechanisms specifically tailored for these new environments.

Part III of this book considers the three above-mentioned areas from the perspective of spatial data management. In particular, Chap. 9 discusses new access control requirements of spatial data by mainly focusing on location-based services (LBSs) and presents GEO-RBAC, a role-based access control mechanism particularly suited to location-aware applications. Then, it discusses a decentralized implementation of GEO-RBAC.

Chapter 10 addresses the security issues arising when geographical data are outsourced to a third party instead of being managed by the data owner. In this case, additional security requirements arise, because data must be protected not only against unauthorized access by users but also against malicious misuse by the party to which they are outsourced. In the chapter, a solution based on non-conventional signature and encryption techniques is presented and an application of the proposed framework to geographical data is discussed.

Finally, Chap. 11 addresses the issue of information hiding for spatial and geographical data. Information hiding can be applied for the protection of intellectual property as well as to convey secret information to authorized users. After introducing watermarking and stenography, which are the most widely used information hiding techniques, the chapter presents a watermarking and stenography model tailored to digital maps. Additionally, the chapter illustrates some experimental results that demonstrate the robustness of the proposed methods.

1.4 Innovative Applications for Mobile Devices

In recent years, there has been a technological evolution of geographic information systems, mobile devices (PDA, smart phones, cellular phones), wireless communication systems (GPRS, UMTS, Wi-Fi), and positioning systems, which are now commonplace (GPS, EGNOS). This evolution made possible the emergence of LBSs in the mobile-geographical information system (GIS) [25]. Such systems have many applications in many fields, such as people safety (e.g. management of natural or industrial crises), in which localization of the event and fast acting are priorities. Besides, LBS has become a part of everyday life in applications like traffic forecasting or mobile phone services and also in the extension of satellite navigation systems toward "context-aware" systems [4], in which user position and context are used to provide information and services. In developing mobile-GIS applications, the user position is the starting point to develop services and it is necessary to study a model of the space surrounding the user [1, 23]. There is a very important distinction to be made between user's position and location [21]: the first one is just the absolute x,y coordinates of the user in the reference space, while the second one is the "surrounding" area of the position, the shape and the dimension of which is established by the application context.

Even if the natural application field of LBSs is the mobile and wireless world, significant benefits could be obtained by applying LBSs to desktop systems, for instance, in trip planning or warehouse asset management [8]. Technological support is given by emerging standards of the OGC. OGC's aim is to advance the standardization process to guarantee an interoperability in every field including LBSs [8]. OGC has created the OGC Web services (OWS) [14], and the latter has developed initiatives relevant to LBSs such as the OpenGIS location service (OpenLS) [15], which is devoted to the development of architectures to support emerging LBSs. In order to present to the user the information in the best way, the capabilities of the mobile device have to be considered.

Part IV consists of two chapters presenting innovative applications for mobile devices. Chapter 12 deals with the problem of data visualization in mobile devices: the main limitation is the small size and resolution of the display, besides the limited processing power and input peripherals. The difficult working conditions of a mobile user are also an issue. Some LBSs are not only dependent on the instantaneous user position, but rely on the path followed by the user, e.g. traffic jam information.

Chapter 13 discusses various techniques to predict the future positions of moving objects. These techniques are based on a server-side representation of past user positions; more often the moving object communicates its position to the server, more accurate can be the prediction of future positions. A challenge is how to predict future positions with a certain level of accuracy while also reducing client–server communication.

References

1. Bartie P J, Mackaness W A (2006) Development of a Speech-Based Augmented Reality System to Support Exploration of Cityscape. Transactions in GIS, 10(1):63–86
2. Bertino E, Sandhu R S (2006) Database Security-Concepts, Approaches, and Challenges. IEEE Trans. on Dependable Secure Computing 2(1):2–19
3. Bishop M (2005) Introduction to Computer Security. Addison-Wesley Professional
4. Cheverst K, Davies N, et al. (2002) Exploring Context-aware Information Push. Personal and Ubiquitous Computing 6:276–281
5. Dasarathy B (2001) Information Fusion - What, Where, Why, When, and How? Information Fusion, 2(2):75–76
6. Ferrari E, Thuraisingham B (2005) Web and Information Security. IRM Press
7. Gruber T (1993) A Translation Approach to Portable Ontology Specifications. Knowledge Acquisition, 5(2):199–220
8. Intergraph (2004) Open Location-Based Services, A White Paper. Mapping and Geospatial Solutions Series
9. ISO/TC 211 Geographic information/Geomatics (2002) 19107 Geographic information - Spatial schema, text for FDIS, N. 1324
10. ISO/TC 211 Geographic information/Geomatics (2003) 19109 Geographic information - Rules for application schema, text for FDIS, N. 1538
11. ISO/TC 211 Geographic Information/Geomatics (2002) 19113 Geographic information - Quality principles
12. ISO/TC 211 Geographic Information/Geomatics (2003) 19114 Geographic information - Quality evaluation procedures
13. Gueting R H, Schneider M (2005) Moving Objects Databases. Morgan Kaufmann

14. Open Geospatial Consortium (2003) OGC Web Services, Phase 2. http://www.opengeospatial.org/ projects/initiatives/ows-2
15. Open Geospatial Consortium (2005) OpenGIS Location Service (OpenLS) Implementation Specification: Core Services
16. Open Geospatial Consortium (2006) Geography Markup Language (GML). http://www.opengeospatial.org/standards/gml
17. Open Geospatial Consortium (2006) Web Feature Service (WFS). http://www.opengeospatial.org/standards/wfs
18. Rahm E, Bernstein P (2001) A Survey of Approaches to Automatic Schema Matching. The VLDB Journal, 10:334–350
19. Rigaux P, Scholl M, Voisard A (2001) Spatial Databases: With Application to GIS. Morgan Kaufmann
20. Shekhar S, Chawla S (2002) Spatial Databases: A Tour. Prentice Hall
21. Tryfona N, Pfoser D (2005) Data Semantics in Location-Based Services. In: Spaccapietra S, Zimanyi E (eds) Journal on Data Semantics III. Lecture Notes in Computer Science, Springer, Berlin Heidelberg New York, 3534:168–195
22. Worboys M F (1995) GIS: a Computing Perspective. Taylor & Francis
23. Ware J M, Anad S, et al. (2006) Automated Production of Schematic Maps for Mobile Applications. Transactions in GIS, 10(1):25–42
24. Wiederhold G (1992) Mediators in the Architecture of Future Information Systems. IEEE Computer, 25(3):38–49
25. Winter S, Tomko M (2004) Shifting the Focus in Mobile Maps. In: Morita T (ed.) Joint Workshop on Ubiquitous, Pervasive and Internet Mapping UPIMap2004, Tokyo pp. 153–165.
26. World Wide Web Consortium. Extensible Markup Language (XML) 1.0, W3C Recommendation. http://www.w3.org/XML/

References containing URLs are validated as of September 1st, 2006.

Models for Representing Spatial Semistructured, Multiresolution, and Multiscale Data

GeoMTGM: A Graphical Data Model
for Semistructured, Geographical, and Temporal Data

Carlo Combi, Sara Migliorini, Barbara Oliboni, and Alberto Belussi

University of Verona, Verona (Italy)

2.1 Introduction

A few decades ago, paper maps were the principal means to synthesize and represent geographical information. Manipulating this information was limited to a manual, non-interactive process. Since then, the rapid development of new technologies for collecting and digitizing geographical data, together with an increasing demand for both interactive manipulation and analysis of this data, has generated a need for dedicated software, namely geographical information systems (GISs) [23].

GISs have made available a large amount of geographical data to several applications, and as a consequence, spatial data have been included in different automatic tools (like, many decision support systems), providing a new dimension of analysis, represented by the common reference space: the Earth surface. Different kinds of problems can benefit of the new spatial dimension: from geomarketing to epidemiological crises.

Recently, with the diffusion of XML [24] as the language for exchanging data over the Web, it has become more and more common to manage complex geographical information represented as *semistructured data*, that is, data having an irregular or incomplete structure that can change over time [2]. For this reason, semistructured data are often explained as "schema-less" or "self-describing," meaning that there is no separate description of the structure of data.

In this chapter, we propose a new model for semistructured data that allows one to represent multimedia and temporal aspects of geographical data described as a set of geographical objects.

With the term *geographical object* (or feature in the Open GeoSpatial Consortium terminology), we mean an abstraction of a real-world phenomenon associated with a location on the Earth's surface [21]. A geographical object may be characterized by the following components:

- One or more *spatial attributes*. They describe the location, shape, orientation, and size of the object in 2D or 3D space.
- A set of *descriptive or thematic attributes*. These attributes provide a description of the object, for example, the population of a country.

- A set of *spatial relations*. This set describes the existing relationships between geographical objects (i.e. topological relations such as disjoint, touch, in, contains, equal, cross, and overlaps [7]).

The above-mentioned components could be only partially specified or they could be defined at different levels of detail, making it difficult to represent the related information within a (pre)fixed schema. A further aspect to deal with is related to the fact that geographical information is often time varying, for example, the borders of a town can change and their evolution must be suitably managed.

The model we propose in this chapter is named geographical multimedia temporal graphical model (GeoMTGM). A GeoMTGM graph is a directed, labeled graph with a single root; it is based on the multimedia temporal graphical model (MTGM) presented in [10] and allows us to represent geographical and semistructured data, with particular attention to their temporal aspects. In particular, GeoMTGM deals with the *Valid Time* (VT) (i.e. the time at which a fact is true in the considered domain) [9] of both objects and their relationships.

The GeoMTGM is a new proposal in the context of semistructured multimedia spatiotemporal data. In Sect. 2.2, we give a brief description of the main issues related to this context.

This work is structured into three main sections. In Sect. 2.2, we present some basic concepts that are useful for understanding the model and the rest of the work. In Sect. 2.3, we go on with the explanation of the GeoMTGM model, in particular, its main characteristics and structure, its formal definition, and its translation into XML. Finally, in Sect. 2.4, we present a realized Web application based on the J2EE [15] and the Struts [1] framework for the definition, manipulation, and translation of GeoMTGM graphs.

2.2 Related Work

In this section we briefly recall semistructured data features, aspects of multimedia data modeling, and some basic notions related to geographical information. Then, we give a brief description of geography markup language (GML) [13], an XML-based language for geographical information, and finally, introduce the data model proposed in [14], which is, to the best of our knowledge, the unique work in the literature addressing issues related to spatiotemporal semistructured data, proposing an XML-based representation.

2.2.1 Semistructured Data

Semistructured data are data having some structure that may be irregular or incomplete and does not necessarily conform to a fixed schema [3]. The structure of semistructured data can change over time; for this reason, for this type of information, it is important to consider the temporal aspect of data.

In this context, the models proposed in the literature allow one to represent and query changes in semistructured data. Among these, we cite here the delta object exchange model (DOEM) [6], a temporal data model representing the *transaction time* [18] (i.e. the time at which an item of information is stored in the database), and the temporal graphical model (TGM) [22], a graph-based data model that is able to represent static and dynamic aspects of semistructured data, considering the VT [18] (i.e. the time at which a fact is true in the considered domain).

It is usual for semistructured data models to represent data as labeled graphs, where nodes and edges suitably represent different (related) data items. In particular, a TGM graph is a labeled graph with two types of nodes: *complex nodes* that describe the objects of the represented reality and *atomic nodes* that represent properties of the objects. The relationships between objects in the reality are represented by labeled relational edges in the graph.

2.2.2 Multimedia Temporal Information

In the field of multimedia database systems, topics related to the integration of several media objects (with their temporal aspects) have been considered in several papers. Among them, we distinguish two different research directions: the first one focuses on data modeling and querying issues [5, 8, 12], while the second one is more specifically devoted to proposals of models for multimedia presentations [4, 19]. An example of the first research direction is the work presented in [5], where a unified data model is proposed for multimedia types, such as images, sounds, videos, and long text data. For the second research direction, a multimedia presentation can be defined as the composition of a set of multimedia objects that have to satisfy some given spatiotemporal constraints during their playback; several proposals deal with the specification of synchronization constraints between different multimedia data items [4]. For example, in [19] the authors present a methodological approach for checking the temporal integrity of interactive multimedia documents (IMDs), focusing on temporal properties of presentations.

2.2.3 Geographical Information and GML

In the literature, there are two approaches to the representation of geographical information [23] : the *Entity-based Model* and the *Field-based Model*. In the entity-based model, geographical information is represented by several geographical objects that are distinguished by an identifier and described by their attributes. The entity-based model properly represents the typical content of a traditional map, where objects are embedded in a reference space, for example, a map can be obtained by composing the road network, the buildings, the administrative units, and so on. Each geographical object can be viewed as an entity, like in traditional information systems, and each of their attributes as a property of this entity.

In the field-based model, each point of the reference space is associated to one or several attribute values, defined as continuous functions in x and y. The altitude above sea level is an example of a function defined over x and y, whose result is the value of a variable h for any point in the 2D space. Measures for several phenomena can be collected as attribute values varying with the location in space, for example, precipitation, temperature, and pollution values. This view of space as a continuous field is in contrast with the view of the entity-based model that describes the geographical objects that populate the space.

Any effective language for representing geographical information should be able to model data in both the above described approaches. The GML [13, 20] satisfies this requirement. It is an XML grammar written in XML Schema that is used to model, exchange, and store geographical information, including both the spatial and non-spatial properties of geographical features.

GML provides different kinds of objects for describing geography including features, coordinate reference systems, geometry, topology, time, units of measure, and generalized values. GML is the only language we found in the literature that supports the representation of semistructured data for geographical information.

In GML, there are essentially three types of geographical object: (1) *basic GML feature* is the type used to represent a geographical entity, that is, any meaningful object in the selected domain of discourse such as a Road, River, Person, or Administrative Boundary; (2) *coverage* is a type of feature that includes a function with a spatial domain and a value representing a set of homogeneous two- to n-dimensional tuples (i.e. it describes the spatial distribution of the Earth phenomena); (3) *observation* is considered to be a GML feature with the time at which the observation took place and with a value for the observation.

GML is designed to support interoperability through the provision of basic geometry tags, a common data model (features-properties), and a mechanism for creating and sharing application schema.

2.2.4 An XML-based Representation of Semistructured Spatiotemporal Information

In [14], the authors explore how information about location and time can be integrated and represented in XML. In particular, they consider how to represent spatiotemporal types through XML Schema. The authors propose several data types in XML Schema to represent both basic temporal elements (i.e. a sequence of disjoint time intervals) and histories of spatial objects. The VT of a spatial object is represented by a temporal element and spatial histories are modeled as temporal sequences of spatial objects, through the parametric XML schema type *Time-Series*. The XML-STQ spatiotemporal query language allows the user to express queries containing both temporal and spatial objects; as an example, it is possible to detect when a geographical area has some specific spatial relation with another spatial object.

This proposal deals with the representation of spatiotemporal information in the XML context, while the data model we present in this chapter is able to manage spatiotemporal information in the semistructured data context. Moreover, in [14], the authors propose the representation of histories by means of sequences (with respect to time) of spatial objects, while our new data model allows one to represent the VT dimension of each piece of geographical information.

2.3 Geographical Multimedia Temporal Graphical Model

The geographical multimedia temporal graphical model (GeoMTGM) is a logical model for the representation of geographic information. More precisely, it enriches MTGM with the constructs necessary for representing information with both the entity-based and the field-based models.

2.3.1 GeoMTGM at Work

In this section, we describe the main features of GeoMTGM through the use of an example.

A GeoMTGM Database

As an example to illustrate the characteristics of GeoMTGM, we propose to model the history of a highway using semistructured data. In particular, we consider the Highway A1 of the Italian highway network, which has the following general features:

- It starts in Milan and ends in Naples.
- It is 754 km long and has been built in several phases: first, the route from Milan to Parma was opened in 1954, then in 1964 the route from Florence to Rome became operative. Finally, the segment from Rome to Naples was added and opened in 1966.
- It has a set of toll stations, where each toll station has a number of payment points and a number of telepass points.
- It has several intersections with other highways.
- Sometimes, accidents occur at several points on the highway route and the highway can eventually be blocked or suffer reduced traffic flow; finally, maintenance work can be present along the route of the highway with consequent traffic flow reduction.

All these features can be represented in a GeoMTGM graph as shown in Fig. 2.1. In particular, the temporal evolution of some of the properties of the highway can also be represented, for example, the evolution of its route that has been built in three separate phases or the different events that can occur along the highway, such as accidents, maintenance work, strikes, and so on.

Figure 2.1 shows the schema of a GeoMTGM graph containing information about the A1 Highway, while Figs. 2.2 and 2.3, show a possible instance of it. As usual in semistructured data models, (part of) the schema and (part of) the instance of a database are represented in a similar way through graphs.

In GeoMTGM the entities of the modeled reality are represented through *complex nodes* (depicted as rectangles). In particular, there is a special kind of complex node, named *geographical complex node* (*geoComplex node* in short), that represents geographical entities (depicted as dashed rectangles). For example, in Fig. 2.1 *Owner*

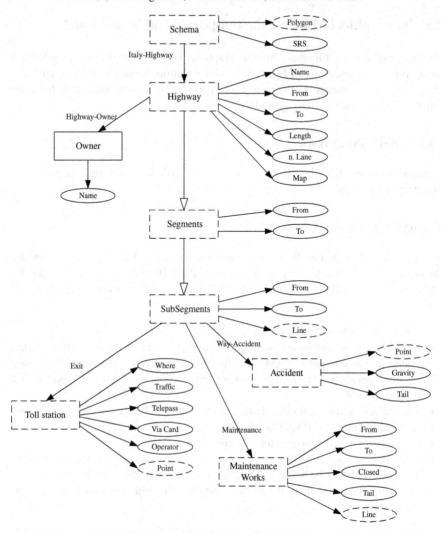

Fig. 2.1. A portion of the GeoMTGM database schema

is a complex node, while *Highway, Segment, SubSegment, Maintenance Work, Toll Station,* and *Accident* are geographical complex nodes.

Each entity has some textual attributes that are represented by *atomic nodes* (depicted as ovals). In Fig. 2.1, *Name, Length, n.Lane,* and *Gravity* are examples of atomic nodes. Moreover, entities can have some multimedia attributes that are represented through a special kind of atomic node, named *stream node*. Stream nodes contain multimedia information as unstructured texts, movies, and sounds; they are depicted as thick ovals in database instances. In Fig. 2.1, the nodes *Traffic* and *Map* are examples of stream nodes that refer to the files that encode the movie of the current traffic on the specified toll station and the image of the Highway map, respectively.

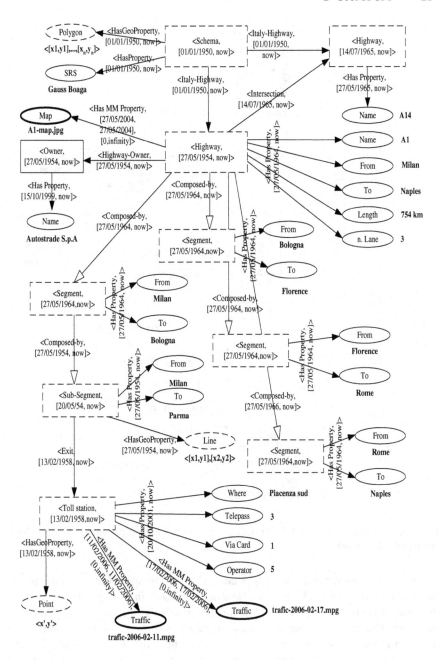

Fig. 2.2. A portion of the GeoMTGM database instance: highway with its segments

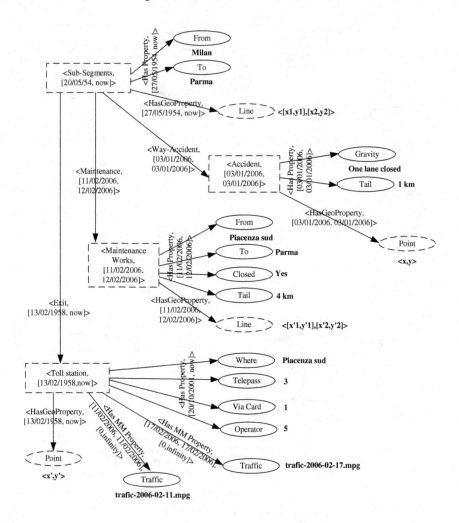

Fig. 2.3. A portion of the GeoMTGM database instance: a highway segment with accidents and maintenance work

For every geographical entity it is necessary also to specify its spatial components that are represented by *geographical atomic nodes* (in short *geoAtomic nodes*), depicted as dashed ovals. The geographical atomic node stands for a general spatial component and has some specializations: a *point node* represents a point, a *line node* represents a line, and a *polygon node* represents a region in the space. For example, in Fig. 2.1 the geographical complex node *Sub-segment* has a geographical attribute of type line. A geographical atomic node is different from an atomic or stream node mainly for two reasons: (1) its name can be only *point*, *line*, or *polygon* (it is distinguished from the others by its identifier); (2) its value is not a string but a set of coordinates like (x, y) in 2D space, or (x, y, z) in 3D space.

As regards the root of the GeoMTGM graph, as shown in Fig. 2.2, it is represented by a geographical complex node named *"Schema"* whose spatial component covers the geographical area described by the graph. The root has a particular attribute named *"SRS"* whose value indicates the spatial reference system used in the graph. If the spatial component of a particular geographical object is represented with respect to a different spatial reference system, then this information is added in the label containing the coordinate values.

Nodes are connected through labeled edges having a direction (graphically specified by an arrow), called "relational edges." In particular, relational edges connecting complex nodes to their atomic nodes have labels containing the name *"HasProperty,"* while those connecting complex nodes to their stream nodes have labels containing the name *"HasMMProperty"*; finally, those connecting complex nodes to their geographical atomic nodes have labels containing the name *"HasGeoProperty."*

The VT [18] is explicitly managed by GeoMTGM both for nodes and edges. The VT of a (geographical) complex node is represented in its label, while the VT of an atomic (stream) or geographical atomic node is represented in the label of the edge between it and its parent. For example, as shown in Fig. 2.2, the VT of the complex node *Highway* is [27/05/1954, now] where *"now"* indicates that the represented fact is currently true, while, in Fig. 2.3, the VT of the atomic node *Gravity* is [03/01/2006, 03/01/2006]. Moreover, the VT of a relational edge is reported in its label, for example, the relational edge *Segment-Accident* between a *SubSegment* and an *Accident* complex node has label ⟨Segment-Accident, [27/05/1954, now]⟩. In general, the VT of a complex node and those of its ingoing relational edges can be different.

As regards a relational edge through a complex node and its stream node, besides the name *"HasMMProperty,"* the label of the edge contains both the VT, the time during which the subpart of the movie has been recorded, and the specific subpart of the stream object to which the complex node is related. For example, in Fig. 2.3 the label of the edge between *Toll-station* and *Traffic* is ⟨HasMMProperty, [11/02/2006,11/02/2006], [0,infinity]⟩.

Moreover, GeoMTGM allows one to represent shared spatial components, in particular it admits that the spatial attribute of a geographical entity is composed by performing the union of the spatial attributes of other geographical entities, without duplicate geometry. This can be done by means of a new kind of edge, named *"composed-by"* edge, that is defined between two geographical complex nodes and is depicted with a big white arrow, as reported in Fig. 2.2 between *Highway* and *Segment* or between *Segment* and *SubSegment* entities in Fig. 2.3.

Defining Geographical Properties

As previously outlined, GeoMTGM allows one to represent geographical information according to both the entity-based model and the field-based model. The first information model is supported by the constructs explained in the previous section. In this section, we present the construct we propose to use for GeoMTGM according to the field-based model.

In a GeoMTGM database, several field-based geographical properties can be stored, so we need to identify each of them in a unique way. For this purpose, we introduce a special complex node named *"GeoField"* node that represents a particular geographical property and has an atomic node named *"GeoFieldName,"* representing its unique name with respect to the considered GeoMTGM graph.

The *GeoField* node is connected by several edges, named *geographical edges* (depicted as dashed lines), to the geographical complex nodes on which the property has been measured. The label of a geographical edge contains the name of the property, the value measured, and the VT (i.e. the time during which the property was measured). The geometric union of the geographical complex nodes, to which a *GeoField* is connected, represents the reference space of the measured property; this space is partitioned by the geographical complex nodes in cells, which can be of different granularities, from the administrative units to cells of a few square meters.

Figure 2.4 shows an example of the use of these constructs. In particular, we store information about the traffic in a particular time period in the various subsegments of the A1 highway.

In real applications, GeoMTGM can be applied effectively in the cases where data have a non-uniform structure, and it is important to describe the temporal evolutions of spatial and non-spatial properties. Moreover, the ability to represent any type of partition of the space is a key feature of the model, in particular for field-based data where the partition blocks can be represented also by pointing to a node of the graph that models an entity-based instance. Traditional geographical information can also be represented using a GeoMTGM graph; however, the implementation of the graph in XML leads to verbose structures with respect to

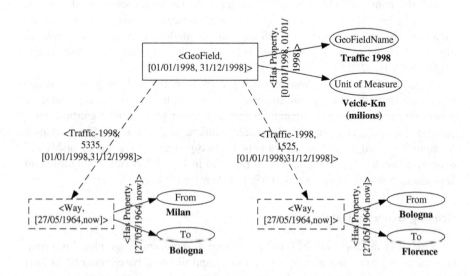

Fig. 2.4. Representation of geographical property through the GeoField model

the corresponding representations in grid-based or vector-based structures usually adopted in spatial database systems.

2.3.2 GeoMTGM Formal Definition

In this section, we provide a formal description of the GeoMTGM. This definition is based on the formal definition of MTGM presented in [10], so we omit the multimedia aspect and constraints about the multimedia part of the model and explain in detail only the geographical part.

Definition 2.1 (GeoMTGM Graph). *A geographical multimedia temporal graph (or GeoMTGM graph) is a directed labeled graph $\langle N, E, \ell \rangle$ with a single root where:*

Nodes
N is a finite set of nodes.

Edges
E is a set of labeled edges of the form $\langle m, label, n \rangle$, with $m, n \in N$, and label is the string label associated to a specific edge, in particular, it has the form

$$label \in (\mathcal{T}_e \times \mathcal{L}_e) \times (\mathcal{L}_{MMProp} \cup \{\bot\}) \times (\mathcal{F} \cup \{\bot\}) \times (\mathcal{V} \cup \{\bot\}))$$

where:

- $\mathcal{T}_e = \{relational, geographical, composed\text{-}by\}$ *is the set of types for edges.*
- $\mathcal{L}_e$ *is the set of strings to be used as name for edges.*
- $\mathcal{L}_{MMProp} = \{[frame\_s, frame\_e]\} \cup \{[time\_s, time\_e]\} \cup \{[x, y, width, height]\}$ *is the set of descriptions to define a subpart of a stream object. In particular, [frame_s, frame_e] stands for a pair of real numbers that describe the start and end frame of a video, [time_s, time_e] is a pair of real numbers that define the limit of an audio stream, and [x, y, width, height] are four real numbers that define a subpart of an image object.*
- $\mathcal{F}$ *is the set of strings to be used as value for the value of a geographical property represented in the edges that start from a GeoField node.*
- $\mathcal{V}$ *is a set of temporal elements of the form: $[t_1, t_2) \cup [t_3, t_4) \cup \cdots \cup [t_{n-1}, t_n)$, with $n > 1$. We use a temporal element to keep track of the time intervals during which an edge exists in the reality.*

Node Labels
ℓ is a function that associates a unique label to each node.

$$\ell : N \longrightarrow (\mathcal{T}_n \cup \{\bot\}) \times (\mathcal{L}_n \cup \{\bot\}) \times (\mathcal{S} \cup \{\bot\}) \times (\mathcal{G} \cup \{\bot\}) \times (\mathcal{V} \cup \{\bot\})$$

where:

- $\mathcal{T}_n = \{complex, geoComplex, atomic, stream, geoAtomic\}$ *is the set of types for nodes.*
- $\mathcal{L}_n$ *is the set of strings to be used as name for all nodes of the graph.*

- S is a set of strings to be used as atomic values.
- $G = G_2 \cup G_3$ where G_2 is the set of all possible pairs (x, y) of real numbers that represents coordinates in the 2D space, and G_3 is the set of all possible triples (x, y, z) of real numbers that represents coordinates in the 3D space. In particular, the possible value for a spatial component can be a single point or a list of coordinates that defines a simple polyline, or a set of coordinates that defines a closed polyline that identifies a simple polygon.
- V is a set of temporal elements, and as for edges, they are used to keep track of the time intervals during which a node exists in reality.

The function ℓ can be seen as composed of the following five single-valued functions:

$$\ell_{T_n}, \ell_{L_n}, \ell_S, \ell_V, \ell_G.$$

$\perp$ means 'undefined,' and in the following, when the context is clear, if an element of a label is mapped in $\perp$, it will be omitted.   □

Each GeoMTGM graph must satisfy the following integrity constraints in order to correctly represent real data. In particular, some of them ensure the correctness of the general structure of the graph (constraints 1–7), some of them ensure the consistency of the temporal elements specified on atomic and complex nodes (constraints 8 and 9), and finally, constraints 10–15 ensure the correctness of the spatial data in geoAtomic and GeoField nodes, and of the geographical and composed-by edges.

Definition 2.2 (Integrity Constraints of a GeoMTGM graph).
The following conditions must be satisfied by each GeoMTGM graph:

1. *Root.* The root of the graph is a geographical complex node with the name "Schema" whose spatial attribute is a polygon containing all the territory managed by the graph. It is connected to an atomic node named "*SRS*" that indicates the considered spatial reference system.

2. *Values.* Values are associated to *atomic* and *stream* nodes only.

$$\forall x \in N(\quad \ell_{T_n}(x) = complex \ \lor$$
$$\ell_{T_n}(x) = geoComplex \ \lor$$
$$\ell_{T_n}(x) = geoAtomic) \rightarrow \ell_S(x) = \perp)$$

3. *General Relational Edges.* Relational edges can be defined only between two *(geographical) complex nodes* or between a *complex node* and its *atomic* or *stream* nodes, or between a *geographical complex node* and its *atomic, stream* nodes, or *geoAtomic* nodes.

$$\forall \langle m, \langle relational, RelName, I \rangle, n \rangle \in E$$
$$((\ell_{T_n}(m) = \ell_{T_n}(n) = complex) \ \lor$$
$$(\ell_{T_n}(m) = \ell_{T_n}(n) = geoComplex) \ \lor$$
$$(\ell_{T_n}(m) = complex \ \land$$
$$(\ell_{T_n}(n) = atomic \ \lor \ \ell_{T_n}(n) = stream \ )) \ \lor$$

$$(\ell_{T_n}(m) = geoComplex \ \wedge$$
$$(\ell_{T_n}(n) = atomic \ \vee \ \ell_{T_n}(n) = stream \ \vee \ \ell_{T_n}(n) = geoAtomic)))$$

4. *"Has Property" Edges.* Each *atomic* node is connected to its parent by an edge labeled *"Has Property"* and with temporal element $\mathcal{I}$.

$$\forall \langle m, label, n \rangle \ \in E$$
$$((\ell_{T_n}(m) = complex \ \vee$$
$$\ell_{T_n}(m) = geoComplex) \ \wedge$$
$$\ell_{T_n}(n) = atomic) \ \rightarrow$$
$$label = \langle relational, "HasProperty," \perp, \perp, \mathcal{I} \rangle$$

5. *"HasMMProperty" Edges.* Each *stream* node is connected to its parent by a relational edge with label formed by the name *"HasMMProperty,"* the temporal element $\mathcal{I}$, and the value $value_p$ that specifies the subpart of the media object.

$$\forall \langle m, label, n \rangle \ \in E$$
$$((\ell_{T_n}(m) = complex \ \vee$$
$$\ell_{T_n}(m) = geoComplex) \ \wedge$$
$$\ell_{T_n}(n) = stream)$$
$$\rightarrow label = \langle relational, "HasMMProperty," value_p, \perp, \mathcal{I} \rangle$$

6. *"HasGeoProperty" Edges.* Each *geographical atomic* node is connected to its parent by an edge labeled *"HasGeoProperty"* and with temporal element $\mathcal{I}$.

$$\forall \langle m, label, n \rangle \ \in E$$
$$(\ell_{T_n}(m) = geoComplex) \ \wedge$$
$$\ell_{T_n}(n) = geoAtomic) \ \rightarrow$$
$$label = \langle relational, "HasGeoProperty," \perp, \perp, \mathcal{I} \rangle$$

7. *Uniqueness of Edges.* At a specific time instant, at most one edge with a specific name *RelName* can exist between two nodes. This is due to the fact that an edge represents a relationship between two nodes; thus it makes no sense to represent the same relationship with two edges.

$$\forall \langle m, \langle relational, RelName, \mathcal{I}_1 \rangle, n \rangle \in E$$
$$(\nexists \langle m, \langle relational, RelName, \mathcal{I}_2 \rangle, n \rangle \in E \ (\mathcal{I}_1 \cap \mathcal{I}_2 \neq \emptyset))$$

8. *Temporal Elements.* Temporal elements are not associated to *atomic, stream,* and *geoAtomic* nodes, because in the database the temporal element of a value coincides with the temporal element of its ingoing edge.

$$\forall x \in N((\ \ell_{T_n}(x) = atomic \ \vee$$
$$\ell_{T_n}(x) = stream \ \vee$$
$$\ell_{T_n}(x) = geoAtomic) \rightarrow \ell_V(x) = \perp)$$

9. *Valid Time.* The VT of an edge between a complex node and an atomic node must be related to the VT of the complex node. Intuitively, the relation between a complex node and an atomic node cannot survive the complex node, thus the VT of the edge cannot contain time points that do not belong also to the VT of the complex node. This is due to the fact that we suppose that a complex node is related to its properties (atomic nodes) while it is valid.

$$\forall \langle m, \langle relational, RelName, \mathcal{I}\rangle, n\rangle \in E \triangleright$$
$$((\ell_{\mathcal{T}_n}(n) = atomic \ \lor \ \ell_{\mathcal{T}_n}(n) = stream \ \lor$$
$$\ell_{\mathcal{T}_n}(n) = geoAtomic) \to \ \ell_{\mathcal{V}_n}(m) \supseteq \mathcal{I})$$

10. *Coordinates.* Coordinates are associated only to *geoAtomic* nodes.

$$\forall x \in N((\ \ell_{\mathcal{T}_n}(x) = complex \ \lor$$
$$\ell_{\mathcal{T}_n}(x) = geoComplex \ \lor$$
$$\ell_{\mathcal{T}_n}(x) = atomic \ \lor$$
$$\ell_{\mathcal{T}_n}(x) = stream) \to \ell_{\mathcal{G}}(x) = \bot)$$

11. *GeoAtomic Nodes.* The name of a *geoAtomic* node can be only: *point* if the node represents a point in the space, *line* if the node represents a simple polyline, or *polygon* if the node represents a simple polygon defined by a closed polyline.

12. *Edges Between Geographical Complex Nodes.* Each relational edge between two *(geographical) complex* nodes has label $\langle relational, RelName, \bot, \bot, \mathcal{I}\rangle$, where *RelName* is the name associated to the relation and $\mathcal{I}$ is the temporal element.

$$\forall \langle m, label, n\rangle \in E$$
$$(\ell_{\mathcal{T}_n}(m) = geoComplex \ \land$$
$$\ell_{\mathcal{T}_n}(n) = geoAtomic) \to$$
$$label = \langle relational, RelName, \bot, \bot, \mathcal{I}\rangle$$

13. *Spatial Components.* Each *geographical complex* node must have a *geographical atomic* property or a *composed-by* edge to other geographical complex nodes. In particular, there cannot be any cycles in the graph formed by *geoComplex* nodes and *composed-by* edges.

14. *GeoField Nodes.* Each geographical property that is modeled by following the *field-based* approach is represented in a GeoMTGM graph by using a complex initial node $n \in N$ with label: $\langle complex, GeoField, \bot, \mathcal{I}\rangle$. Such *GeoField* node has a unique name described by the value of its atomic node *GeoFieldName*.

$$\forall m \in N((\ell_{\mathcal{T}_n}(m) = complex \land \ell_{\mathcal{L}_n}(m) = \text{``}GeoField\text{''}) \to$$
$$(\exists! \ n \in N(\exists\langle m, \langle relational, \text{``}HasProperty\text{''}, \mathcal{I}\rangle, n\rangle \in E$$
$$(\ell_{\mathcal{T}_n}(n) = atomic \land \ell_{\mathcal{L}_n}(n) = \text{``}GeoFieldName\text{''}))))$$

15. *Gographical Edges.* Each complex node *GeoField* has a set of outgoing edges named *geographical edges* that are connected to the geographical complex nodes

on which the geographical property has been measured. The label of these edges is ⟨*geographical, GeoFieldName, GeoFieldValue, I*⟩, where *GeoFieldName* is the name associated to the GeoField node, *GeoFieldValue* is the value measured on the particular geographical complex node, and I is the temporal element on which the property has been measured.   □

2.3.3 Translating GeoMTGM Graphs into XML Documents

We used GML [20] as a starting point for the definition of the rules to translate the geographical information, represented by means of GeoMTGM graphs, into XML documents. In particular, some tags and conventions have been extended and adopted to implement the translation rules defined in [10] for the MTGM data model.

GML elements are used for translating geoAtomic nodes into XML elements and are characterized by the following notation: `<gml:tagName>`, as depicted in Fig. 2.5.

In particular, point, line, and polygon nodes are translated into XML elements that extend the `PointType`, `LineStringType`, and `AbstractSurfaceType` defined in GML, respectively. In Fig. 2.5 the XML Schema for point, line, and polygon are shown; only the `PointWithTimeType` is described in detail, the other two types are very similar to this one.

In GML the coordinates of any geographical object are encoded either as a sequence of `coord` elements that encapsulate tuple components, or as a single string contained within a `coordinates` element. We use a `coord` element instead of a `coordinates` element, because it allows us to perform basic type checking and enforce constraints on the number of tuples that appear in a particular geometric instance.

The coordinates of a geometry are defined within some Spatial Reference System (SRS). In GML for all geometry elements, the SRS is specified by the attribute `srsName`; instead, in GeoMTGM this attribute must be specified only for the geometric atomic node nested in the root element, and it is optional for any other geometric entity. Moreover, unlike GML, the `id` attribute is required and not optional.

In the same way, for translating the VT information we have defined an element type `ValidTimeType` that is composed of a sequence of specialized GML elements `TimePrimitivePropertyType`, named `ValidIntervalType`, as shown in Fig. 2.6.

The overall main ideas underlying the designed translation technique can be summarized as follows:

- *(Geographical) Complex nodes* are translated into complex elements (i.e. elements that contain other elements); in particular, they have a (nested) element for the corresponding VT and an element for each outgoing edge.
- *Atomic nodes* are translated into mixed elements (i.e. elements containing both string values and other elements); in particular, they contain the string representing their value and an element for their VT (which is contained in the label of the ingoing relational edge).
- *Geographical atomic nodes* are translated into complex elements; in particular, they have some (nested) elements for the coordinates and an element for their VT (as for atomic nodes it is contained in the label of the ingoing relational edge).

```
. . . . . . . .
<xs:element name="Point" type="PointWithTimeType"/>
<xs:element name="Line" type="LineWithTimeType"/>
<xs:element name="Polygon" type="PolygonWithTimeType"/>

<xs:complexType name="PointWithTimeType">
    <xs:complexContent>
        <xs:restriction base="gml:PointType">
            <xs:sequence>
                <xs:element ref="gml:coord"/>
                <xs:element ref="ValidTime" maxOccurs="unbounded"/>
            </xs:sequence>
            <xs:attribute name="gid" use="required"/>
            <xs:attribute name="type" type="xs:string" use="required"/>
        </xs:restriction>
    </xs:complexContent>
</xs:complexType>

<xs:complexType name="LineWithTimeType">
    <xs:complexContent>
        <xs:restriction base="gml:LineStringType">
            <xs:sequence>
                <xs:element ref="gml:coord" minOccurs="2" maxOccurs="unbounded"/>
                <xs:element ref="ValidTime" maxOccurs="unbounded"/>
            </xs:sequence>
            <xs:attribute name="gid" use="required"/>
            <xs:attribute name="type" type="xs:string"/>
        </xs:restriction>
    </xs:complexContent>
</xs:complexType>

<xs:complexType name="PolygonWithTimeType">
    <xs:complexContent>
        <xs:restriction base="gml:AbstractSurfaceType">
            <xs:sequence>
                <xs:element ref="gml:coord" minOccurs="3" maxOccurs="unbounded"/>
                <xs:element ref="ValidTime" maxOccurs="unbounded"/>
            </xs:sequence>
            <xs:attribute name="gid" use="required"/>
            <xs:attribute name="type" type="xs:string"/>
        </xs:restriction>
    </xs:complexContent>
</xs:complexType>
. . . . . . . .
```

Fig. 2.5. The XML Schema for the GeoAtomic nodes

- *Edges between complex nodes* are represented through complex elements nested into the element corresponding to the complex node from which the edge originates. The element corresponding to the node the edge points to is referred through a suitable attribute in the element representing the edge.
- *Edges between a complex node and an atomic (stream) node* are not translated into an XML element but are represented by nesting the atomic (stream node) within the complex one.
- *Compound labels* are managed by introducing suitable subelements (i.e. nested elements), as for representing VTs of nodes and edges.

```
.........
<xs:element name="ValidInterval" type="ValidIntervalType"/>
<xs:element name="ValidTime" type="ValidTimeType"/>

<xs:complexType name="ValidIntervalType">
  <xs:complexContent>
    <xs:restriction base="gml:TimePeriodType">
      <xs:sequence minOccurs="1" maxOccurs="1">
        <xs:element name="vt_start" type="gml:TimeInstantPropertyType"
                    minOccurs="1" maxOccurs="1"/>
        <xs:element name="vt_end" type="gml:TimeInstantPropertyType"
                    minOccurs="1" maxOccurs="1"/>
      </xs:sequence>
    </xs:restriction>
  </xs:complexContent>
</xs:complexType>

<xs:complexType name="ValidTimeType">
  <xs:complexContent>
    <xs:restriction base="gml:TimePrimitivePropertyType">
      <xs:sequence minOccurs="1" maxOccurs="unbounded">
        <xs:element ref="ValidInterval"/>
      </xs:sequence>
    </xs:restriction>
  </xs:complexContent>
</xs:complexType>
.........
```

Fig. 2.6. The XML Schema for the VT information

- Elements corresponding to GeoMTGM nodes have an attribute (of type ID), which allows one to refer to them in an unambiguous way.
- A root element for the XML document is explicitly created; it contains a single element corresponding to the root of the GeoMTGM graph.

We have decided to use GML elements only for representing some constructs of the model. Actually, we could represent the whole graph using GML. However, considering the semistructural nature of GeoMTGM graphs and the impossibility of fixing a common schema for them, the trivial GML translation has to create a generic GML dynamic feature element for all the (geographical) complex nodes of the graph. Then add to it an element for representing the VT and a sequence of dynamic properties, representing generic atomic, stream, and geoAtomic nodes. Doing so, the name of the node becomes an attribute and not the name of the resulting XML element. In our opinion this translation is not satisfactory, because the resulting XML document reflects the graph structure more than the semantic information contained in it.

In the following, we will describe in detail the translation rules defined for a GeoMTGM graph.

Translation of a Complex Node with an Atomic Property

Let us suppose we have a (geographical) complex node c_i, with name $name\_c_i$ and VT $[t'_1, t'_2) \cup \cdots \cup [t'_i, t'_{i+1})$. c_i has an atomic node a_i, with name $name\_a_i$ and value $value\_a_i$ that describes a property of c_i. The edge between c_i and a_i has label

$$\langle HasProperty, [t_1, t_2) \cup \cdots \cup [t_n, t_{n+1}) \rangle.$$

The translation of c_i is shown in Fig. 2.7; in particular the edge between the complex node and its atomic node is translated by nesting the element representing the atomic node within the element representing the complex node.

Translation of a Complex Node with a Stream Node

Let us suppose we have a complex node c_i, with name $name\_c_i$ and VT $[t'_1, t'_2) \cup \cdots \cup [t'_i, t'_{i+1})$. c_i has a stream node a_i, with name $name\_a_i$ and value $value\_a_i$. The edge between c_i and a_i has label

$$\langle HasMMProperty, [t_1, t_2) \cup \cdots \cup [t_n, t_{n+1}), [Int\_start, Int\_end] \rangle.$$

The translation of c_i is shown in Fig. 2.8. Note that the edge between the complex node and its stream node is not translated into an XML element, adopting the same nesting strategy as the previous case.

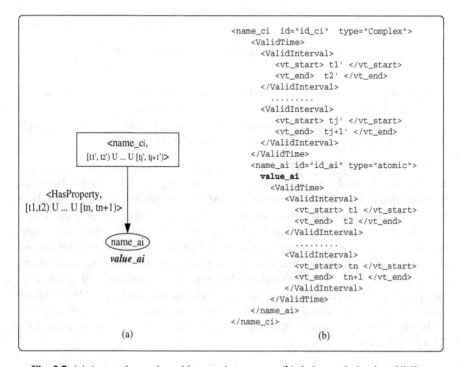

Fig. 2.7. (a) A complex node and its atomic property; (b) their translation into XML

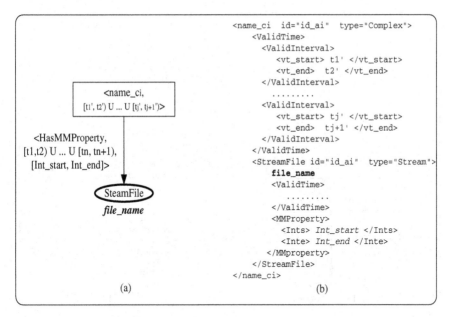

Fig. 2.8. (a) The complex node and its stream node; (b) their translation into XML

Translation of a Geographical Complex Node with a Geographical Atomic Node

Let us suppose we have a geographical complex node c_i, with name $name\_c_i$ and VT $[t_1, t_2) \cup \cdots \cup [t_i, t_{i+1})$. c_i has a geographical atomic node with value formed by a set of n coordinates $\langle (x_i, y_i) \cup \cdots \cup (x_n, y_n) \rangle$. The edge between c_i and a_i has label

$$\langle HasGeoProperty, [t1, t2) \cup \ldots \cup [t_i, t_{i+1}) \rangle$$

The translation of c_i is shown in Fig. 2.9; in particular the edge between the geographical complex node and its geographical atomic node is translated by nesting the element representing the geoAtomic node within the element representing the geoComplex node.

Translation of a Relational Edge Between Two Complex Nodes

Let us suppose we have two complex nodes c_i and c_j, with names $name\_c_i$ and $name\_c_j$, VT $[t'_1, t'_2) \cup \cdots \cup [t'_i, t'_{i+1})$ and $[t''_1, t''_2) \cup \cdots \cup [t''_j, t''_{j+1})$, respectively. The edge from c_i to c_j has label

$$\langle RelName, [t_1, t_2) \cup \ldots \cup [t_n, t_{n+1}) \rangle.$$

The translation of c_i and c_j is shown in Fig. 2.10. In this case, the two complex nodes are translated into XML elements as described above, and c_j is not nested into c_i. Moreover, the relational edge between the two complex nodes is translated into

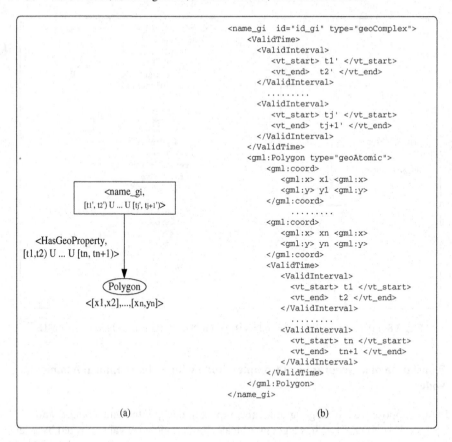

Fig. 2.9. (a) A geographical complex node and its geographical atomic property; **(b)** their translation into XML

an XML element with name *RelName*. In particular, the relational edge is nested into the XML element related to the complex node c_i.

Translation of a Geographical Edge Between a Complex Node and a Geographical Complex Node

We suppose to have a complex node that represents a geographical property with *name_p* and VT $[t_1, t_2) \cup \ldots \cup [t_i, t_{i+1})$, in the *field-based* model, and a geographical complex node c_j with name *name_c_i* and valid time $[t'_1, t'_2) \cup \ldots \cup [t'_i, t'_{i+1})$.

The geographical edge that connects the *GeoField* and c_i nodes has label

$$\langle geographical, name_p, value, [t1, t2) \cup \ldots \cup [t_i, t_{i+1}) \rangle$$

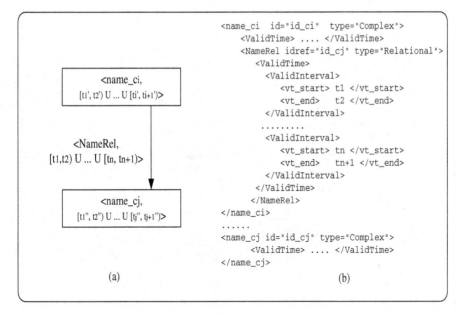

Fig. 2.10. (a) Two complex nodes and their relational edge; and **(b)** their translation into XML

and is translated into an XML element with name equal to the name of the geographical property and nested into the XML element related to the complex node *GeoField*. We specify the attribute *idref* to refer to the element c_i.

Figure 2.11 shows the translation of the complex node *GeoField* into XML code and its related "geographical edge" into XML elements.

Translation of a Composed-by Edge Between Two Geographical Complex Nodes

Let us suppose we have two geographical complex nodes g_i and g_j with names *name_g_i* and *name_g_j*, VT $[t'_1, t'_2) \cup \ldots \cup [t'_i, t'_{i+1})$ and $[t''_1, t''_2) \cup \ldots \cup [t''_j, t''_{j+1})$, respectively. The composed-by edge from g_i to g_j has label

$$\langle Composed - by, [t_1, t_2) \cup \ldots \cup [t_n, t_{n+1}) \rangle$$

The translation of g_i and g_j is shown in Fig. 2.12. In this case, the two complex nodes are translated into XML elements as described above, and g_j is not nested into g_i. Moreover, the composed-by edge between the two geographical complex nodes is translated into an XML element with name *Composed-by* and nested into the XML element related to the complex node g_i.

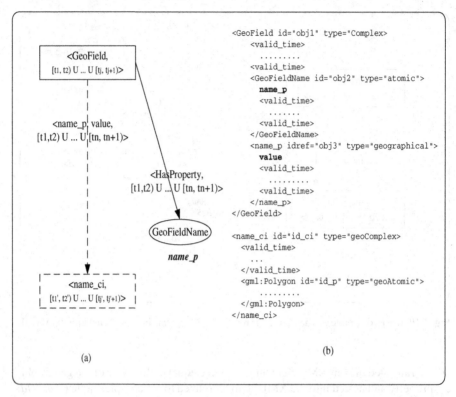

Fig. 2.11. (a) A GeoField complex node, its atomic property GeoFieldName, and the geographical property between it and a geographical complex node; **(b)** their translation into XML

2.4 The GeoMTGM Web Application

In this section we present a Web application, named GeoMtgmApp, that allows us to manage GeoMTGM graphs (create a new graph, load, and modify an existing one) giving a textual or graphical representation of them, and finally it allows us to translate GeoMTGM graphs into XML documents and store these into an XML native database.

2.4.1 The GeoMTGM Actions

The GeoMtgmApp Web application has been developed on the Java 2 Enterprise Edition platform [15] with the use of the Struts framework [1]. In particular, we have used the ActionServlet and the FrontController provided by Struts and have defined a custom `Action` object for every business operation we would perform on the model. All these custom `Action` classes are contained in the package `it.univr.geomtgm.actions`.

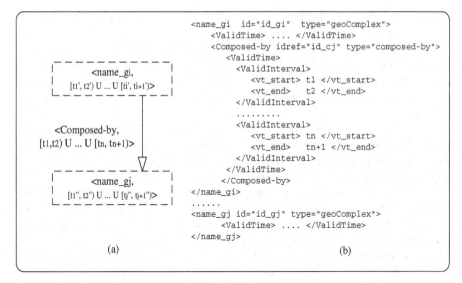

```
                          <name_gi  id="id_gi"  type="geoComplex">
                              <ValidTime> .... </ValidTime>
                              <Composed-by idref="id_cj" type="composed-by">
                                  <ValidTime>
                                      <ValidInterval>
                                          <vt_start> t1 </vt_start>
                                          <vt_end>   t2 </vt_end>
                                      </ValidInterval>
                                      .........
                                      <ValidInterval>
                                          <vt_start> tn </vt_start>
                                          <vt_end>   tn+1 </vt_end>
                                      </ValidInterval>
                                  </ValidTime>
                              </Composed-by>
                          </name_gi>
                          ......
                          <name_gj id="id_gj" type="geoComplex">
                              <ValidTime> .... </ValidTime>
                          </name_gj>
```

(a)   (b)

Fig. 2.12. (a) Two geographical complex nodes, their composed-by edge; **(b)** their translation into XML

The business operations provided by this application deal with the insertion and deletion of nodes and edges, the change of VT, atomic value or geographical coordinates, the visualization of graph representation, the translation from and into XML document, and finally the communication with the database.

In general, a GeoMTGM action deals with the retrieval of the data inserted into an HTML form, the population of a particular JavaBean, and the invocation of several service methods. In particular, the GeoService class represents the service that really performs the operation on the model, while the DatabaseService class provides some methods to load or store XML documents into the eXist [11] XML database.

The GeoMTGM graph is maintained in the current session and every operation on it is performed and stored in the session itself. The database is used only to load a previously saved database or to save the graph permanently.

The HTML form is defined with the use of the Struts html taglib and the data here inserted are retrieved via some ActionForm objects that transfer them to the Action class. In particular, we have defined a custom ActionForm object for every needed HTML form.

2.4.2 Representation of the GeoMTGM Graph

For the representation of a GeoMTGM graph we have used the JGraphT [17] library. In particular, we have defined some specialized node, edge, and graph objects that are contained into the it.univr.geomtgm.model.-node, it.univr.geomtgm.model.edge and it.univr.geomtgm.model.graph packages, respectively. In Fig. 2.13 a class diagram for the main package `it.univr.geomtgm.model` is shown.

The `it.univr.geomtgm.model.graph` package contains the `GeoMtgm-Concrete` class that defines some methods for creating and populating a GeoMTGM graph. In particular, these methods allow one to define the root element, add nodes, establish new relations between them, delete, and modify nodes and edges.

This class extends the `DirectMultigraph` class offered by the JGraphT library. The use of a multigraph allows the definition of several directed edges between the same pair of nodes to represent into a unique graph both relational, geographical and multimedia information.

The package `it.univr.geomtgm.model.node` contains a set of classes for the definition of the various node types that appear in a GeoMTGM graph. In particular, the class `Node` represents the base class for all other types of nodes and is characterized by the attributes and methods shared by all nodes of the graph.

The atomic nodes are represented by the `AtomicNode` class from which the `StreamNode` class for the representation of stream nodes derives.

The geographical atomic nodes have been implemented by a hierarchy of types. In particular, the `GeoAtomicNode` class defines a generic geographical atomic node from which the `PointNode`, `LineNode`, and `PolygonNode` classes for the representation of geographical point, line, and polygon object, respectively, derive.

As regards the complex nodes, they are represented by the `ComplexNode` class from which the `GeoComplexNode` class for the representation of geographical complex nodes derives. This class shares the same attributes of the parent class, but it has been distinguished in order to define some constraints for the edge definition. In

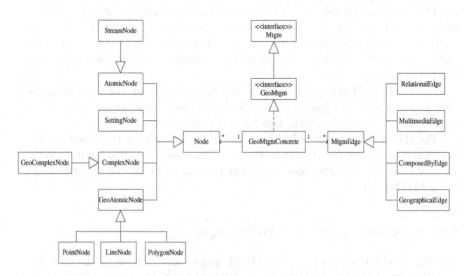

Fig. 2.13. The class diagram of the package `it.univr.geomtgm.model`

particular, only geographical complex nodes can have relational edges to geographical atomic nodes.

The it.univr.geomtgm.model.edge package contains a set of classes for the definition of directed labeled edges of a GeoMTGM graph. In particular, the MtgmEdge defines a generic edge and it inherits from the DirectedEdge class of the JGraphT library.

From this class the RelationalEdge, MultimediaEdge, ComposedBy Edge, and GeographicalEdge classes for the definition of relational, multimedia, composed-by, and geographical edges, respectively, derive.

2.4.3 Translation From and Into XML and the Application Layout

The it.univr.geomtgm.parser package contains a set of classes that offer static methods for the translation of a GeoMTGM graph into an XML document and for the construction of a graph from an XML document, according to the translation rules defined in the previous section.

The parsing is performed by the use of the JDOM [16] interface; in particular a GeoMTGM graph is first converted into a JDOM Document object and then translated into an XML document. Conversely from an XML file is first constructed a JDOM Document object and then the graph is rebuilt.

We have organized and assembled the content and the layout of JSP pages via the use of the Tiles framework, included with the Struts distribution. This framework provides a templating mechanism that allows us to separate the responsibilities of

Fig. 2.14. The GeoMtgmApp home page

layout from those of content. In particular, instead of being hardcoded within the layout page, the content is supplied or "passed" as parameters to the layout page at runtime.

In Fig. 2.14 the home page of the Web application from which it is possible to create a new graph or load an existing one is shown.

2.5 Conclusions

In this work we have proposed a new data model for representing spatiotemporal semistructured data considering the VT of the represented information. It is called GeoMTGM; it is an extension of the MTGM model and is based on a formalism using the graph notation.

We also show a possible representation of a GeoMTGM graph in XML by using an extension of the XML elements proposed by GML 3.0; in particular, we extended GML elements for adding VT to points, lines, and polygons. Moreover, we show by an example the effectiveness of this model in specifying the spatiotemporal data that describe the state of a transport network that considers the Highway A1 of the Italian highway network. Finally, we implemented a tool for storing GeoMTGM graphs in a XML-native database and for managing graphs with a Web application.

Future work will regard the introduction of transaction time in the data model, and the design and implementation of a query language that allows one to specify queries on GeoMTGM graphs, considering both the expressiveness and the syntax of widely known languages, such as XPath, XQuery and OQL, for XML data and objects.

Acknowledgments

This work has been partially supported by the Department of Computer Science, University of Verona, Italy.

References

1. The Apache Software Foundation (Apache Struts Web Application Framework). http://struts.apache.org/
2. Abiteboul S. (1997) Querying Semi-Structured Data. In: Afrati F N Kolaitis P (eds.) Proc. 6th International Conference on Database Theory (ICDT'97), Delphi, Greece, Lecture Notes in Computer Science Vol 1186, Springer, 1–18
3. Abiteboul S, Buneman P, Suciu D (1999) Data on the Web. From Relations to Semistrucutred Data and XML. Morgan Kaufmann Publishers, Los Altos, USA
4. Bertino E, Ferrari E (1998) Temporal Synchronization Models for Multimedia Data. IEEE Transactions on Knowledge and Data Engineering 10:612–631

5. Cárdenas AF, Dionisio JDN (1998) A Unified Data Model for Representing Multimedia, Timeline and Simulation Data. IEEE Transactions on Knowledge and Data Engeneering 10:746–767

6. Chawathe S, Abiteboul H, Widom J (1999) Managing historical semistructured data. Theory and Practice of Object Systems 5:143–162

7. Clementini E, di Felice P, van Oosterom P (1993) A Small Set of Formal Topological Relationships Suitable for End-User Interaction. In: Proc. of SSD'93, Lecture Notes in Computer Science Vol 692, Springer, 277–295

8. Combi C (2000) Modeling Temporal Aspects of Visual and Textual Objects in Multimedia Databases. In: Proc. 7th International Workshop on Temporal Representation and Reasoning (TIME 2000), IEEE Computer Society Press, 59–68

9. Combi C, Montanari A (2001) Data Models with Multiple Temporal Dimensions: Completing the Picture. In: Proc. of CAiSE'01, Lecture Notes in Computer Science Vol 2068, Springer, 187–202

10. Combi C, Oliboni B, Rossato R (2005) Merging Multimedia Presentations and Semistructured Temporal Data: a Graph-based Model and its Application to Clinical Information. Artificial Intelligence in Medicine 34:89–112

11. exist. http://exist.sourceforge.net/.

12. Gibbs S, Breiteneder C, Tsichritzis D (1994) Data Modeling of Time-Based Media. In: Snodgrass R T, Winslett M (eds) ACM Special Interest Group on Management of Data (SIGMOD 1994), Vol 23, Minneapolis, USA, ACM Press, 91–102

13. Geography Markup Language. World Wide Web Consortium W3C. www.w3.org/Mobile/posdep/Presentations/GMLIntro/

14. Huang B, Yi S, Chan WT (2004) Spatio-temporal information integration in xml. Future Generation Comp. Syst. 20:1157–1170

15. Java 2 Platform Enterprise Edition. http://java.sun.com/j2ee/

16. Hunter J, McLaughlin B JDom. http://www.jdom.org

17. Naveh B JGraphT a free Java graph library. http://jgrapht.sourceforge.net

18. Jensen C, Snodgrass R (1999) Temporal Data Management. IEEE Transactions on Knowledge and Data Engineering 11:36–44

19. Mirbel I, Pernici B, Sellis T, Tserkezoglou S, Vazirgiannis M (2000) Checking the Temporal Integrity of Interactive Multimedia Documents. VLDB Journal: Very Large Data Bases 9:111–130

20. OpenGeoSpatial Consortium (2004), OpenGIS, Geography Markup Language (GML) Implementation Specification. Version 3.1.1, http://www.opengeospatial.org/docs/03-105r1.pdf

21. OpenGeoSpatial Consortium (1999) OpenGIS Simple Features Specification for SQL. Technical Report OGC 99-049

22. Oliboni B, Quintarelli E, Tanca L (2001) Temporal aspects of semistructured data. In: Proc. 8th International Symposium on Temporal Representation and Reasoning (TIME-01), Cividale del Friuli, Italy, IEEE Computer Society Press, 119–127

23. Worboys MF (1995) GIS: a computing perspective. Taylor & Francis

24. Wide Web Consortium. XML http://www.w3.org/xml/1.0. http://www.w3.org/xml/

References containing URLs are validated as of September 1st, 2006.

3

Out-of-core Multiresolution Terrain Modeling

Emanuele Danovaro[1,2], Leila De Floriani[1,2], Enrico Puppo[1], and Hanan Samet[2]

[1] University of Genoa, Genoa (Italy)
[2] University of Maryland, College Park (USA)

3.1 Introduction

In this chapter we discuss issues about level of detail (LOD) representations for digital terrain models and, especially, we describe how to deal with very large terrain data sets through out-of-core techniques that explicitly manage I/O operations between levels of memory. LOD modeling in the related context of geographical maps is discussed in Chaps. 4 and 5.

A data set describing a terrain consists of a set of elevation measurements taken at a finite number of locations over a planar or a spherical domain. In a digital terrain model, elevation is extended to the whole domain of interest by averaging or interpolating the available measurements. Of course, the resulting model is affected by some approximation error, and, in general, the higher the density of the samples, the smaller the error. The same arguments can be used for more general two-dimensional scalar or vector fields (e.g. generated by simulation), defined over a manifold domain, and measured through some sampling process.

Available terrain data sets are becoming larger and larger, and processing them at their full resolution often exhibits prohibitive computational costs, even for high-end workstations. Simplification algorithms and multiresolution models proposed in the literature may improve efficiency, by adapting resolution on-the-fly, according to the needs of a specific application [32]. Data at high resolution are preprocessed once to build a multiresolution model that can be queried online by the application. The multiresolution model acts as a black box that provides simplified representations, where resolution is focused on the region of interest and at the LOD required by the application. A simplified representation is generally affected by some approximation error that is usually associated with either the vertices or the cells of the simplified mesh.

Since current data sets often exceed the size of the main memory, I/O operations between levels of memory are often the bottleneck in computation. A disk access is about one million times slower than an access to main memory. A naive management of external memory, for example, with standard caching and virtual memory policies, may thus highly degrade the algorithm performance. Indeed, some computations are inherently non-local and require large numbers of I/O operations. Out-of-core algorithms and data structures explicitly control how data are loaded and how they are stored. Here, we review methods and models proposed in the literature for simplification and multiresolution management of huge datasets that cannot be handled in main memory. We consider methods that are suitable to manage terrain data, some of which have been developed for more general kinds of data (e.g. triangle meshes describing the boundary of 3D objects).

The rest of this chapter is organized as follows. In Sect. 3.2, we introduce the necessary background about digital terrain models, focusing our attention on triangulated irregular networks (TINs). In Sect. 3.3, we review out-of-core techniques for simplification of triangle meshes and discuss their application to terrain data to produce approximated representations. In Sect. 3.4, we review out-of-core multiresolution models specific for regularly distributed data, while in Sect. 3.5, we describe more general out-of-core multiresolution models that can manage irregularly distributed data. In Sect. 3.6, we draw some conclusions and discuss open research issues including extensions to out-of-core simplification and multiresolution modeling of scalar fields in three and higher dimensions, to deal, for instance, with geological data.

3.2 Digital Terrain Models

There exist two major classes of digital terrain models, namely digital elevation models (DEMs) and triangulated irregular networks (TINs).

A DEM consists of a regular tiling of a planar domain into rectangular cells, called "pixels" with elevation values attached to pixels, that provide a piecewise constant approximation of terrain surface. DEMs are most often defined over rectangular domains encoded just as two-dimensional arrays of elevation values, which are easily geo-referenced by defining an origin, an orientation, and the extent of the cells in the tiling.

A TIN consists of a subdivision of the domain into triangles (i.e. a triangle mesh). Triangles do not overlap, and any two triangles are either disjoint or may share exactly either one vertex or one edge. Elevation samples are attached to vertices of triangles, thus each triangle in the mesh corresponds to a triangular terrain patch interpolating measured elevation at its vertices. Linear interpolation is usually adopted at each triangle, hence the resulting surface is piecewise-linear. More sophisticated interpolation models can also be used without changing the underlying domain subdivision. Data structures used to encode TINs are more sophisticated and expensive than those used for DEMs, but TINs have the advantage of being adaptive. A high density of samples may be adopted over more irregular portions of a terrain, while other portions that are relatively flat may be represented at the same accuracy with a small number of samples.

Multiresolution DEMs usually consist of a collection of grids at different resolutions. Well-known multiresolution representations for DEMs are provided by region quadtrees or pyramids [40]. One recent example in computer graphics is provided in the work of Losasso and Hoppe [31], where fast rendering of a large area of terrain is supported through a compressed pyramid of images that can be handled in main memory. Fast decompression algorithms and an effective use of a graphics processing unit (GPU) allow traversing the pyramid, by extracting data at the appropriate resolution, and rendering them on the fly.

In most cases, however, multiresolution models require use of adaptive representations in the form of TINs, even when full-resolution data come in the form of a DEM. For this reason, we consider here techniques that process and produce TINs at all intermediate levels of resolution. If a DEM at high resolution is given as input, most of the techniques developed in the literature consider a TIN built as follows. The DEM is interpreted as a regular grid where elevation values are attached to the vertices (usually placed at the center of pixels) and each rectangular cell is subdivided into two triangles through one diagonal. This model carries the same information as the DEM but is fully compatible with the structure of a TIN. Lower resolution models are TINs obtained adaptively from the full-resolution model through some terrain simplification procedure, as described in the following section.

3.3 Out-of-core Terrain Simplification

Dealing with huge amounts of data is often a difficult challenge. An obvious workaround is to reduce the data set to a more manageable size. Simplification algorithms take a terrain model as input and produce a simplified version of it, which provides a coarser (approximated) representation of the same terrain, based on a smaller data set.

Many simplification algorithms have been proposed in the literature (see, e.g. [16] for a survey of specific methods for terrain, and [32] for a survey of more general methods for polygonal 3D models). Most methods are based on the iterated or simultaneous application of local operators that simplify small portions of the mesh by reducing the number of vertices. Most popular techniques are based on the following:

- *Clustering.* A group of vertices that lie close in space, and the submesh that they define, are collapsed to a single vertex, and the portion of mesh surrounding it is warped accordingly.
- *Vertex decimation.* A vertex is eliminated, together with all its incident triangles, and the hole is filled by a new set of triangles.
- *Edge collapse.* An edge is contracted so that its two vertices collapse to a single point (which may be either one of them or a point at a new position computed to reduce approximation error), and its two incident triangles collapse to edges incident at that point; the portion of mesh surrounding such two triangles is warped accordingly.

Classical techniques require that the model is completely loaded in main memory. For this reason, such techniques cannot be applied to huge data sets directly. Simplification of huge meshes requires a technique that is able to load and process at each step a subset of the data that can fit in main memory. Existing out-of-core techniques have been developed mainly for simplification of triangle meshes bounding 3D objects, and can easily be adapted in order to simplify TINs that are usually simpler to handle. Out-of-core simplification algorithms can be roughly subdivided into the following three major classes:

- Methods based on vertex clustering
- Methods based on space partition
- Streaming techniques

In the following subsections, we review the main contributions in each class.

3.3.1 Algorithms Based on Clustering

Rossignac and Borrel [39] proposed an in-core method for the simplification of triangle meshes embedded in the three-dimensional Euclidean space. Their algorithm subdivides the portion of 3D space on which the mesh is embedded into buckets by using a uniform grid and collapses all vertices inside each bucket to a new vertex, thus modifying the mesh accordingly. In [27], Lindstrom proposes an out-of-core version of the above method that has a lower time and space complexity, and improves mesh quality. The input mesh is kept in secondary memory, while only the portion of the mesh within a single bucket is loaded in main memory. However, the whole output mesh is assumed to fit in main memory. In [28], Lindstrom and Silva propose further improvements over the method in [27] that increase the quality of approximation further and reduce memory requirements. The new algorithm removes the constraint of having enough memory to hold the simplified mesh, thus supporting simplification of really huge models. The work in [28] also improves the quality of the mesh, preserving surface boundaries and optimizing the position of the representative vertex of a grid cell.

Another extension of Rossignac and Borrel's approach is the algorithm by Shaffer and Garland [42]. This algorithm makes two passes over the input mesh. During the first pass, the mesh is analyzed and an adaptive space partitioning, based on a BSP tree, is performed. Using this approach, a larger number of samples can be allocated to more detailed portions of the surface. However, their algorithm requires more RAM than the algorithm by Lindstrom to maintain a BSP tree and additional information in-core.

Garland and Shaffer in [17] present a technique that combines vertex clustering and iterative edge collapse. This approach works in two steps: the first step performs a uniform vertex clustering, as in the methods described above. During the second step, edge collapse operations are performed iteratively to simplify the mesh further, according to an error-driven criterion. The assumption is that the mesh obtained after the first step is small enough to perform the second step in main memory.

Note that all these methods have been developed for simplifying triangle meshes representing objects in 3D space. Adaptation to terrain data is straightforward; it is sufficient to simplify the triangle mesh subdividing the domain that describes the structure of the TIN, while elevation values are used just to perform error computation. Therefore, for terrains, it is sufficient to partition the domain of the TIN with a 2D grid subdividing its domain into buckets.

Methods based on clustering are fast, but they are not progressive. The resolution of the simplified mesh is a priori determined by the resolution of the regular grid of buckets and no intermediate representations are produced during simplification. For this reason, the algorithms in this class are not suitable to support the construction of multiresolution models that will be discussed in Sect. 3.5.

3.3.2 Algorithms Based on Space Partitioning

The approach to simplification based on space partitioning consists of subdividing the mesh into patches, each of which can fit into main memory, and then simplifying each patch with standard techniques. Attention must be paid to patch boundaries to maintain the topological consistency of the simplified mesh.

The method proposed by Hoppe [20] starts from a regular grid and partitions the domain by using a PR quadtree subdivision based on the data points. The quadtree is built in such a way that data points in each leaf node fit in main memory. Simplified TINs are built bottom-up as described below. A full-resolution TIN is built by connecting the vertices of the input grid inside each leaf, and this is simplified iteratively by applying edge collapse. Only internal edges can be collapsed, while edges on the boundary of each patch are left unchanged. Once the meshes corresponding to the four siblings of a node A in the quadtree are reduced to a manageable size, they are merged into a mesh that will be associated with node A, and the simplification process is repeated recursively. The fact that edges on the boundaries of the quadtree blocks are frozen leads to a somehow unbalanced simplification, because data along the boundaries between adjacent blocks are maintained at full resolution even in the upper levels of the quadtree.

This technique was later generalized by Prince [37] to arbitrary TINs, whose vertices do not necessarily lie on a regular grid. While conceptually simple, the time and space overhead of partitioning the TIN and of later stitching the various pieces together leads to an expensive in-core simplification process, making such method less suitable for simplifying very large meshes.

El-Sana and Chiang [12] propose an algorithm that works on irregularly distributed data. They partition the mesh at full resolution into patches, where each patch is bounded by chains of edges of the triangle mesh. Patches are sized in such a way that a few of them can be loaded in main memory if necessary. Simplification of a single patch is performed by iteratively collapsing the shortest internal edge of the corresponding triangle mesh. Simplification of a patch is interrupted when its shortest edge lies on its boundary to preserve matching between adjacent patches. Once all patches have been simplified independently, the shortest edge of the mesh lies on the common boundary between two patches. Two such patches are merged in main memory and edge collapse is restarted. In this way, the result of simplification is consistent with that obtained in-core by collapsing at each step the shortest edge of the mesh. This method has been used to build a multiresolution model as described in Sect. 3.5.

Magillo and Bertocci [33] present a method specific to a TIN. It subdivides the TIN into patches that are small enough to be simplified individually in-core. They take a different approach to preserve interpatch boundaries. The skeleton of edges that defines the boundary of the patches in the decomposition is simplified first through an algorithm for line simplification based on vertex removal. This maintains the consistency of patch boundaries through different levels of detail. The history of simplification of each chain of edges forming a boundary line is maintained. Then the interior of each patch is also simplified independently through vertex removal. Removal of internal and boundary vertices can be interleaved to obtain a more uniform simplification process.

Cignoni et al. [3] propose a simplification method for triangle meshes in 3D space based on a hierarchical partition of the embedding space through the use of octrees. Octree subdivision stops when the set of triangles associated with a leaf fits in a disk page. The portion of the triangle mesh contained in each octree leaf is independently simplified through iterative edge collapse. Once the leaves are simplified, they can be merged and simplified further. The problem caused by avoiding performing edge collapses on the boundary of the patches, as in [19], is overcome. Vertices and edges of the mesh do not lie on the faces of quadtree blocks, while only edges that cross the boundary between adjacent blocks may exist. Such interblock edges cannot be collapsed while independently simplifying the patches, but they can be either stretched or identified in pairs because of the other edge collapses occurring inside the patches. Thus, the independent simplification of the interior of one patch is interleaved with some simplification on the strip of triangles joining it to its adjacent patches, thus preserving interpatch consistency. This method can be easily adapted to the special case of terrain data by using quadtrees instead of octrees to partition the domain of the TIN. Note that the input data need not be regularly distributed. In general, the vertices of the mesh will not lie on the boundary of quadrants of the quadtree.

3.3.3 Streaming Algorithms

The philosophy underlying streaming techniques is that a strictly sequential processing order is followed, where each datum is loaded only once to main memory such that the result is written to secondary memory as soon as possible.

Isenburg et al. [23] present a simplification technique for triangle meshes that takes a sequential indexed representation of the mesh as input in which vertices and triangles are suitably interleaved. This representation may also be built from more common mesh data structures, such as triangle soups or indexed data structures, through preprocessing techniques also based on streaming [22]. The algorithm streams very large meshes through main memory, but at each step, only a small portion of the mesh is kept in-core. Mesh access is restricted to a fixed traversal order, but full connectivity and geometry information is available for the active elements of the traversal. The simplification step is performed only on the portion of the mesh loaded in main memory.

Note that streaming algorithms, as those based on clustering, are not progressive and cannot be used to build multiresolution models.

Table 3.1. Comparison among simplification algorithms for triangle meshes

	data set	partitioning	simplification	multires	space
Lindstrom [27]	tri mesh	regular grid	vertex clustering	no	$O(out)$
Lindstrom et al. [28]	tri mesh	regular grid	vertex clustering	no	$O(1)$
Shaffer et al. [42]	tri mesh	BSP tree	vertex clustering	no	$O(out)$
Garland et al. [17]	tri mesh	regular grid	clust.+collapse	yes	$O(out)$
Hoppe [20]	DEM	space part.	edge collapse	yes	$O(1)$
Prince [37]	TIN	space part.	edge collapse	no	$O(1)$
El-Sana et al. [12]	tri mesh	greedy dec.	edge collapse	yes	$O(1)$
Magillo et al. [33]	tri mesh	user defined	vertex removal	yes	$O(1)$
Cignoni et al. [3]	tri mesh	space part.	edge collapse	yes	$O(1)$
Isenburg et al. [23]	tri mesh	streaming	various	no	

3.3.4 Comparison

Table 3.1 summarizes the main features of out-of-core simplification techniques. For each method, we list the following: the type of input (*Data set*); the type of space partition adopted (*Partitioning*); the type of approach (*Simplification*); the possibility to build a multiresolution model based on the specific simplification technique (*Multires*); and the space requirements in main memory (*Space*). The algorithms by Hoppe [20], Prince [37], and Magillo and Bertocci [33] are designed for either regular or irregular terrain data, while all the other techniques can handle arbitrary triangle meshes describing the boundary of 3D objects. When applied to terrain simplification, algorithms based on space partitioning should be modified by replacing the octree that partitions 3D space with a simpler quadtree partition of the 2D domain of the TIN. Note that only methods based on space partitioning are suitable to build multiresolution models.

3.4 Out-of-core Representation of Regular Multiresolution Models

Multiresolution terrain models that work on data regularly distributed on a grid have been proposed in the literature [10, 13, 18, 25, 26, 35]. Such models are based on a nested subdivision that starts from a simple regular tiling of the terrain domain into regular triangles and is generated through a refinement process defined by the uniform subdivision of a triangle into scaled copies of it. The two most common refinement operators used for generating regular multiresolution models are *triangle quadrisection* and *triangle bisection*.

The quadrisection of a triangle t consists of inserting a new vertex on each edge of t in such a way that the original triangle t is split into four subtriangles (see Fig. 3.1 (a,b)). The resulting (nested) hierarchy of triangles is encoded as a quadtree, called a "triangle quadtree" (see Fig. 3.1 (c)). A triangle quadtree is used for terrain

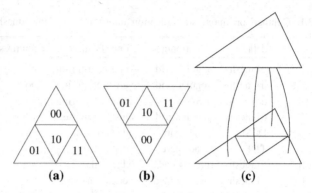

Fig. 3.1. The quadrisection of a triangle oriented: (a) tip-up; (b) tip-down; (c) an example of a triangle quadtree

data distributed on the plane or on the sphere. In this latter case, the idea is to model (i.e. to approximate) the sphere with a regular polyhedron, namely one of the five Platonic solids (i.e. tetrahedron, hexahedron, octahedron, dodecahedron, and icosahedron that have 4, 6, 8, 12, and 20 faces, respectively), and then to subdivide the surface of the polyhedron using regular decomposition. In the case of the tetrahedron, octahedron, and icosahedron, the individual faces of the solid are triangles and they are in turn represented by a triangle quadtree that provides a representation that has both a variable and a multiple resolution variant. Clearly, the fact that the icosahedron has the most faces of the Platonic solids means that it provides the best approximation to a sphere and consequently has been studied the most (e.g. [14, 15, 25]). The goal of these studies has been primarily to enable a way to rapidly navigate between adjacent elements of the surface (termed "neighbor finding"). However, the methods by Lee and Samet [25] are not limited to the icosahedron and, in fact, are also applicable to the tetrahedron and octahedron. In particular, neighbor finding can be performed in worst-case constant time on triangle quadtrees.

Most of the regular multiresolution terrain models are based on *triangle bisection*. The square domain is initially subdivided in two right triangles. The bisection rule subdivides a triangle t into two similar triangles by splitting t at the midpoint of its longest edge (see Fig. 3.2 (a)). The recursive application of this splitting rule to the subdivided square domain defines a binary tree of right triangles in which the children of a triangle t are the two triangles obtained by splitting t. The multiresolution model generated by this subdivision rule is described by a forest of triangles, called a "triangle bintree." Each node in a triangle bintree represents a triangle t generated in the recursive subdivision, while the children of node t describe the two triangles arising from the subdivision of t (see Fig. 3.2 (b)).

If vertices are available at all nodes of the supporting regular grid, then a triangle bintree consists of two full trees that can be represented implicitly as two arrays [10, 13]. On the contrary, if data are available at different resolutions over different parts of the domain, then the binary forest of triangles is not complete, and

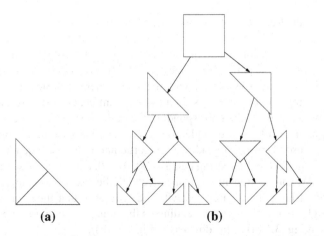

Fig. 3.2. (a) The bisection of a triangle; (b) An example of a triangle bintree

dependencies must be represented explicitly, thus resulting in a more verbose data structure [18].

When extracting adaptive TINs from a triangle bintree, we need to guarantee that whenever a triangle t is split, the triangle adjacent to t along its longest edge is also split at the same time. In this way, the extracted mesh will be conforming that is, it will not contain cracks. This can be achieved through the application of neighbor finding techniques to the hierarchy. In [13], an algorithm for neighbor finding is proposed that works in worst-case constant time. An alternative approach consists of using an error saturation technique, which through manipulation of the errors associated with the triangles in the hierarchy, allows for extracting conforming meshes (but not with a minimal number of triangles) without the need for neighbor finding [34]. Other representations for multiresolution models generated through triangle bisection have been proposed that encode the model as a directed acyclic graph (DAG) of atomic updates, each of which is called a "diamond," formed by pairs of triangles that need to be split at the same time [26, 35, 38].

The space requirements of a hierarchical representation can be reduced through the use of pointerless tree representations, also known as "compressed representations." There are a number of alternative compressed representations for a binary tree, and, in general, for a tree with fanout equal to 2^d (i.e. a quadtree for $d = 2$ and an octree for $d = 3$). One simple method, known as a "DF expression" [24], makes use of a list consisting of the traversal of the tree's constituent nodes, where a one bit code is used to denote if the corresponding node is a non-leaf or a leaf node. This method is very space-efficient but does not provide for random access to nodes without traversing the list from the start for each query. Thus, most implementations make use of structures that are based on finding a mapping from the cells of the domain decomposition to a subset of the integers (i.e. to one dimension).

In the case of a triangle bintree or triangle quadtree, this mapping is constructed by associating a *location code* with each node that corresponds to a triangle element (i.e. noblock) in the tree. A location code for a node implicitly encodes the node as a bit string consisting of the path from the root of the tree to that node, where the path is really the binary (1 bit) or quaternary (2 bits) representation of the transitions that have been made along the path. If the node is at level (i.e. depth) i in the tree, a string of length d_i binary digits is associated with the node, where each step in the descent from the root is represented by d bits. In a triangle bintree, each step is encoded as 0(1) depending on whether the corresponding arc in the tree leads to the left (right) child of its parent, while in a triangle quadtree a labeling scheme is applied that extends the one used for region quadtrees. In particular, in a region quadtree where the blocks are square, the 2-bit string patterns 00, 01, 10, and 11 are associated with the NW, NE, SW, and SE transitions, respectively. In the case of a triangle quadtree, the same 2-bit string patterns are used with the difference that they correspond to different triangles in the hierarchy depending on the triangle orientation (i.e. whether it is tip-up as in Fig. 3.1 (a) or tip-down as in Fig. 3.1 (b)).

Note that the location codes in a square quadtree are equivalent to the Z order (or Morton order) (e.g. see [41]), which is an ordering of the underlying space in which the result is a mapping from the coordinate values of the upper-left corner u of each square quadtree block to the integers. The Morton order mapping consists of concatenating the result of interleaving the binary representations of the coordinate values of the upper-left corner (e.g. (a, b) in two dimensions) and i of each block of size 2^i so that i is at the right. In the case of a triangle quadtree, the analog of a Z or Morton order can still be constructed but there is no interpretation in terms of bit interleaving as can be seen by examining the three-level labeling of the triangle quadtree in Fig. 3.3 that is three levels deep. If we record the depth of the tree at which the node is found and append it to the right of the number corresponding to the path from the root to the node thereby forming a more complex location code, then the result of sorting the resulting location codes of all the nodes in increasing order yields the equivalent of a depth-first traversal of the tree. If we vary the format of the resulting location codes so that we record the depth of the tree at which the node is found on the left (instead of on the right) and the number corresponding to the path from the root to the node is on the right, then the result of sorting the resulting location codes of all the nodes in increasing order yields the equivalent of a breadth-first traversal of the tree access structure [2]. These depth-first and breadth-first traversal characterizations are also applicable to triangle quadtrees.

Location codes are used for performing neighbor finding efficiently as well as to retrieve the vertices of a triangle, the value of the field associated with a vertex, and so on. An efficient implementation involving arithmetic manipulation and a few bit operations allows performing such computations in constant time [13, 25].

Out-of-core representations of triangle quadtrees and bintrees are based on encoding the location codes of both internal and leaf nodes in an external memory index such as a B-tree, as done for encoding a quadtree or an octree in external memory.

In [18], Gerstner presents a compressed representation of a triangle bintree that works in main memory, but it could be implemented to provide an effective out-of-core representation. One of the most interesting features of this approach is

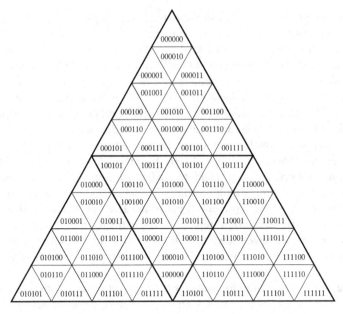

Fig. 3.3. Labeling of a triangle quadtree that is three levels deep

that the triangle bintree is not necessarily a forest of two full trees as in the implicit in-core representations commonly used for triangle bintrees [10, 13].

In [29], Lindstrom and Pascucci present an implementation of a triangle bintree in external memory with the purpose of visualizing huge sets of terrain data at uniform and variable resolutions. The triangle bintree is represented by using two interleaved quadtrees that appear at alternating levels. The first quadtree is aligned with the domain boundary, while the other one is rotated to 45°. The two quadtrees are encoded level-by-level, and the resolution increases with the level. Then, the out-of-core handling of data is left to the file paging system of the operating system (data are simply loaded through an mmap system call). Two different data layouts are proposed: the first one corresponds to a row-by-row traversal, while the second one is based on the Z-order. The latter appears to be more efficient even though it is more complex. In this approach, a version of the triangle bintree with error saturation is encoded to avoid neighbor finding to extract topologically consistent TINs.

3.5 Out-of-core Multiresolution Models Based on Irregular Meshes

Many multiresolution models based on irregular triangle meshes have been proposed for both terrain and arbitrary 3D shapes (see [9, 32] for surveys). The basic elements of a multiresolution model are the following: a *base mesh* that defines the coarsest available representation; a set of local *updates* U refining it; and a *dependency relation* among updates [38]. In general, an *update* u defines two sets of triangles u^- and u^+, representing the same portion of a surface at a lower and a higher LOD, respectively. An update can be applied locally either to refine or to coarsen a mesh. The *direct dependency* relation is defined as follows. An update u_2 depends on an update u_1 if it removes some cells inserted by u_1, that is, if the intersection $u_2^- \cap u_1^+$ is not empty. The transitive closure of the dependency relation is a partial order, and the direct dependency relation can be represented as a directed acyclic graph (DAG). Any subset of updates, which is closed with respect to the partial order, that is, which defines a cut of the DAG, can be applied to the base mesh in any total order extending the partial one, and gives a mesh at an intermediate (uniform or variable) resolution. In Fig. 3.4 we show a simple multiresolution model composed of a base mesh and three updates, the gray line represents a cut of the DAG encoding the dependency relation and the mesh on the right represents the extracted mesh, associated to the depicted cut. In [8], we have shown that all the existing multiresolution models are captured by this framework, which can be extended to higher dimensions and to cell complexes.

In some cases, when a multiresolution model is built through some specific local modification operators (such as vertex removal, edge collapse, or vertex-pair contraction), an implicit, procedural encoding of the updates can be used and the dependency relation can be encoded in a more compact form than a DAG, such as the view-dependent tree proposed in [11].

Various techniques have recently been proposed in the literature for out-of-core multiresolution modeling of large irregular datasets. The general strategy in the design of out-of-core data structure consists of organizing information in disk pages in such a way that the number of page loads/swaps is minimized when traversing the model. In this respect, the basic queries on a multiresolution model are instances of *selective refinement* that consists of extracting adaptive meshes of minimal size according to application-dependent requirements. The idea is to select and apply to

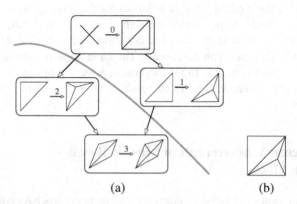

(a)   (b)

Fig. 3.4. (a) A simple multiresolution model composed of a base mesh and three updates, bold gray line represents a cut; (b) the mesh corresponding to the front represented by the bold gray line

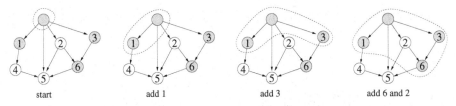

start   add 1   add 3   add 6 and 2

Fig. 3.5. Selective refinement with a top-down depth-first approach

the base mesh all and only those updates that are necessary to achieve a given, user-defined LOD in a user-defined region of interest. Figure 3.5 shows the behavior of selective refinement with a top-down depth-first approach. The update without label denotes the dummy modification corresponding to creating the base mesh. White updates satisfy the user criterion, gray updates do not. The dashed line encloses the current set of updates.

There are two main strategies to organize data into clusters that fit disk pages. The first approach consists of clustering nodes of the hierarchy, corresponding to atomic updates. The second approach consists of clustering data by spatial proximity. We will follow this classification to describe the various methods in the following subsections.

3.5.1 Methods Based on Clustering of Atomic Updates

In [12], El-Sana and Chiang build an out-of-core multiresolution model based on edge collapse, which is generated through the simplification algorithm described in Sect. 3.3, and targeted to support view-dependent rendering. The dependency relation among updates is represented as a binary forest of vertices, called a "view-dependent tree." If an edge $e = (v_1, v_2)$ is collapsed into a vertex v, then the node corresponding to v will be the parent of the two nodes corresponding to vertices v_1 and v_2, respectively. There is a one-to-one relation between vertices in the forest and nodes in the DAG of the general model. Arcs of the binary tree are not sufficient to encode all dependencies in the DAG. However, a vertex enumeration mechanism is associated with the vertices in the binary forest to correctly represent the dependency relation among updates, as described in [11].

The binary vertex forest is clustered in subtrees of height h, and thus, each subtree contains at most $2h-1$ nodes of the tree. The value h is selected to maximize disk page filling. The selective refinement algorithm keeps in memory the set of subtrees that store the vertices belonging to the extracted mesh, and keeps in cache their parents and children also. If prefetched subtrees exceed cache size, the first prefetched page, not used by the extracted mesh, is removed from the cache. Figure 3.6 shows a simple binary vertex forest clustered in subtrees of height 2.

Pajarola in [36] proposes a compact representation for multiresolution models built through half-edge collapse, that is, by contracting an edge $e = (v_1, v_2)$ to one of its extreme vertices, let us say v_1. The multiresolution model encodes a binary forest of half-edge collapses. This forest is obtained from the forest of binary vertices by

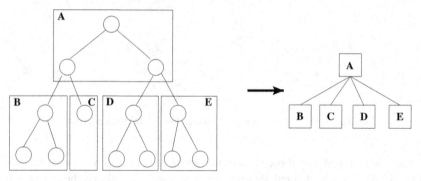

Fig. 3.6. Clustering of a binary vertex forest in subtrees of height 2

replacing the subtree formed by vertices v_1, v_2, and by their parent, which will be again v_1 for a half-edge collapse with a pointer to the corresponding half-edge e in the data structure describing the currently extracted mesh. In [7], DeCoro and Pajarola propose an out-of-core representation of the same model to support view-dependent rendering of large 3D meshes. The proposed technique starts from the binary forest and computes almost balanced subtrees that can fit in a disk page. Pointers to empty children are removed, thus reducing the storage cost by 25%. This, together with a compact encoding of the information associated with the nodes of the binary forest of edges, results in a compact data structure that requires less disk accesses. On the other hand, computation of the disk blocks requires two depth-first traversals of the binary forest. The objective is to perform selective refinement out-of-core efficiently, while both the simplification step and the construction of the multiresolution model are assumed to be performed in-core.

The strategies described above apply only to models based on edge collapse in which the dependency relation is represented as a binary forest. In [6] we propose and analyze clustering techniques that work on the full DAG of dependency relations. Such techniques are general and can be applied to any multiresolution model, regardless of the way it is generated. An important assumption is that the updates in the multiresolution model are atomic, that is, each update involves a relatively small number of triangles. Conversely, the DAG may have a very large number of nodes.

On the basis of an analysis of selective refinement queries and of the shape of the DAG describing the MT, we have defined and implemented the following techniques for grouping the updates in an MT according to sorting criteria:

- *Approximation error* (Err)
- *Layer*: shortest path from the root (Lyr)
- *Level*: longest path from the root (Lev)
- *Distance*: average path length from the root (Ly2)
- *Depth-first* (DFS) and *breadth-first* (BFS) DAG traversal
- *Multiresolution depth-first traversal (GrD) and multiresolution breadth-first traversal* (GrB): similar to depth-first or breadth-first DAG traversal, respectively, but before performing an update u on the currently extracted mesh, all

ancestors of u are visited recursively if they have not been visited before. These criteria simulate the strategies used in a selective refinement algorithm.

We have also defined and implemented two spatial grouping strategies. The first strategy is based on the *R*-tree* [1], while the second strategy is based on a *Point Region k-d (PR k-d) tree* for partitioning in space combined with a *PK-tree* [44] as a grouping mechanism.

Finally, we have designed and implemented a class of strategies that combines space grouping with DAG traversals (according to depth). At a lower resolution, we are interested in having clusters that span the whole domain, while, as the resolution increases, we are looking for clusters associated with finer space subdivisions. In order to achieve this goal, we have developed techniques that interleave the effect of a sorting rule and a space partitioning rule similar to a PR k-d tree. A description of such techniques can be found in [6].

In all our experiments, the GrB outperforms the other clustering techniques. It is interesting to note that even with a small cache (about 1% the size of the whole model), a clustering technique based on GrB exhibits a very limited overhead, compared to loading the whole model just once in main memory.

3.5.2 Methods Based on Domain Partitioning

In [20], Hoppe describes an out-of-core multiresolution model generated through edge collapse, called a progressive mesh (PM) quadtree, which is built through the simplification method described in Sect. 3.3. The model works on gridded data (DEM). The input grid is triangulated and then partitioned into square blocks in such a way that the content of each block fits in the main memory. Simplification is performed bottom-up, and a quadtree is built where every quadrant contains a simplified representation of the terrain represented by its four children. The resulting data structure is a sort of pyramid, where each block contains a sequence of vertex splits, that inverts the sequence of edge collapses that produced the mesh associated with that block. When selective refinement is performed, only the blocks of the quadtree that are interested in the query are transferred to the main memory. Also this model was designed specifically for visualization.

In [4], Cignoni et al. propose an out-of-core multiresolution model based on the decomposition of the domain into a nested triangle mesh described as a triangle bintree. Each triangle in the bintree, which we call a *macrotriangle*, contains an irregular mesh, formed by a relatively large number of triangles (typically between 256 and 8k triangles). Leaves of the triangle bintree are associated with portions of the mesh at full resolution, while internal nodes are associated with simplified meshes. This multiresolution representation is created during a fine-to-coarse simplification process, based on Hoppe's method [21] and adopted in order to keep boundary coherence among adjacent macrotriangles. The meshes associated with the macrotriangles are carefully created during the simplification process, so that, when assembled together, they form a conforming triangle mesh. As in Hoppe's approach, this multiresolution model is targeted at out-of-core terrain visualization. To enhance rendering performances, each mesh associated with a macrotriangle is organized into triangle strips.

The out-of-core organization in this case is very simple, thanks to the approach that acts on the different levels of granularity of nodes in the hierarchy. The triangle bin-tree is relatively small, even for huge models, and can easily be kept in main memory. The mesh associated with a macrotriangle is stored in a disk page through a compact data structure.

In [5], the same approach is extended to work on multitessellations, that is, the general model, in which the hierarchy is represented by a DAG. Also in this case, every node of the DAG contains a patch of a few thousands of triangles and is stored in secondary memory, while the DAG describing the dependency relation among updates is small enough to be kept in core. As for the standard multitessellation, this model can be used for both terrains and 3D shapes.

The model is constructed by computing a partition of the input mesh into patches of almost equal size. Such patches are computed by uniformly distributing on the mesh a set of seed points. The number of such points must be proportional to the mesh size. Then, an approximated Voronoi diagram of the seed points is computed, so that the Voronoi cells define the patches of the subdivision. Each patch is then independently simplified, without modifying its boundary. Simplification of a patch usually reduces the number of triangles by a factor of 2. The same process described above is applied again to the simplified mesh by starting with a new, smaller set of seed points. The process is repeated until the resulting mesh satisfies some size constraint. It has been shown that the number of steps is usually logarithmic in the size of the mesh. The DAG describing the dependency relation is built during this process.

During selective refinement, the DAG is traversed, and only those patches that are needed to achieve the resolution required by the user are loaded into the memory. Each patch is stored as a compressed triangle strip to improve rendering performances.

Lindstrom [30] and Shaffer and Garland [43] have proposed two similar out-of-core multiresolution models for massive triangle meshes describing 3D scenes generated through vertex clustering. The purpose is in both cases view-dependent rendering of very large 3D meshes. The multiresolution model is an octree in which the leaves store the vertices of the mesh at full resolution, or of an already simplified mesh (in the case of Lindstrom's approach), while the internal octree cells represent vertices obtained as a result of the vertex clustering process and the triangles are associated with cells containing their vertices.

Lindstrom's approach is based on the out-of-core simplification technique reported in Sect. 3.3. The construction of the multiresolution model is performed in two steps. During the first step, the mesh is regularly sampled. Each side of the sampling grid is composed by 2^n cells, where n is a user-defined quantity. After that, the vertices are sorted in external memory according to their position in the grid. This guarantees local access during the construction of the hierarchy. The second step considers the simplified mesh, composed of the list of vertices, error information, and the list of triangles, and produces an octree having the simplified mesh stored in its leaves. Starting from a group of sibling vertices, position, normal, and error of the parent are computed. Note that a leaf cell stores just the vertex and its normal, while an internal cell stores a vertex v and the triangles are removed from the representation when v is collapsed.

The mesh indexed in the leaves of the resulting multiresolution model is the one generated by the first simplification step. Thus, the original full-resolution mesh cannot be reconstructed. The multiresolution model is completely built on disk. View-dependent refinement is performed by two threads: one extracts the variable-resolution mesh, according to an approximate breadth-first octree traversal of the tree, while the other thread renders the mesh. Disk paging is left to the operating system. This is a reasonable choice, since data are sorted in a cache coherent way, but the technique could be further improved by an explicit paging scheme.

Shaffer and Garland's approach [43] is to develop a design for a data structure that offers explicit access to the original mesh. On the other hand, Lindstrom's method has the benefit of working completely out of core, while Shaffer and Garland's method keeps a hash table that refers only to non-empty cells of the grid in the memory. This could be a problem for very dense meshes filling the space. Moreover, hash keys are stored in 32 bits, and each key is composed of the three vertex coordinates. This bounds the size of the uniform grid to 1024^3. For out-of-core terrain modeling, both approaches can be simplified by using a quadtree to describe the vertex clustering. In this scenario, Lindstrom's approach could be definitely more efficient, since their performances are not affected by the percentage of full cells in the domain decomposition.

In [45], Yoon et al. propose an out-of-core multiresolution model for view-dependent rendering of massive triangle meshes describing 3D scenes. The multiresolution model is called a clustered hierarchy of progressive meshes (CHPMs), and it consists of a hierarchy of clusters that are spatially localized mesh regions and of linear sequences of edge collapses, each associated with a cluster, that simplify the corresponding meshes. Each cluster consists of a mesh formed by a few thousand triangles. The clusters are used to perform coarse-grained view-dependent refinement of the model, while the linear sequences of edge collapses are used for fine-grained local refinement. The cluster hierarchy is enhanced with dependencies among clusters that act as constraints to be able to generate crack-free meshes. A CHPM is computed in three steps. First, the vertices of the mesh at full resolution are organized into clusters containing almost the same number of vertices. A regular grid is superimposed on the set of vertices and a graph $G = (N, A)$ is computed, in which the nodes correspond to the non-empty cells of the grid, while any arc in A connects a node in N and its k-nearest neighboring cells. Graph G is partitioned into clusters, and a new graph is computed in which the nodes are associated with clusters, and two nodes are connected by an arc if the corresponding clusters share vertices or if they are within a threshold distance of each other. Then, the cluster hierarchy is generated top-down by recursively partitioning the cluster graph into halves, thus producing a binary hierarchy. Finally, the triangles of the full-resolution mesh are associated with the clusters in the hierarchy and a mesh simplification process is applied bottom-up on the hierarchy of clusters by performing half-edge collapses. During each pass of simplification only the cluster is simplified and the clusters with which it shares vertices must be resident in the memory. When performing view-dependent refinement,

Table 3.2. Comparison among multiresolution approaches

	approach	data	update size	in RAM
El-Sana and Chiang [12]	clustering dep.	free form	atomic	clusters of C.M.
DeCoro and Pajarola [7]	clustering dep.	free form	atomic	binary forest
Danovaro et al. [6]	clustering dep.	nD	atomic	a few clusters
Hoppe [20]	partitioning	scalar field	atomic	cluster hierarchy
Cignoni et al. [4]	partitioning	scalar field	large	binary tree
Yoon et al. [45]	partitioning	free form	large	cluster hierarchy
Lindstrom [30]	partitioning	free form	medium	a few MB
Shaffer and Garland [43]	partitioning	free form	medium	cluster indices
Cignoni et al. [5]	partitioning	free form	large	cluster hierarchy

the cluster hierarchy is kept in the main memory, while the sequences of edge collapses are fetched from the disk. Also, vertices and triangles corresponding to the active clusters are stored in GPU memory. This approach has been developed for 3D meshes, but can be easily adopted to TINs by using a 2D grid built on the projection of the data points in the plane.

3.5.3 Comparison

Table 3.2 summarizes various multiresolution techniques that we have presented, by highlighting the approach used to organize the out-of-core data structure, the data that can be represented, and the size of updates. Overall, models based on the clustering of nodes in the hierarchy are not only more general, but also more complex to manage. They have the advantage that the meshes extracted from them have the same granularity and thus the same accuracy of the meshes extracted from the corresponding in-core multiresolution models. On the other hand, dealing with large sets of atomic updates can become a bottleneck in some visualization tasks.

On the contrary, methods that use large patches highly simplify the management of secondary memory and result more efficient, but they trade-off this advantage by being coarser-grained, hence less smooth in the transition between different levels of detail. In particular, the method by Yoon et al. [45] seems to be more suitable for large scenes than terrains, and also the dependencies among clusters are not easily managed.

3.6 Conclusions

We have analyzed and compared out-of-core approaches for simplification of triangle meshes and for out-of-core multiresolution modeling of TINs, both for regularly and irregularly distributed data sets. Most of the mesh simplification algorithms and the out-of-core multiresolution representations have been developed for visualization purposes. Most of the techniques for irregular meshes have been developed for triangle meshes describing the boundary of a 3D object or a 3D scene. These techniques, however, can be applied either directly or easily adapted to TINs.

While there are several techniques able to deal with huge triangle meshes, or regular tetrahedral meshes, there is no technique for simplification and multiresolution modeling of tetrahedral meshes. We are currently developing an out-of-core multiresolution modeling system for scalar fields based on the out-of-core multitessellation, able to handle both triangle and tetrahedral meshes as well as simplicial meshes in higher dimensions.

Acknowledgments

This work has been partially supported by the European Network of Excellence AIM@SHAPE under contract number 506766, by the SHALOM project under contract number RBIN04HWR8, by the National Science Foundation under Grants CCF-0541032, IIS-00-91474, and CCF-05-15241.

References

1. Beckmann N, Kriegel HP, Schneider R, Seeger B (1990) The R*-tree: an efficient and robust access method for points and rectangles. In: Proc. of ACM SIGMOD Conference, Atlantic City, NJ, ACM Press, 322–331
2. Bogdanovich P, Samet H (1999) The ATree: a data structure to support very large scientific databases. In: Agouris P, Stefanidis A (eds) Integrated Spatial Databases: Digital Images and GIS, Portland, ME, 235–248, also University of Maryland Computer Science Technical Report TR–3435, March 1995
3. Cignoni P, Montani C, Rocchini C, Scopigno R (2003) External memory management and simplification of huge meshes. IEEE Transactions on Visualization and Computer Graphics 9:525–537
4. Cignoni P, Ganovelli F, Gobbetti E, Marton F, Ponchio F, Scopigno R (2003) Bdam: Batched dynamic adaptive meshes for high performance terrain visualization. Computer Graphics Forum 22:505–514
5. Cignoni P, Ganovelli F, Gobbetti E, Marton F, Ponchi F, Scopigno R (2005) Batched multi triangulation. In: Proc. of IEEE Visualization 2005, IEEE Computer Society, 207–214
6. Danovaro E, De Floriani L, Puppo E, Samet H (2005) Multiresolution out-of-core modeling of terrain and geological data. In: Symposium on Advances in Geographic Information Systems, New York, NY, USA, ACM Press, 143–152
7. De Coro C, Pajarola R (2002) XFastMesh: Fast view-dependent meshing from external memory. In: Proc. of IEEE Visualization 2002, Boston, MA, IEEE Computer Society, 263–270
8. De Floriani L, Magillo P (2002) Multiresolution mesh representation: models and data structures. In: Floater M, Iske A, Quak E, (eds) Principles of Multiresolution Geometric Modeling. Lecture Notes in Mathematics, Berlin, Springer Verlag, 364–418
9. De Floriani L, Kobbelt L, Puppo E (2004) A survey on data structures for level-of-detail models. In: Dodgson N, Floater M, Sabin M, (eds) Multiresolution in Geometric Modeling, Springer Verlag, 49–74

10. Duchaineau M, Wolinsky M, Sigeti DE, Miller MC, Aldrich C (1997) Mineev-Weinstein, M.B.: ROAMing terrain: real-time optimally adapting meshes. In: Yagel R, Hagen H (eds) Proc. of IEEE Visualization 1997, Phoenix, AZ, IEEE Computer Society, 81–88
11. El-Sana J, Varshney A (1999) Generalized view-dependent simplification. Computer Graphics Forum 18:C83–C94
12. El-Sana J, Chiang Y J (2000) External memory view-dependent simplification. Computer Graphics Forum 19:139–150
13. Evans W, Kirkpatrick D, Townsend G (2001) Right-triangulated irregular networks. Algorithmica 30:264–286
14. Fekete G, Davis LS (1984) Property spheres: a new representation for 3-d object recognition. In: Proc. of the Workshop on Computer Vision: Representation and Control, Annapolis, MD (1984), 192–201, also University of Maryland Computer Science Technical Report TR–1355
15. Fekete G (1990) Rendering and managing spherical data with sphere quadtrees. In: Kaufman A (ed) Proc. of IEEE Visualization 1990, San Francisco, 176–186
16. Garland M, Heckbert PS (1997) Surface simplification using quadric error metrics. In: Computer Graphics Proc., Annual Conference Series (SIGGRAPH'97), ACM Press, 209–216
17. Garland M, Shaffer E (2002) A multiphase approach to efficient surface simplification. In: Proc. of Visualization 2002, Boston, Massachusetts, IEEE Computer Society, 117–124
18. Gerstner T (2003) Multiresolution visualization and compression of global topographic data. GeoInformatica 7:7–32
19. Hoppe H (1998) Smooth view-dependent level-of-detail control and its application to terrain rendering. In: Proc. of IEEE Visualization 1998, Research Triangle Park, NC, IEEE Computer Society, 35–42
20. Hoppe H (1998) Efficient implementation of progressive meshes. Computer & Graphics 22:27–36
21. Hoppe H (1999) New quadric metric for simplifying meshes with appearance attributes. In: Proc. of IEEE Visualization 1999, San Francisco, California, United States, IEEE Computer Society Press, 59–66
22. Isenburg M, Gumhold S (2003) Out-of-core compression for gigantic polygon meshes. ACM Transactions on Graphics 22:935–942
23. Isenburg M, Lindstrom P, Gumhold S, Snoeyink J (2003) Large mesh simplification using processing sequences. In: Proc. of Visualization 2003 Conference, IEEE Computer Society Press, 465–472
24. Kawaguchi E, Endo T (1980) On a method of binary picture representation and its application to data compression. IEEE Transactions on Pattern Analysis and Machine Intelligence 2:27–35
25. Lee M, Samet H (2000) Navigating through triangle meshes implemented as linear quadtrees. ACM Transactions on Graphics 19:79–121
26. Lindstrom P, Koller D, Ribarsky W, Hodges LF, Faust N, Turner GA (1996) Real-time continuous level of detail rendering of height fields. In: Proc. of SIGGRAPH 1996, 109–118
27. Lindstrom P (2000) Out-of-core simplification of large polygonal models. In: ACM Computer Graphics Proc., Annual Conference Series (SIGGRAPH'00), New Orleans, LA, USA, ACM Press, 259–270
28. Lindstrom P, Silva CT (2001) A memory insensitive technique for large model simplification. In: IEEE Visualization 2001, IEEE Computer Society, 121–126

29. Lindstrom P, Pascucci V (2002) Terrain simplification simplified: a general framework for view-dependent out-of-core visualization. IEEE Transactions on Visualization and Computer Graphics 8:239–254
30. Lindstrom P (2003) Out-of-core construction and visualization of multiresolution surfaces. In: Proc. of ACM SIGGRAPH 2003 Symposium on Interactive 3D Graphics, Monterey, California, ACM Press, 93–102
31. Losasso F, Hoppe H (2004) Geometry clipmaps: terrain rendering using nested regular grids. ACM Transaction on Graphics 23:769–776
32. Luebke D, Reddy M, Cohen J, Varshney A, Watson B, Huebner R (2002) Level of Detail for 3D Graphics. Morgan-Kaufmann, San Francisco
33. Magillo P, Bertocci V (2000) Managing large terrain data sets with a multiresolution structure. In: Tjoa A M, Wagner R R, Al-Zobaidie A (eds) Proc. 11th Int. Conf. on Database and Expert Systems Applications, Greenwich, UK, IEEE, 894–898
34. Ohlberger M, Rumpf M (1997) Hierarchical and adaptive visualization on nested grids. Computing 56:365–385
35. Pajarola R (1998) Large scale terrain visualization using the restricted quadtree triangulation. In: Ebert D, Hagen H, Rushmeier H (eds) Proc. of IEEE Visualization 1998, Research Triangle Park, NC, IEEE Computer Society, 19–26
36. Pajarola R (2001) FastMesh: efficient view-dependent meshing. In: Proc. of Pacific Graphics 2001, IEEE Computer Society, 22–30
37. Prince C (2000) Progressive Meshes for Large Models of Arbitrary Topology. Master Thesis, Department of Computer Science and Engineering, University of Washington
38. Puppo E (1998) Variable resolution triangulations. Computational Geometry Theory and Applications 11:219–238
39. Rossignac J, Borrel P (1993) Multiresolution 3D approximations for rendering complex scenes. In: Falcidieno B, Kunii TL (eds) Modeling in Computer Graphics, Springer-Verlag, 455–465
40. Samet H (1990) The Design and Analysis of Spatial Data Structures. Addison-Wesley, Reading, MA
41. Samet H (2006) Foundations of Multidimensional and Metric Data Structures. Morgan-Kaufmann, San Francisco
42. Shaffer E, Garland M (2001) Efficient adaptive simplification of massive meshes. In: Proc. of IEEE Visualization 2001, IEEE Computer Society, 127–134
43. Shaffer E, Garland M (2005) A multiresolution representation for massive meshes. IEEE Transactions on Visualization and Computer Graphics 11:139–148
44. Wang W, Yang J, Muntz R (1998) PK-tree: a spatial index structure for high dimensional point data. In: Tanaka K, Ghandeharizadeh S (eds) Proc. 5th International Conference on Foundations of Data Organization and Algorithms (FODO), Kobe, Japan, 27–36
45. Yoon S, Salomon B, Gayle R, Manocha D (2005) Quick-VDR: Out-of-core view-dependent rendering of gigantic models. IEEE Transactions on Visualization and Computer Graphics 11:369–382

Progressive Techniques for Efficient Vector Map Data Transmission: An Overview

Michela Bertolotto

University College Dublin, Dublin (Ireland)

4.1 Introduction

Progressive data transmission techniques are commonly used for data exchange over the Internet: a subset of the data is sent first and then incrementally refined by subsequent stages. The advantages of progressive transmission have been highlighted by many researchers [5–8, 10, 22] and include efficient data transmission (as smaller files are sent), quick response, and, possibly, transmission of only the most relevant detail. Besides being able to perform preliminary operations on temporary versions of the data, during progressive transmission, users can realize that the detail of the currently displayed representation is good enough for their purpose and so can decide to interrupt the downloading of more detailed representations (stored in larger files). Therefore, both time and disk space can be saved. This is particularly important for users in the field trying to download datasets of interest. Even if the fully detailed version is needed, they can start working with a coarser version while more detail is being added progressively.

The first and most successful implementations have been developed for raster images [12, 21, 41]. The data is efficiently compressed with different techniques and sent to the user. The full resolution image is reconstructed on the user's machine by gradually adding detail to coarser versions. The success of progressive raster transmission relies on the availability of effective compression techniques for such data: these techniques provide good compression ratios while causing low information loss. Furthermore, they are efficient and relatively easy to implement.

These mechanisms are applied also in the spatial domain: raster geospatial datasets (including high resolution satellite images, aerial photos, and scanned maps) can be exchanged progressively from a server to a client. This is acceptable for applications that mainly involve visualization. However, a raster version of the data might not be adequate in certain applications, including contexts in which actual object manipulation is involved. In these cases a vector representation of the data is required. Vector data sets consist of collections of spatial entities in the form of points, polylines, and polygons that are related through spatial relations (e.g. topological, metric, and direction relations [20]). Examples include thematic maps, road network maps, city maps, mesh-based digital terrain models, and so on.

As a consequence more recently attention has turned to the progressive transmission of vector data: several approaches based on effective compression mechanisms have been defined and successfully implemented for the particular case of triangular meshes [1, 16, 27–29].

In contrast, the investigation and development of progressive transmission techniques for more general vector spatial data still present many challenges. Our focus is on map data (e.g. city street maps, road networks, hydrography networks, administrative boundary maps, etc.). Compressing this type of data is not straightforward as many factors have to be taken into account. For example, spatial relations must be preserved to guarantee reliable query results.

The level of detail of a vector map can be decreased by applying the so-called map generalization process. However, such a process relies on cartographic principles whose complete formalization is still lacking (see [48] for a survey). Although few automated solutions and algorithms have been proposed for specific cases, they involve complex and time-consuming calculations and therefore cannot be applied for real-time Web mapping.

Recent studies [10, 11] have distinguished between two different approaches for the development of progressive map data transmission: on-the-fly map generation versus pre-computation of multiple map representations. The first approach aims at generating new generalized maps upon user request. The second approach relies on the pre-computation of a fixed sequence of map representations that are stored and transmitted progressively in order of increasing detail. Both approaches have advantages and drawbacks. This chapter provides a critical overview of these issues and of the few prototype systems that have been described and developed for progressive vector map transmission.

The remainder of this chapter is organized as follows. Sect. 4.2 reviews well-established implementations for progressive transmissions of data over the Internet. These include the cases of raster images and triangular meshes. The problem of transmitting geographical map data in vector format is analyzed in the subsequent sections. Section 4.3 discusses the use of progressive transmission techniques for Web mapping. This is followed by an overview of the prototype systems developed and discussed in the literature. These systems rely on techniques that can be differentiated on the basis of the type of changes applied during generalization: geometric versus topological changes. Therefore, in Sect. 4.4 we describe research related to changes in the geometry of the data, while in Sect. 4.5 we focus on approaches based on topological changes. Based on our experience in the development of progressive transmission prototype systems, Sect. 4.6 is dedicated to a critical discussion of the main research and implementation challenges still associated with progressive vector data transmission for Web mapping. Finally in Sect. 4.7 we present a summary and some concluding remarks.

4.2 Progressive Transmission: Established Approaches

In this section we review existing methods commonly employed for progressive transmission of spatial data over the Internet. The first approaches were developed for raster data and include interleaving techniques and other mechanisms based on image compression. The success and application of image compression mechanisms for progressive raster transmission is due to their effectiveness: they provide good compression with low loss of information. Furthermore, they are relatively easy to implement. The main characteristics of these approaches are summarized in Sect. 4.2.1.

In the vector domain the most successful methods for progressive transmission are restricted to the distribution of data in the form of triangular meshes. Some of the proposed approaches are discussed in Sect. 4.2.2.

4.2.1 Progressive Transmission of Raster Images

The first successful implementations of progressive raster transmission rely on interleaving techniques. The simplest approach consists of randomly extracting subsets of pixels from the image and incrementally completing it by adding pixels. If the implicit row/column ordering of images is exploited, pixels can be sampled in a uniform fashion; alternatively, hierarchical structures, such as quadtrees [45], can be used to extract more pixels from parts of the image with higher density of detail.

Sophisticated compression techniques have generated more effective progressive raster transmission methods (see, for example, [41, 47]). The most commonly used image compression method is the JPEG (joint photographic experts group) format. The JPEG method is a transform-based compression method that decomposes the original data before compression (see [23] for a detailed description). As images in JPEG format are segmented into rectangular sub-blocks and each block is transformed independently, at high compression ratios they take on unnatural blocky artifacts.

Alternative techniques are based on wavelet decompositions [33]. Wavelet methods are also transform-based but they act on the entire image. The result is a robust digital representation of a picture, which maintains its natural look even at high compression ratios (see [12] for an overview). Wavelets are well suited to progressive transmission because, besides providing more efficient overall compression, they naturally represent image data as a hierarchy of resolution features and its inverse at each level provides subsampled versions of the original image. Therefore, progressive transmission corresponds to a natural reconstructive mode for a wavelet-based compression algorithm (see, for example, [41]).

Much work on image compression relies on fractal theory [3]. A fractal is a geometrical figure whose local features resemble its global characteristics (self-similarity property). The challenges in fractal-based compression include finding a small number of affine transformations to generate the image as well as subparts of the input image that have self-similarity properties [13, 21]. Hybrid compression methods, combining fractals and wavelets, have also been defined [14, 50].

When transform-based methods are used to compress an image, during transmission over the Internet, instead of the image, the function coefficients are transmitted, and the image is subsequently reconstructed by inverting the transformation. Both transmission and reconstruction are very efficient, as they can be performed simply by means of a single pass through each color band of the image.

The purpose of raster data exchanging over the Internet is very often only visualization. In this case, techniques for progressive raster transmission are usually suitable and efficient. Even when measurements need be performed on the images (e.g. for photogrammetric purposes), common compression mechanisms (such as JPEG) have proved to be effective [30]. However, some applications involve direct object handling and manipulation. In this case a vector representation is required and preservation of topological and metric properties is a major issue. This is discussed in the following sections.

4.2.2 Progressive Transmission of Meshes

Part of the spatial data in vector format is represented by means of triangular meshes. For example, triangulations are used for digital terrain modeling and for real objects surface reconstruction and rendering [1, 16].

As discussed in Chap. 3, several compression methods for triangular meshes have been defined in the literature (see also [15] for a survey). These methods are either based on optimal point decimation techniques or they exploit combinatorial properties of triangulations for efficient encoding. In this section we cite some that have been applied for progressive transmission. Among them, the progressive mesh scheme proposed by Hoppe [27] represents a fundamental milestone. However, this method does not have a satisfactory performance as, like many other traditional methods [19, 53], it is topology preserving. This is important for data consistency but it limits the level of simplification. More recent methods achieve higher compression by slightly modifying the topology of the input mesh [1].

Juenger and Snoeyink [28, 29] defined a parallel point decimation technique for triangulated irregular network (TIN) simplification that allows progressive transmission and rendering of TIN data. Finally, De Floriani and Puppo [15] proposed a general framework for multiresolution hierarchical mesh representation. Such a model has a high storage cost. A more efficient encoding structure has been described in [16] to facilitate progressive transmission. Although these progressive transmission techniques are effective and they have already produced several prototypes [1, 28], they can only be applied to data represented in the form of triangular meshes.

4.3 Progressive Vector Transmission for Web Mapping

The increasing abundance of geographical map data sets on the Internet has recently generated a lot of interest in downloading such data sets for public and commercial use. Although several systems have been developed, Web mapping still suffers from slow transmission and information overload. Indeed, vector maps usually occupy very large data sets due to the inherent complexity of their geometry and topology. Furthermore, they are often represented at high levels of detail (LOD).

The set of spatial objects that need to be represented in a map depends on its level of detail. In this context, the term 'detail' refers both to the amount of entities contained in the data set and to the information stored about such entities, for example their geometry [24]. Therefore, changes of detail can be caused by the addition or elimination of some entities, as well as by the refinement or coarsening of the representation of existing entities (e.g. change of dimension or shape).

Limited bandwidth makes the (one-step) transmission time of large, detailed vector maps unacceptable for real-time applications. However, as downloaded maps may contain more detail than users require, progressive transmission has been identified as a viable solution: a less detailed version can be sent first and then refined upon user request.

A coarser map version can be generated by applying map generalization techniques to a fully detailed map [9, 31, 32, 34, 38, 42, 43, 48]. This may lead to a remarkable reduction of data. If the extracted representation satisfies users' requirements, an improvement is obtained both from a transmission point of view (a smaller file is sent) and from an information overload point of view (only relevant data is received and displayed).

However, there are still several impediments to the online application of map generalization: first, the cartographic principles on which it is based have not been completely formalized yet. Secondly, it is a very complex and time-consuming process that currently still involves interaction between semi-automatic solutions and expert cartographers (see [48] for a survey of map generalization techniques and operators).

Several tools for map generalization have recently been developed, based on the application of object-oriented principles, expert systems, neural networks, and case-based reasoning [31, 42, 49].

A critical issue in map generalization is preservation of consistency (e.g. topological constraints) while decreasing the level of detail. For example, random selection of objects (to be eliminated or simplified) does not usually generate a topologically consistent representation at coarser detail. Consistency is an essential property for data usability (see also Chap. 8): the answer to a query on the generalized map must be consistent with the answer obtained when the same query is applied to the original (fully detailed) map. Most of the proposed generalization algorithms do not guarantee the preservation of this property. Therefore they often require the application of a posteriori checks to rectify introduced inconsistencies.

With these issues in mind, researchers have studied the realization of Web mapping systems based on different types of solutions.

Cecconi and Galanda [11] have summarized the different scenarios in which users may issue a request to a Web mapping system. These include the following:

- *On-the-fly Mapping Request:* The user wants to download a first map for an overview of the requested information. The waiting time should be short and generalization only partially applied.

- *On-demand Mapping Request:* In this case the time component is less critical and the user is willing to wait for the desired map. The generation can thus involve the entire process of generalization to obtain a high-quality map.
- *On-the-fly and On-demand Mapping Request:* In this case the advantages of the other two approaches are combined. The user needs a map with good cartographic quality within a limited time.

For real-time solutions, the application of the map generalization process is unacceptable. Alternative solutions have been proposed based on the pre-computation of multiple map representations at increasing levels of detail for support to real-time Web mapping (see also Chap. 5 for further discussions on the use of multiple representations for Web applications). One such proposal was described in [6] and relies on a simple client–server architecture. A sequence of consistent representations at lower levels of detail are pre-computed and stored on the server site. Such representations are transmitted in order of increasing detail to the client upon request. The user can stop the downloading process when satisfied with the current map version (Fig. 4.1).

An obvious disadvantage of this approach is that it requires pre-computation. Furthermore, it is based on a predefined set of representations thus limiting the flexibility of obtaining a level of detail that completely suits users' requirements. However, as the computations are done offline, this model allows to ensure that consistency is preserved. Note that the architectural organization is completely independent of both the data set used and the technique for generalization employed.

In the following sections we provide an overview of the prototype systems that have been developed following this architectural approach and partially combining it with other methods. These systems differ in the way they generate, store, and transmit multiple map representations. In our survey we subdivide them on the basis of the type of generalization operators they rely on: namely operators that mainly apply changes to the geometry of the data (Sect. 4.4) and operators that apply changes to the underlying topological structure (Sect. 4.5).

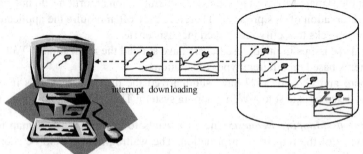

interrupt downloading

Fig. 4.1. Client–server architecture for progressive map transmission

4.4 Approaches Based on Geometric Operators

Much of the work on map generalization and multiple map representations for progressive transmission rely on the application of line simplification algorithms. This is due to the availability of successful algorithms and to the fact that the majority of features in a map are represented by polylines. Line simplification is a generalization operator that generates an approximated (simplified) representation of a polyline by eliminating a subset of its vertices.

The classical Ramer–Douglas–Peucker (RDP) algorithm [18, 40] is the most commonly used line simplification algorithm in both GIS research and commercial applications. The basic idea of the algorithm is as follows: in order to simplify a polyline with respect to a pre-defined tolerance value, it selects the vertex v with maximum distance from the segment connecting the endpoints of the polyline and compares such a distance with the tolerance value; if the distance is greater, then the polyline is subdivided into two polylines incident at v and the process is recursively applied to them; otherwise the polyline is approximated by the segment itself (see Fig. 4.2 for an example). The algorithm is very simple and easy to implement. In addition it provides good visual results as it preserves the overall shape of polylines.

Notwithstanding its popularity, such an algorithm presents a major drawback, which prevents its completely automated application. Indeed, it treats each polyline as an isolated feature and simplifies it individually. As a consequence, inconsistencies may be introduced into the simplified map, including unwanted intersections with other polylines and topologically incorrect positioning of features (e.g. an island jumps inland after simplification). A posteriori checks are required to rectify these. Examples are shown in Fig. 4.3.

Several efforts have been made to improve such an algorithm to preserve topological consistency. In particular, Saalfeld [44] proposed a topologically consistent line simplification method by adding further checks to the terminating condition in the classical RDP algorithm. His improvement is based on the fact that, while

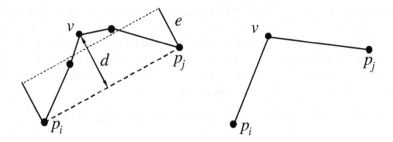

Fig. 4.2. A polyline l with five vertices (*left*) is reduced to a polyline with three vertices (*right*) when the RDP algorithm with tolerance e is applied. First l is split at v into two subpolylines, each of which is then replaced by the segment connecting its endpoints during the subsequent application of the algorithm

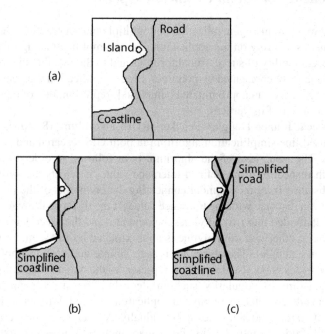

Fig. 4.3. (a) The original map containing a coastline, an island, and a road; **(b)** the island jumps inland after simplification of the coastline; **(c)** the simplified coastline and road intersect

generalizing a polyline, conflicts can only occur with vertices of other polylines (or isolated points) that lie within its convex hull (i.e. the boundary of the smallest polygon containing the vertices of the polyline such that the segments connecting any two of these vertices are completely contained inside the polygon).

Saalfeld's algorithm has been very successful and has been applied by other researchers for the development of progressive transmission of simplified maps [8, 22, 51, 52].

A prototype implementation developed by Buttenfield [8] applies Saalfeld's modified RDP algorithm to pre-computed series of generalized maps that are stored in hierarchical structures and can be transmitted progressively upon request. The building procedure consists of the following three steps:

1. Store different thematic layers (e.g. hydrography and transportation) in separate files.
2. Establish an ordering of the features within a theme for transmission purposes.
3. Subdivide each polyline (represented as a set of arcs) iteratively using the RDP algorithm and store it in a hierarchical strip tree [2].

A pointer-based encoding structure for the tree is provided and used to visit the tree during progressive transmission and for reconstruction of each level. However, full support for flexible navigation and browsing across map levels in the form of vertical links is not provided.

A major drawback of the system proposed in [8] is that it does not efficiently manage massive amounts of data as it stores vector map data in files without appropriate indexing mechanisms. This causes problems also for progressive transmission within a multi-user Internet environment. The pre-processing phase also represents a limitation. However, using pre-computations allows for progressive real-time analysis of generalized data that might not be possible if online simplification was applied upon request. Another limitation relates to the fact that each line is simplified, stored, and transmitted separately. This does not guarantee preservation of map consistency. Full experimental results on the performance of the progressive transmission system have not been reported.

Another system based on online simplification and developed for application to progressive transmission has been presented by Boujou et al. [22]. The approach proposed in [22] attempts to overcome some of the limitations of other systems using multiple map representations for progressive transmission. The idea is to try and link pre-existing map sequences (generated applying line simplification) vertically in order to easily access different versions of the same feature and generate variable LOD maps based on the pre-computed map representations. This system was developed for deployment in mobile environments. The emphasis is then on reducing to the minimum the amount of data to be transferred. Therefore it is paramount to allow reuse of already transferred objects at previous LOD. The authors identify three data transfer models between client and server:

1. *Simple Communication:* All queries are executed on the server and the entire query result is transmitted to the client.
2. *Two-step Communication:* All queries are executed on the server but the client maintains data cache and can reuse already received objects.
3. *Pre-computed Answer:* The client can execute some queries locally without connection to the server.

Depending on the current scenario, each of these models represents advantages over the others. Experiments conducted on three LOD generated for the transportation network of La Rochelle (France) have shown interesting and promising results for the gain in data transfer. Evaluation with other data sets is still ongoing.

In the following paragraphs we describe our work on progressive exchange of simplified maps [51, 52]. The system we have developed relies on the application of Saalfeld's modified RDP algorithm. Several map representations are generated and stored in a (Oracle^{TM} 9i [37]) Spatial Database. The different map versions are organized into a hierarchical structure that avoids redundancy and favors efficient transmission.

Indeed, an interesting property of some techniques for progressive transmission of raster data (e.g. interleaving techniques) is the fact that they only transmit increments (i.e. sets of new pixels) at each step. Such increments are added to the currently displayed image to improve its resolution without requiring the transmission or downloading of another complete image version.

Since vector files can be very large, it would be useful to speed up the transmission in a similar way and thus reduce network traffic. We have tried to incorporate this property into our proposed data structure. Instead of storing and transmitting the entire intermediate map representations, only the coarser map version is fully stored while subsequent levels just store increments (i.e. in this case, newly introduced entities and refined representations of entities present at previous LOD). An example is illustrated in Fig. 4.4. An important aspect of our data structure is that vertical links between different representations of the same polyline are explicitly encoded and transmitted (see [51] for more details).

Note that as intermediate levels are not fully stored in the database, they have to be reconstructed on the client site by merging increments into the coarser fully stored map version. However, increments between two consecutive levels of detail of a vector data set are not simply sets of pixels. They can be complex sets of entities that are added to the level at finer detail to refine the representation of a set of entities at the lower level. The integration of such increments with the currently downloaded or displayed representation might be a non-trivial task if topological consistency between different representations must be preserved.

In the case of sequences of maps generated by a topologically consistent line simplification algorithm (such as Saalfeld's algorithm), the reconstruction process is relatively simple as topological relations between different entities do not change (the endpoints of polylines remain unchanged, only their shape varies).

We have performed experiments on the progressive transmission of these multiple map representations using real data sets [52]. Results showed that transmitting incrementally the entire sequence takes approximately (with a difference of very few seconds on average) the same amount of time as sending the fully detailed map in one step. An example is reported in Table 4.1, where a hierarchy with three different levels is considered. The specific data set that produced these results contains over 30,000 lines represented by over 70,000 points (see [52] for more details).

The experiments were performed using a server with 1.6 GHz CPU processor and 2 GB memory and a client with 2.4 GHz CPU processor and 512 MB memory. We measured the average response time for transmission via LANs at the speed of 100 Mbps.

Table 4.1 shows that, when the map is transmitted progressively, users can start working with a coarser representation downloaded from the server within half of the time required to download the full resolution map; such a representation is then gradually refined until an acceptable version is obtained. Even if the user is interested in the fully detailed map, the response time is not increased significantly (one additional second in this case). However, we expect that frequently the time of transmission would be even further reduced, as in general users would not need to download all

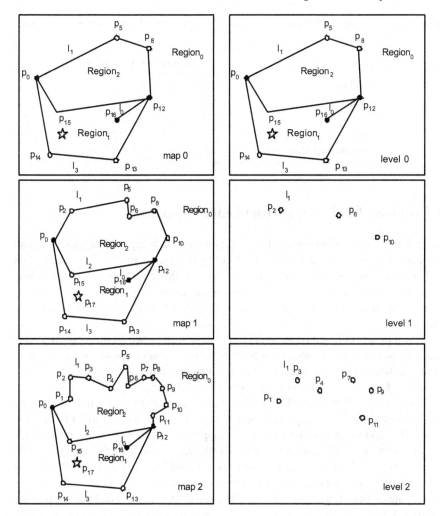

Fig. 4.4. (*Left*) Sequence of simplified maps: map 0, map 1, map 2; (*right*) corresponding representation levels stored in the data structure: level 0, level 1, level 2 (*for simplicity vertical links are not drawn here*)

levels in the sequence available on the server to obtain the requested information. Note that this approach decreases risk of information loss due to network failure: if a network failure occurs during data transmission, users need request to re-download a smaller file rather than the fully detailed map. This is particularly beneficial in mobile environments where network connections are unreliable.

Table 4.1. Comparing transmission time of real data with and without progressive transmission

	Average response time (s)
level 0	10
level 1	6
level 2	9
total	25
one step transmission	24

4.5 Approaches Based on Topological Operators

Several operators commonly used in map generalization apply changes to the map topology. For example, a geographical entity can be represented as a region at higher detail, and as a point at a coarser level. In this case a region-to-point contraction operation can be applied to obtain the generalized version. The dimension of the entity, and consequently its spatial relations with surrounding entities, changes.

The approaches based on geometric operators seen in Sect. 4.4 do not take these changes into account. However, in real applications, these operations are needed. In this section we discuss some progressive transmission prototypes that have been developed based on operators that apply topological changes.

The framework proposed by Bertolotto and Egenhofer [5, 6] relies on a model that was initially presented by Puppo and Dettori [17, 39] and then further formalized in [4]. The model consists of a sequence of map representations of a given area, each corresponding to a different LOD. Such a sequence is stored on the server and sent progressively to the client upon request (as described in Sect. 4.3). In [4] a set of generalization operators have been defined and shown to be minimal and sufficient to generate consistent generalized maps. The set of operators defined is the following (see Fig. 4.5):

- *Point Abstraction:* elimination of an isolated point inside a region
- *Line Abstraction:* elimination of a line inside a region
- *Line-to-point Contraction:* contraction of an open line (including its endpoints) to a point
- *Line Merge:* fusion of two lines sharing an endpoint into a single line
- *Region-to-point Contraction:* contraction of a simply connected region (and its boundary) to a point
- *Region Merge:* fusion of two regions sharing a boundary line into a single region
- *Region-to-line Contraction:* reduction of a region (and its bounding lines) to a line

These operators are defined as functions between map representations. They are called atomic as they perform atomic changes on the map. Complex generalization

Operator	Description	Example
point abstraction	A point p_0 is immersed into the region r containing it.	
line abstraction	A polyline l_0 is immersed into the region r containing it.	
line-to-point contraction	A simple polyline l and its endpoints p and p' reduce to a point p_0	
line merge	Two polylines l and l' sharing an endpoint p are immersed into a polyline l_0.	
region-to-point contraction	A simply connected region r and its boundary reduce to a point p_0	
region merge	Two regions r and r' sharing a bounding line l are immersed into a region r_0.	
region-to-line contraction	A simply connected region r and its boundary reduce to a polyline l_0.	

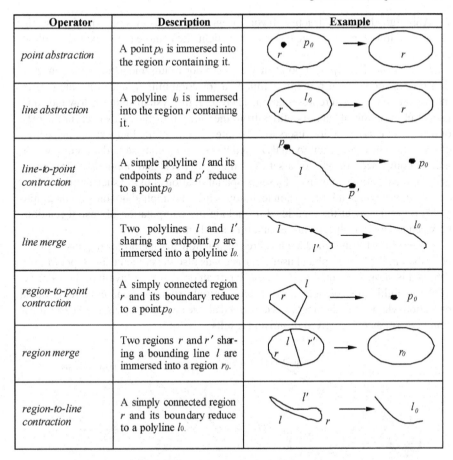

Fig. 4.5. Atomic topological operators

operations can be obtained by composition of these functions (more details can be found in [4]).

Note that only topological changes (e.g. changes of dimension, complexity, etc.) are possible by applying such operators: geometric and semantic changes are not being taken into account. Furthermore, although this model has the important advantage of being implicitly consistent, some commonly used generalization operations, such as aggregation [48], are not supported.

In [25], we defined a hierarchical data structure for efficiently storing and transmitting a sequence of multiple map representations generated by applying this set of operators. Such a data structure is based on the same principle as the data structure described in [51]: only the coarser map representation is completely stored while intermediate levels just store increments. Vertical links are also explicitly stored.

Although this data structure favors progressive transmission, in this case the reconstruction of intermediate levels on the client site is more complex: the topology needs be reconstructed.

From a purely graphical point of view, a major requirement for the reconstruction process consists of providing a method for combining a map with increments into a unique graphical representation. Therefore, the system should allow the addition or elimination of sets of entities from the current display. However, in the case of progressive transmission, users may be interested in more than a visual inspection of the data. They may want to analyze and work with a consistent data set. A vector map is composed not only of a set of points, polylines, and polygons representing geographical entities, but also of a set of spatial relations linking such entities. Thus, suitable overlaying and integration techniques must be applied not only at the graphical level, but also at the data level to include the computation of spatial relations between newly introduced entities and preserved entities.

Algorithms for intermediate level reconstruction and hierarchical querying were described in [25]. A graphical user interface was also developed using OpenMap^{TM} [36] and is shown in Fig. 4.6. As OpenMap is an open source Java-based GIS package, it is highly portable and customizable. Initial experiments on the effect of data reduction when using the operators described above and on the progressive transmission performance have given promising results.

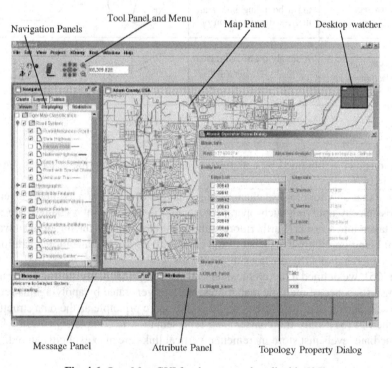

Fig. 4.6. OpenMap GUI for the system described in [25]

Cecconi and Weibel [10] describe a framework for map generalization to be applied for on-demand Web mapping. Rather than structure spatial data with respect to scale-dependent transmission and display, they suggest adapting existing generalization algorithms for on-the-fly generalization and develop new ones to provide for missing generalization functionality.

In [11] an adaptive zooming method for Web mapping is presented. The database is organized as a number of LOD and groups of selected feature classes. The LOD are derived from pre-selected scales, while the groups of selected feature classes are generalized on-the-fly. Given a user zooming request with a specific target scale, some objects at the corresponding LOD are retrieved from the database. The remaining objects are dynamically created by on-the-fly generalization. Different generalization operations (including those involving changes in the map topology) are considered. This results in vector maps with proven cartographic quality. Since on-the-fly generation applies only to a small number of objects instead of the entire map, this approach is applicable in real time. However, it requires the a priori selection of which features are to be stored in the database and which are to be generalized on-the-fly.

Finally, recently a few Web mapping systems have been developed using object-oriented databases and multi-agent technology. One such system, called LAMPS2 [26] relies on an object-oriented multiscale database and applies multi-agent technology to support automated map generalization (including operations that change the map topology). Although the system shows to be promising, in its current version it does not represent a solution to real-time online generalization. In addition, neither a progressive data delivery approach nor the problem of avoiding redundancy caused by multiple geometry objects storage have been discussed in relation to the LAMPS2 system.

4.6 Research and Implementation Issues

As seen in previous sections, research on progressive vector map transmission mostly relates to map generalization techniques that allow for coarser representations to be generated, each one at different LOD.

Although map generalization and multiple map representations have been studied for several years by the GIS community, several research and implementation issues remain unsolved for the successful online application of progressive vector map data transmission.

More specifically, based on our experience in the development of progressive transmission systems and on a thorough analysis of the literature, we have identified some main research priorities that require much investigation. These are summarized below:

- *Real-time Generalization and Data Compression*: A complete formalization of map generalization principles and practice is still lacking; more efforts are required to develop new algorithms for support to automated generalization.

- *Geometry and Topology Integration*: Approaches based on line simplification are applicable only for linear data sets and do not take into account other types of generalization operators (such as those involving changes in topology); vice versa, approaches based on topological operators do not support geometric changes. There is a need for the development of approaches that fully support the integration of topological and geometric operators. Semantic aspects should also be integrated.
- *Data Consistency Control*: Many existing algorithms do not implicitly preserve consistency and still require a posteriori checks; automated mechanisms to guarantee this property must be studied (see also Chap. 8 for further discussions about consistency between vector maps).
- *Data Storage and Multiple Representations*: Efficient hierarchical data structures for storing and manipulating different map versions are needed to facilitate transmission. These should avoid duplication and re-transmission of data and include explicit links for effective navigation and querying across levels. Another important challenge when using multiple representations consists of providing a suitable answer to the question: "how many and what levels should be pre-computed?" The choice of what representations need to be stored is often both application- and data-dependent. This problem poses limitations and should be further studied (see also Chap. 5 for further discussions on the use of multiple representations for Web applications).
- *Efficient Spatial Indexing Mechanisms*: With available vector spatial data sets becoming larger and larger, there is a need to efficiently index this data for fast access and retrieval. Many of the system prototypes proposed (e.g. [8]) did not consider this aspect. Others, that rely on commercially available spatial application servers (e.g. [51]) exploit the indexing functionality provided. However, a full evaluation of how these indexes influence the performance has not been conducted and constitutes an interesting research direction.
- *Accuracy Control*: A full study on evaluating the accuracy of the data exchanged and how essential it is in the different application domains would also represent an interesting study in the context of progressive data exchange.

From a more practical (implementation-oriented) point of view, further developments and decisions are required in relation to the following aspects:

- Definition of a Transmission Technique: This should be efficient (reduce users' waiting time). Studies should be conducted on deciding, for example, whether to segment transmitted data sets so as to transmit smaller files. However, this might further complicate operations for topology reconstruction.
- Development of algorithms to efficiently compile transmitted increments into a topologically consistent format for the reconstruction of intermediate levels of detail. This should be facilitated by the use of vertical links.
- Development of graphic interfaces and dynamic visualization tools for displaying different map versions corresponding to different levels of detail (during transmission).

- Concurrent Accesses: As with all online systems, a progressive transmission system should be able to handle many concurrent accesses. As the systems proposed so far were mainly prototypes this aspect has not been fully analyzed.
- Thin vs Thick Client Solutions: While the server is provided with methods for building, manipulating, and transmitting a sequence of map representations at different levels of detail, depending on the device capabilities and bandwidth, a client could be simply provided with either visualization functionality or a set of operations for updating and integrating the transmitted levels as well as for analyzing and querying. Different caching techniques should also be investigated.
- Interoperability and Openness: So far mainly application-specific prototype implementations, often not complying with interoperability specifications [35, 46], have been developed.

Finally progressive transmission systems should make available, depending on users' purposes, a combination of raster and vector formats. Full comparisons on the transmission of vector map versions and their rasterized counterparts have not been conducted. It would be interesting to evaluate performance versus users' requirements in different scenarios.

4.7 Conclusions

In this chapter, we have reviewed existing approaches to progressive transmission of spatial data over the Internet. Successful and commonly used approaches are limited to the transmission of raster images and triangular meshes. For these types of data, compression mechanisms are effective and efficient. For more general spatial data in vector format (e.g. map data), several challenges and open issues still remain that have thus far prevented the successful implementation of effective progressive data exchange. In this chapter, we have discussed these critical aspects and identified research areas that still need much investigation.

The recent diffusion and popularity of mobile devices has given rise to a lot of research into mobile GIS applications. Due to screen, bandwidth, and processing capabilities limitations, progressive vector transmission would be extremely beneficial within this context. However, the same conceptual problems remain, and from an implementation point of view, mobile device limitations impose even more impediments. Studies in this direction are still at an early stage.

References

1. Bajaj CL, Pascucci V, Zhuang G (1999) Single resolution compression of arbitrary triangular meshes with properties. In: Proc. of IEEE Data Compression Conference, 247–256
2. Ballard DH (1981) Strip Trees: A Hierarchical Representation for Curves. Communications of the ACM, 24(5):310–321
3. Barnsley M (1989) Fractals Everywhere. Academic Press, San Diego

4. Bertolotto M (1998) Geometric Modeling of Spatial Entities at Multiple Levels of Resolution. Ph.D. Thesis, Department of Computer and Information Sciences, University of Genova, Italy
5. Bertolotto M, Egenhofer MJ (1999) Progressive Vector Transmission. In: Proc. of ACMGIS'99, Kansas City, MO, 152–157
6. Bertolotto M, Egenhofer MJ (2001) Progressive Transmission of Vector Map Data over the World Wide Web. GeoInformatica, Kluwer Academic Publishers, 5(4):345–373
7. Buttenfield B (1999) Progressive Transmission of Vector Data on the Internet: A Cartographic Solution. In: Proc. 18th International Cartographic Conference, Ottawa, Canada
8. Buttenfield B (2002) Transmitting Vector Geospatial Data across the Internet. In: Proc. of GIScience2002, Boulder, Colorado, USA, 51–64
9. Buttenfield B, McMaster R (1991) Map Generalization: Making Rules for Knowledge Representation. Longman, London
10. Cecconi A, Weibel R (2000) Map Generalization for On-demand Web Mapping. In: Proc. of GIScience2000, Savannah, Georgia, USA, 302–304
11. Cecconi A, Galanda M (2002) Adaptive Zooming in Web Cartography. In: Proc. of SVG Open 2002, Zurich, Switzerland
12. Davis G, Nosratinia A (1998) Wavelet-Based Image Coding: An Overview. Applied and Computational Control, Signals, and Circuits, 1(1)
13. Davis G (1996) Implicit Image Models in Fractal Image Compression. In: Proc. of SPIE Conference on Wavelet Applications in Signal and Image Processing IV, Denver, TX
14. Davis G (1998) A Wavelet-based Analysis of Fractal Image Compression. IEEE Transactions on Image Processing, 7(2):141–154
15. De Floriani L, Puppo E (1995) Hierarchical Triangulation for Multiresolution Surface Description. ACM Transaction on Graphics, 14(4):363–411
16. De Floriani L, Magillo P, Puppo E (1998) Efficient Implementation of Multi-Triangulations. In: Proc. of IEEE Visualization '98, Research Triangle Park, NC, USA, 43–50
17. Dettori G, Puppo E (1996) How generalization interacts with the topological and geometric structure of maps. In: Proc. of Spatial Data Handling '96, Delft, The Netherlands
18. Douglas DH, Peucker TK (1973) Algorithms for the reduction of the number of points required to represent a digitized line or its caricature. The Canadian Cartographer, 10(2):112–122
19. Eck M, DeRose T, Duchamp T, Hoppe H (1995) Multiresolution analysis of arbitrary meshes. In: Proc. of SIGGRAPH '95, 173–182
20. Egenhofer MJ, Franzosa R (1991) Point-set topological spatial relations. International Journal of Geographic Information Systems, Taylor and Francis, 5(2):161–174
21. Fisher Y (ed) (1995) Fractal Image Compression: Theory and Application to Digital Images. Springer, Berlin Heidelberg New York
22. Follin J M, Boujou A, Bertrand F, Stockus A (2005) An Increment Based Model for Multiresolution GeoData Management in a Mobile System. In: Proc. of W2GIS'05, Lausanne, Switzerland, Lecture Notes in Computer Science, Springer, Berlin Heidelberg New York, 42–53
23. Gonzales R C, Woods R E (1993) Digital Image Processing. Addison-Wesley
24. Goodchild MF, Proctor J (1997) Scale in Digital Geographic World. Geographical & Environmental Modelling, 1(1):5–23
25. Han Q, Bertolotto M (2004) A Multi-Level Data Structure for Vector Maps. In: Proc. of ACMGIS'04, Washington, DC, USA, 214–221

26. Hardy P (2000) Multi-Scale Database Generalisation for Topographic Mapping, Hydrography and Web-Mapping, Using Active Object Techniques. In: International Society for Photogrammetry and Remote Sensing (IAPRS), Vol. XXXIII, Amsterdam

27. Hoppe H (1996) Progressive Meshes. In: Proc. of SIGGRAPH'96, 99–108

28. Junger B, Snoeyink J (1998) Importance Measures for TIN simplification by parallel decimation. In: Proc. of Spatial Data Handling '98, Vancouver, Canada, 637–646

29. Junger B, Snoeyink J (1998) Selecting Independent Vertices For Terrain Simplification. In: Proc. of WSCG '98, Plzen, Czech Republic

30. Kern P, Carswell JD (1994) An Investigation into the Use of JPEG Image Compression for Digital Photogrammetry: Does the Compression of Images Affect Measurement Accuracy. In: Proc. of EGIS'94, Paris, France

31. Lagrange J, Ruas A (1994) Geographic Information Modelling: GIS and Generalisation. In: Proc. of Spatial Data Handling '94, Edinburgh, Scotland, 1099–1117

32. McMaster RB, Shea KS (1992) Generalization in Digital Cartography. Association of American Geographers Washington DC

33. Morlet J, Grossman A (1984) Decomposition of Hardy Functions into Square Integrable Wavelets of Constant Shape. Siam J Math Anal, 15(4):723–736

34. Müller JC, Weibel R, Lagrange JP, Salgé F (1995) Generalization: State of the Art and Issues. In: Müller JC, Lagrange JP, Weibel R (eds) GIS and Generalization: Methodology and Practice, Taylor and Francis, 3–7

35. OGC (OpenGeoSpatial Consortium) http://www.opengeosptial.org

36. OpenMap^{TM}, BBN Technologies, http://openmap.bbn.com

37. Oracle^{TM} www.oracle.com

38. Parent C, Spaccapietra S, Zimanyi E, Donini P, Planzanet C, Vangenot C (1998) Modeling Spatial Data in the MADS Conceptual Model. In: Proc. of Spatial Data Handling '98, Vancouver, Canada

39. Puppo E, Dettori G (1995) Towards a formal model for multiresolution spatial maps. In: Egenhofer MJ, Herring JR (eds) Advances in Spatial Databases, Lecture Notes in Computer Science, Springer, Berlin Heidelberg New York, 152–169

40. Ramer U (1972) An interactive procedure for the polygonal approximation of plane curves. Computer Vision, Graphics and Image Processing, 1:244–256

41. Rauschenbach U, Schumann H (1999) Demand-driven Image Transmission with Levels of Detail and Regions of Interest. Computers and Graphics, 23(6):857–866

42. Robinson G, Lee F (1994) An automatic generalisation system for large-scale topographic maps. In: Worboys MF (ed) Innovations in GIS-1, Taylor and Francis, London

43. Robinson AH, Sale RD, Morrison JL, Muehrcke PC (1984) Elements of Cartography. John Wiley, New York, 5th edition

44. Saalfeld A (1999) Topologically Consistent Line Simplification with the Douglas-Peucker Algorithm. Cartography and Geographic Information Science, 26(1):7–18

45. Samet H (1990) The Design and Analysis of Spatial Data Structures. Addison-Wesley, Reading, MA

46. Sondheim M, Gardels K, Buehler K (1999) GIS Interoperability. In: Longley P, Goodchild MF, Maguire DJ, Rhind DW (eds) Geographical Information Systems: Principles, Techniques, Management and Applications, Second Edition, Cambridge, GeoInformation International, 347–358

47. Srinivas BS, Ladner R, Azizoglu M, Riskin EA (1999) Progressive Transmission of Images using MAP Detection over Channels with Memory. IEEE Transactions on Image Processing, 8(4):462–75

48. Weibel R, Dutton G (1999) Generalising spatial data and dealing with multiple representations. In: Longley P, Goodchild M F, Maguire D J, Rhind D W (eds) Geographical Information Systems: Principles, Techniques, Management and Applications, Second Edition, Cambridge, GeoInformation International, 125–155

49. Werschlein T, Weibel R (1994) Use of Neural Networks in Line Generalisation. In: Proc. of EGIS '94, Paris, 76–85

50. Zhao Y, Yuan B (1996) A hybrid image compression scheme combining block-based fractal coding and DCT. Signal Processing: Image Communication, 8(2):73–78

51. Zhou M, Bertolotto M (2004) Exchanging generalized maps across the Internet. In: Proc. of KES'04, Wellington, New Zealand, Lecture Notes in Artificial Intelligence, Springer, Berlin Heidelberg New York, 425–431

52. Zhou M, Bertolotto M (2005) Efficiently generating multiple representations for web mapping. In: Proc. of W2GIS'05, Lausanne, Switzerland, Lecture Notes in Computer Science, Springer, Berlin Heidelberg New York, 54–65

53. Zorin D, Shroder P, Sweldens W (1997) Interactive multiresolution mesh editing. In: Proc. of SIGGRAPH '97, 259–268

References containing URLs are validated as of June 1st, 2006.

5

Model Generalization and Methods
for Effective Query Processing and Visualization
in a Web Service/Client Architecture

Marian de Vries and Peter van Oosterom

Delft University of Technology, Delft (The Netherlands)

5.1 Introduction

Generalization is a long-standing research subject in geo-science. How to derive 1:50,000 scale maps from 1:10,000 scale maps, for example, has been an issue also in the pre-Internet age. But now, because of distributed geo-processing over the Web (Web mapping and Internet GIS), research into automated generalization of geo-data received a new impulse. One of the research issues is how to enable 'zoom-level dependent' retrieval and visualization of geo-data in Web service/client contexts.

In a Web context, to be able to display the right amount of detail on a map in a Web client is important for a number of reasons. First of all there is the 'classic' requirement of cartographic quality. Too much information will clutter the map image and make it unreadable and too little detail is not very useful either. Secondly, especially because of the Web service/client context, there is the requirement of response time: the LOD of the geo-information that is displayed in the Web client will influence both the amount of bytes transferred over the network and the visualization (rendering) time in the client. Generalization at the Web-server side, before the information is sent to the client, has therefore two advantages: the user gets the LOD that is needed at that zoom level, and response time as experienced by the user will be reduced.

Generation and display geo-information at the right LOD can be tackled in different ways. One approach is to use multiscale/multirepresentation databases that are created in an offline generalization process, and then switch to the appropriate representation for a certain zoom level when the user zooms in or out (e.g. [7]). Another approach is to use on-the-fly generalization, where, for example, line simplification is carried out in real time as part of the visualization process (in the Web client or by a specialized generalization Web service, see [3]).

In this chapter we present a third approach, based on a variable scale data structure complex: the topological GAP structure (generalized area partitioning structure), or tGAP structure for short. The purpose of this structure is to store the geometry only once, at the original, most detailed, resolution. There is an offline generalization process, which builds a hierarchical, topological structure of faces (the GAP-face tree) and edges (the GAP-edge forest). Faces and edges are assigned an importance range (LOD range) for which they are valid. This set of tree-like structures is used when the data is requested by a Web client to derive the geometry on-the-fly at the right LOD, based on the importance values that were assigned during the generalization process.

In the first part of this chapter we give a short overview of the tGAP structure. In Sect. 5.2 the basic principles are described, followed by an explanation in Sect. 5.3 of how the generalization process builds a tGAP data set in a succession of steps. In the second part of the chapter we explore how the tGAP data structures can be used in a Web service/client environment. The focus is on two aspects: how the tGAP structure can support progressive transfer of vector data from Web service to client, and how adaptive zooming can be realized, preferably in small steps ('smooth' zooming). The relevant standards and protocols for vector data Web services are discussed in Sect. 5.4: web feature service (WFS) and geography markup language (GML). Section 5.5 explains how the tGAP structure can be used for progressive data transfer and smooth zooming and proposes some necessary extensions to the current standards (WFS and GML) in order to support vario-scale geo-information. Finally, Sect. 5.6 concludes this chapter with a summary of the most important findings and suggestions for further research.

5.1.1 Related Work

Vario-scale

Data structures supporting variable scale data sets are still very rare. There are a number of data structures available for multiscale databases based on multiple representations (MRDBs) fixed number of scale (or resolution) intervals [7, 9, 25]. These multiple representation data structures try to explicitly relate the objects at the different scale levels, in order to offer consistency during the use of the data. Most data structures are intended to be used during the pan and zoom (in and out) operations of a user, and in that sense multiscale data structures are already a serious improvement for interactive use as they do speed-up interaction and give a reasonable representation for a given LOD (scale). Drawbacks of the multiple representation data structures are that they store redundant data (same coordinates, originating from the same source) and that they support only a limited number of scale intervals.

Progressive Transfer

Another drawback of the multiple representation data structures is that they are not suitable for progressive data transfer, as each scale interval requires its own (independent) graphic representation to be transferred. In a Web service/client context progressive transfer can be very useful, because it reduces the waiting time as experienced by the end-user.

Nice examples of progressive data transfer are raster images, which are first presented relatively fast in a coarse manner and then refined when the user waits a little longer. These raster structures can be based on simple (raster data pyramid) [17] or more advanced (wavelet compression) principles [8, 10, 16]. For example, JPEG2000 (wavelet-based) allows both compression and progressive data transfer from the server to the end-user. Similar effects are more difficult to obtain with vector data and require more advanced data structures, though recently there are a number of interesting initiatives [2, 4, 9, 25]. (See Chap. 4 of this book for an extensive overview of issues and possibilities in the field of progressive transfer of vector data.)

5.2 Generalized Area Partitioning: The (t)GAP-tree

The tGAP structure can be considered the topological successor of the original GAP-tree [22]. The idea of the GAP-tree was based on first drawing the larger and more important polygons (area objects), which then results in a generalized representation. After this first step, the map can be further refined through the additional drawing of the smaller and less important polygons on top of the existing polygons (based on the Painters algorithm). If one keeps track of which polygon refines which other polygon, then the result is a tree structure, as a refining polygon completely falls within one parent polygon and there is only one 'root' polygon (covering the whole area). The tree structure is built during the generalization process by thinking the other way around: starting with the most detailed representation, find the least important object (child) and assign this to the most compatible neighbor (parent). This process is then repeated until only one single polygon remains, the root (see Fig. 5.1).

Drawbacks of this original GAP-tree were as follows: (i) redundancy as boundaries between neighbor polygons are stored twice at a given scale (and redundancy

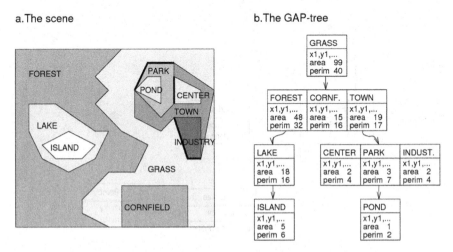

Fig. 5.1. The original GAP-tree [22]

of the boundaries shared between scales; and (ii) boundaries are always drawn at the highest detail level (all points are used, even if it would not be meaningful for a given scale). Therefore, the tGAP structure was developed: a topological structure with faces and edges and no duplicate geometries, but using the same principle as the original GAP-tree – coarse representations of an area are refined by more detailed ones. A more in-depth explanation and motivation of the steps taken in the development of the tGAP structure can be found in [23].

The parent–child relationships between the faces form a tree structure with a single root: the GAP-face tree. Also the parent–child relationships between edges at the different scale (detail) levels form tree structures, but now with multiple roots; therefore this is called the GAP-edge forest. In order to take care of the line simplification, the edges are not stored as simple polylines, but as binary line generalization trees: BLG-trees [20]. The BLG-tree is a binary tree for a variable scale representation of a polyline based on the Douglas–Peucker [6] line generalization algorithm. The BLG-tree will be explained in more detail in Sect. 5.3.3.

When two edges are combined to form one edge at a higher importance level (i.e. lower LOD), then their BLG-trees are joined, but no redundant geometry is stored.

Figure 5.2 shows the conceptual model of the tGAP structure complex. Only the Point class has a geometry property. This is where the actual (vertex) coordinates are stored. The Face class and the BLG-tree class have methods to construct the geometry at a certain LOD: constructPolygon() (to derive the polygons) and constructPolyline() (to derive the polylines).

In order to efficiently select the requested faces and edges for a given area (spatial extent) and scale (importance) during query and visualization of the data, a specific index structure is proposed: the Reactive-tree [21]. As this type of index is not

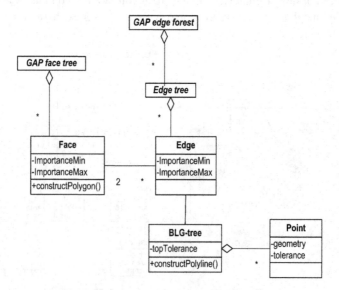

Fig. 5.2. Abstract model of the tGAP structure

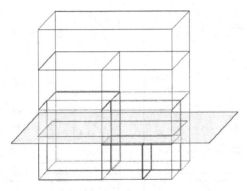

Fig. 5.3. Importance levels represented by the third dimension (*at the most detailed level, bottom, there are several objects while at the most coarse level, top, there is only one object*); the hatched plane represents a requested level of detail and the intersection with the symbolic '3D volumes' then gives the faces

available in most systems, an alternative is to use a 3D R-tree to index the data: the first two dimensions are used for the spatial domain and the third dimension is used for the scale (or more precisely for the importance level as this is called in the context of the tGAP structure).

Together, the GAP-face tree, the GAP-edge forest, the BLG-trees, and the Reactive-tree are called the tGAP structure. The tGAP structure can be used in two different ways (see Fig. 5.3): to produce a representation at an arbitrary scale (a single map) and to produce a range of representations from rough to detailed representation. Both ways of using the tGAP structure are useful, but it will be clear that in the context of progressive transfer (and smooth zooming) the second way must be applied.

5.3 Building the tGAP-tree

Throughout this section an example illustrates the steps in the generalization process that results in the tGAP structure (see Figs. 5.4 and 5.5). Different subsections will explain additional details. Starting with the source data (at the most detailed resolution) the generalization steps are carried out until the complete topological GAP structure is computed. Figures 5.4 and 5.5 top left show the edges and top right shows the faces, each with a color according to their classification. The faces have a number as id and a computed importance value (shown in a smaller font). The edges have a letter as id (just for illustration purposes, in a normal implementation edge id's will also be numbers). Note that all edges are directed, as is normally the case in a topological structure. So the edges have a left- and a right-hand side.

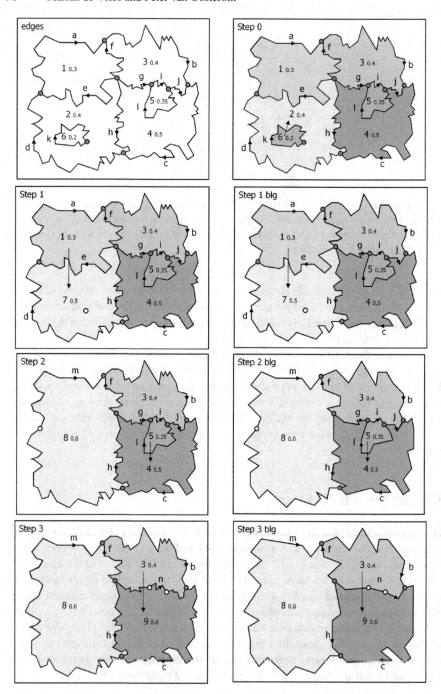

Fig. 5.4. Generalization example in five steps (*see Fig. 5.5 for the last two steps*) from detailed to coarse (*left side shows the effect of merging faces, right side shows the effect of also simplifying the boundaries*). Nodes are depicted in gray and removed nodes are shown for only one next step in white

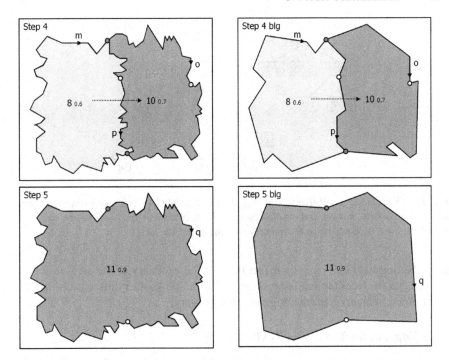

Fig. 5.5. Generalization example in five steps from detailed to coarse (*see Fig. 5.4 for the first three steps*)

5.3.1 Faces in the Topological GAP-tree

In the example above, in every step the least important object is removed and its area is assigned to the most compatible neighbor (as in the original GAP-tree). In the first step in Fig. 5.4, the least important object, face '6,' has been added to its most compatible neighbor, face '2.' This process is continued until there is only one face left. A little difference with the original GAP-tree is that a new id is assigned to the face, which is enlarged, that is the more important face. The enlarged version of face '2' (with face '6') is called face '7,' and faces '2' and '6' are not used at this detail level. However, this face keeps its classification as indicated by the color of the faces. In the example, face '7' has the same classification as face '2,' but the importance of face '7' is recomputed: as it becomes larger the importance increases from '0.4' to '0.5.' In the next step, face '1' (the least important with importance value '0.3') is added to face '7' (the most compatible neighbor) and the result is called face '8' (and its importance raises from '0.5' to '0.6'). Then face '5' (with importance '0.35') is added to face '4' (best neighbor) and the result is called face '9' (with increased importance from '0.5' to '0.6'). This process continues until one large area object is left, in the example this is face '11.'

From the conceptual point of view the generalization process for the faces is the same as with the original, non-topological GAP-tree (Fig. 5.6). Quite different

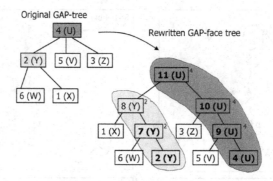

Fig. 5.6. The 'classic' GAP-tree rewritten to the topological GAP-face tree (*with new object id whenever a face changes and the old object id shown in a small font to the upper-right of a node*); the class is shown between brackets after the object id

from the original GAP-tree is the fact that not the geometry of the area objects is merged, but the faces (and the edges, see next section). Another difference is that the tGAP-tree is a binary tree (the original tree was n-ary).

5.3.2 Edges in the Topological GAP-tree

When faces are merged during the generalization process, three different things may happen with the edges:

1. An edge is removed; for example, edge 'e' in step 2 (see Fig. 5.4);
2. Two or three edges are merged into one edge; for example, edges 'a' and 'd' into new edge 'm' in step 2 or edges 'g,' 'i,' and 'j' into new edge 'n' in step 3;
3. Only edge references are changed; for example, the reference to the right face of edge 'h' is changed from '7' to '8' in step 2.

The parent–child relationships between the edges at the different importance levels again form a tree structure, as with the faces (Fig. 5.7). However, there is no single root for the edges (as there is for the faces in the GAP-face tree) due to the edges that are removed (scenario 1). When an edge is removed, it turns into a 'local' root, at the top of its own tree of descendants. So, there is not one GAP-edge tree, but a collection of trees: the GAP-edge forest.

Similar to the faces, which have an associated importance range, the edges are also assigned an importance range, indicating when they are valid. These edge importance ranges are shown in Fig. 5.7. Note that only the top edge (in this case 'q') has no upper value for the importance range, all other nodes have both a lower and a upper importance value associated with the edge.

5.3.3 BLG-trees in the Topological GAP-tree

The edges are not represented by polylines but by BLG-trees, enabling line simplification later on, when the data is retrieved and visualized in the Web client. Figure 5.8

Edge GAP-forest

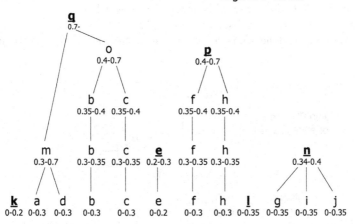

Fig. 5.7. GAP-edge forest (*with importance ranges*); note that the edges shown in bold and underlined font (*k, q, e, p, l, and n*) are the roots of the different GAP-edge trees

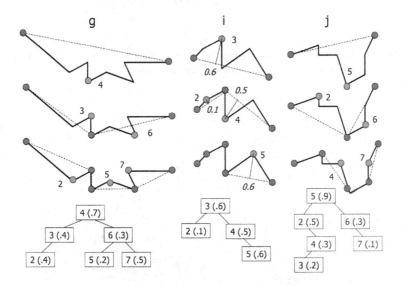

Fig. 5.8. Three examples of BLG-trees of edges 'g,' 'i,' and 'j'; the nodes in the BLG-tree show the id and between brackets the tolerance value of each point (*vertex*)

shows three different edges (top) with their corresponding BLG-trees (below the edges); nodes in the BLG-trees indicate point number and tolerance values.

The BLG-trees can be used very well in the tGAP structure to represent the edges. When edges are merged into longer edges (as a consequence of the merging of faces), the BLG-trees of these edges are joined to form a new BLG-tree. If three edges, for example 'g,' 'i,' and 'j,' are merged, this is done in two steps: first 'i' and 'j' are merged (see Fig. 5.9), then edge 'g' is merged (see Fig. 5.10).

When two BLG-trees are joined a new top-level tolerance value has to be assigned. This can be done using a 'worst-case' estimation or by computing the

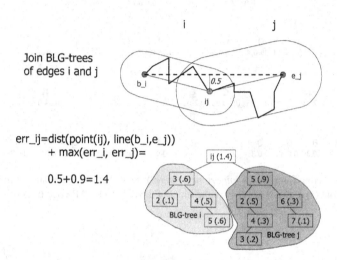

Fig. 5.9. First step in the merging of edges 'g,' 'i,' and 'j': the BLG-trees of 'i' and 'j' are joined (*worst-case estimation for new top-level tolerance value is used – '1.4'*)

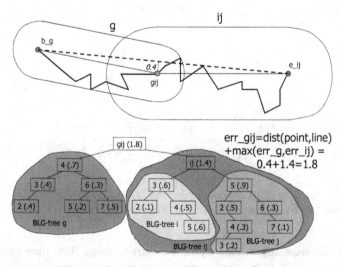

Fig. 5.10. Second step in the merging of edges 'g,' 'i,' and 'j': the BLG-trees of 'ij' and 'g' are joined (*note that for BLG-tree 'ij' the worst-case estimation for the tolerance value '1.4' was used and therefore the tolerance band does not touch the polyline in the middle as might be expected*)

exact new tolerance value. The worst-case estimation of the new top tolerance value 'err_ij,' according to the formula given in [20] page 92 and reformulated in Fig. 5.9 as 'err_ij = dist(point(ij), line(b_i, e_j) + max(err_i, err_j)' only uses the top-level information of the two participating trees.

An improvement, which keeps the structure of the merged BLG-tree unaffected (so the lower level BLG-tree can be reused again), is to compute the exact tolerance value ('err_ij_exact') of the new approximated line, which is less than or equal to the estimated worst-case ('err_ij'). In Fig. 5.9, this would be the distance from point 5 of edge 'j' to the dashed line. This tolerance value would be '1.1,' which is less than the worst-case estimate of '1.4' for the tolerance. The drawback is that one has to descend in the lower level BLG-tree for the computation (this may be a recursion) and this will take more time (but normally has to be done only once: during the generalization process when the tGAP structure is built). The advantage is that during the use of the structure (which happens more often than creation) one has a better estimate and will not descend the BLG-tree unneeded. For example, assume one needs a tolerance of '1.2,' then with the worst-case estimate one has to descend into the two child BLG-trees. This is not necessary when the top-level tolerance value for the joined BLG-tree is computed instead of estimated.

When the data set is accessed and the faces and edges are selected based on their importance range, the corresponding BLG-trees are used for line simplification. Depending on the requested tolerance value the (joined) BLG-tree is traversed in order to produce the appropriate detail level. Note that the relationship between the tolerance value and map scale is quite direct (e.g. one could use the size of a pixel on the display screen as the tolerance value).

5.4 Current Internet GIS Protocols

One of the main features of the proposed structure is that it supports progressive transfer and smooth zooming in a Web service/client context. Therefore in this section we will look at the most relevant Web service Protocols for vector data. We also point out the main characteristics of a Web service/client architecture.

5.4.1 Geo-Web Services

For Web services that provide access to geo-information the standardization efforts of the open geospatial consortium (OGC) are very important, therefore some background is presented here first.

The OGC was founded in 1994 by a number of software companies, large data providers, database vendors, and research institutions. It can be considered an industry-wide discussion and standardization forum for the geo-application domain. One of the goals of OGC is to enhance interoperability between software of different vendors. With this purpose also a number of Web service interface specifications have been developed, the first two were the Web map service (WMS) and the Web feature service (WFS) protocols.

An OGC Web Map Service is an example of a Portrayal service. It has raster images as output, which can be displayed without further processing in the Web client. Some WMS implementations also offer scalable vector graphics (SVG) as output format. A disadvantage of a WMS service that serves raster images is that every zoom or pan action of the user leads to a new map request to the server, which slows down the interaction. A WMS is also limited in its selection possibilities: only a selection on spatial extent (bounding box) and on complete map layer is possible [14].

For our purpose we need a Web service that will send vector data to the client. Only in this way are the progressive refinement and gradual changes in the level of detail possible without continuous requests to the service every time the user changes the zoom level. For that reason, a WFS is a better choice. A WFS is primarily a data service. It produces vector geo-data in geography markup language (GML) that still has to be processed into a graphic map in the client. A Transactional WFS service (WFS-T) can also be used for editing data sets [13]. In the next paragraphs we will give some more information about the WFS protocol, and in Sect. 5.5 discuss whether extensions to that protocol are needed to enable progressive refinement using the tGAP data structures.

5.4.2 WFS Basic Principles

The WFS requests and responses are sent between client and server using the common Internet HTTP protocol. There are two methods of encoding WFS requests. The first uses keyword-value pairs (KVP) to specify the various request parameters as arguments in a URL string (Example 1 below), the second uses XML as the encoding language (Example 2). In both cases (XML and KVP) the response to a request (or the exception report in case of an error) is identical. KVP requests are sent using HTTP GET, while XML requests have to be sent with HTTP POST. With WFS different types of selection queries are possible:

1. Get all feature instances (objects) of a certain feature type (object class) (no selection). For this the following HTTP GET request can be used:

   ```
   http://www.someserver.org/servlet/wfs?REQUEST=
   GetFeature&VERSION=1.0.0&TYPENAME=gdmc:parcel
   ```

2. Select a subset of the features, by specifying a selection filter. Both spatial and non-spatial properties of the feature (attributes of the class) can be used in these filter expressions. Using a filter implies the use of HTTP POST, for sending an XML-encoded request to the WFS service.

   ```
   <wfs:GetFeature service="WFS" version="1.0.0"
       outputFormat="GML2" >
   <wfs:Query typeName='gdmc:parcel'>
     <ogc:Filter>
       <ogc:PropertyIsEqualTo>
         <ogc:PropertyName>gdmc:municip</ogc:PropertyName>
         <ogc:Literal>GDA01</ogc:Literal>
   ```

```
        </ogc:PropertyIsEqualTo>
      </ogc:Filter>
    </wfs:Query>
  </wfs:GetFeature>
```

3. Get one specific feature, identified with a feature id. Here there is again a choice: either the filter encoding can be used (see above) or the KVP 'featureid=...' in a HTTP GET 'GetFeature' request, as in the URL string:

> http://www.someserver.org/servlet/wfs?REQUEST=GetFeature
> &VERSION=1.0.0&TYPENAME=gdmc:parcel&featureid=parcel.93

5.4.3 Geography Markup Language

The output of a WFS service is a GML data stream. GML is one of the other important interoperability initiatives of OGC. Its current version is 3.1.1 [12]. GML is a domain-specific XML vocabulary meant for the exchange of geo-data. Until version 3, GML only contained the basic geometry types (point, polygon, line string). Version 3 introduced more complex geometry types and also topology. Apart from its grammar and syntax, GML also has a meta-model (or conceptual model). At first (GML 1 and GML 2) this was the Simple Features Specification of OGC itself. From GML 3 the conceptual model is based on ISO 19107. The core of GML consists of a large number of geometry and topology data types (primitives, aggregates, and complexes) to encode the spatial properties of geographical objects. Apart from the core data types, the GML specification also contains rules how to derive other, domain-specific spatial and non-spatial data types (using XML Schema).

5.4.4 Interoperability and Getcapabilities

The basic goal of the Web service specifications of OGC is that it must be possible for service software and client software of different vendors to communicate without having 'inside knowledge' of that other component. This is achieved first of all by writing down in an interface specification (or protocol) how to access that type of Web service, which requests are possible, what are the parameters, which of these parameters are optional. What is also specified is what the response of the Web service to the requests should be (in the case of a WFS GetFeature request always GML must be offered by the service, other formats are optional).

The second method to enhance interoperability is by letting an OGC Web service be 'self-describing': for that purpose, all OGC Web services have a GetCapabilities request. The GetCapabilities response of the service contains meta-information about

- the requests that are possible on that particular Web service and the URLs of those requests;
- the responses that are supported (the output that can be expected);

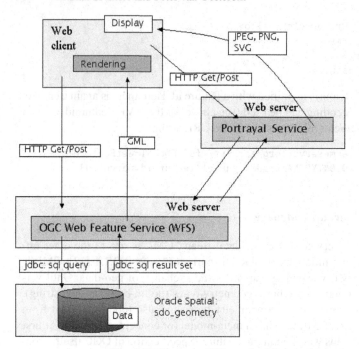

Fig. 5.11. Three-tier Web service/client architecture

- the data that is published by the service and the way this is done: map layers (WMS), feature types (WFS), bounding box of the area, what data is offered in what coordinate system, in the case of a WMS also visualization styles that are available, in the case of a WFS what selection (filter) operators will work, and finally also the output formats that are supported.

Because an OGC compliant Web service of a certain type only has to offer the 'must have' requests and possibilities, the capabilities document is a kind of ingredients label for that particular service.

5.4.5 Service/Client Architecture

A basic Web service/client architecture consists of three layers: data layer, application layer (Web server with one or more Web services), and presentation layer (Web client). This is no different in the case of the OGC Web services like the Web feature service (WFS).

A WFS service runs on a Web server, for example as a Java Servlet application, JSP, or .NET application. This can be open source software like GeoServer or UMN MapServer, or software packages of vendors like Ionic Software, Intergraph, ESRI, Autodesk, MapInfo. (See [15] for a list of WFS products.)

For the WFS client-side software there are a number of options. WFS clients can either be

- HTML-based Web clients that can be run in an Internet browser like Mozilla or Internet Explorer;
- desktop applications that have to be installed first on the user's computer. Also in this case there is a choice between open source clients (uDig, Gaia) and WFS-enabled desktop GIS or CAD software of companies like Intergraph (GeoMedia); or
- other Web services, for example a WMS service that plays the role of Portrayal service for the GML output of the WFS service (see Fig. 5.11).

In this three-tier architecture the geo-database software or data-storage formats that are used in the data layer are not 'visible' for the WFS service–client interaction. The WFS client does not have to now the implementation details of the data layer, because it only communicates with the WFS service. So in the next section, we will concentrate on the WFS service–client interaction.

5.5 Server–Client Set-up and Progressive Refinement

Progressive refinement of data being received by a client could be implemented in the following way. The server starts by sending the most important nodes in tGAP structure (including top levels of associated edge BLG-trees) in a certain search rectangle. The client builds a partial copy of the data in tGAP-structure, which can then be used to display the coarse impression of the data. Every (x) second(s) this structure is displayed and the polygons are shown at the then available level of detail. The server keeps on sending more data, and the tGAP-structure at the client side is growing (and the next time it is displayed with more detail). Several stop criteria can be imagined: (a) 1000 objects (meaningful information density), (b) required importance level is reached (with associated error tolerance value), or (c) the user interrupts the client.

In this section, we will look in some detail how the tGAP structure can be used in a Web service/client setting for progressive transfer and for refinement during zooming. As shown in Fig. 5.11, the WFS service acts as middle layer between Web client and geo-database. So we have to establish two things: (1) what queries are necessary from WFS to database to retrieve the vario-scale data, enabling progressive transfer and smooth zooming later on (Sect. 5.5.1), and (2) what does this mean for the requests and responses between Web client and WFS service (Sect. 5.5.2). We will start with these two questions. In the last paragraph of Sect. 5.5.3, we will shortly discuss the requirements of progressive transfer and refinement for the software used for geo-database, WFS server, and WFS client.

5.5.1 Queries to the Database

From a functional perspective there are two situations in the Web service–client interaction: (1) the initial request to the WFS to get features for the first time during the session; and (2) additional requests when zooming in, zooming out, and panning.

Initial Requests

When the Web client requests the data at one specific importance level (e.g. 101) the query to the database would be something like this:

```
select face_id as id, '101' as impLevel,
    RETURN_POLYGON(face_id, 101) as geom
from tgapface
where imp_low <= 101 and 101 < imp_high
```

This query will retrieve polygons from the database (constructed dynamically by the RETURN_POLYGON function). The result is one 'slice' of the data set at a specific importance level. The disadvantage is that all objects have to be visualized at once (to get a complete map without holes), progressive transfer is not possible, and the GAP-tree structure is not 'rebuilt' at the client for smooth zooming later on. The next query will also retrieve polygons from the database, but now not one slice, but a range of objects from the most important level until a certain importance level (e.g. 90), sorted in order of importance (descending from high to low):

```
select face_id as id, imp_low, imp_high,
    RETURN_POLYGON(face_id,imp_high-1) as geom
from tgapface
where imp_high > 90
order by imp_high desc
```

Note no upper boundary of the required importance is specified, only a lower boundary of '90.' This means that everything starting above 90 and up to the root importance will be selected. When the WFS service receives the results from the data layer the data is already in the right importance order (from high importance to low importance) and can be passed to the client (as GML) in that same order. When the client receives the GML, there are three possibilities that differ with respect to the moment that the data is visualized as map:

a. Rendering in small steps. The client software already starts visualizing parts of the incoming GML data before the whole data stream is received. Here we have an example of 'true' progressive transfer. The objects are received and visualized in sets of two objects which replace one parent at a time (see Sect. 5.2, binary GAP-face tree). The map is progressively rendered in small steps. The visualization is very smooth: the user first sees the contours and then slowly the details are filled in (see screenshots in Fig. 5.12).

b. Visualize the incoming GML in two or more larger steps, refreshing the complete spatial extent that is displayed in the map after a certain time interval (of 2 seconds for example). In this case the client will at given times visualize the level of the latest objects it has received. Here also the data is in the right importance order so that progressive refinement/generalization in the later stages is possible, only the refresh is done for the whole area.

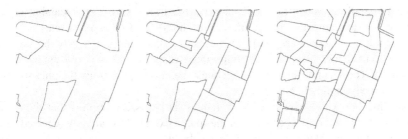

Fig. 5.12. Progressive rendering in the Web client in small steps (*method 1*)

c. No progressive transfer and rendering (the common situation with current WFS client software): the client waits until the complete data stream is received and then visualizes the data at the appropriate level all at once. The data is still vario-scale and in order of importance, so the basis for smooth zooming during user interaction is also there.

Not using the topology (at client side) in the above scenarios is a limiting factor. When the client receives server-side constructed polygons and there is no line simplification, the Painters algorithm works fine: it takes care of hiding the coarser objects when the more detailed objects are received. However, in the case of line generalization the Painters algorithm does not suffice because with line simplification the shape of the derived polygons will change, so the more detailed lines will not always 'hide' the coarser lines. This means that in the case of line simplification the approach will only work when the topology-to-polygon reconstruction is carried out in the Web client and not in the database. The GAP-edges, BLG-trees, and nodes with their geometry will now be streamed to the client; the polylines can already be visualized during retrieval, so there is progressive transfer, but instead of the Painters algorithm other methods are needed to hide the previously received, coarser lines.

Zooming and Panning

In the case of a WFS service (providing vector data) zooming and panning by the user can often be handled in the client itself, without having to send new GetFeature requests to the server. But when only a part of the objects for that spatial extent is already in the client, new requests might be necessary: for example when zooming in, more detailed data could be needed. With panning beyond the original spatial extent the situation is a bit more complicated because it depends on the characteristics of the new spatial extent (the density of objects there) what needs to happen.

When an additional GetFeature request is not necessary, zooming in will mean hiding coarser objects and making visible more detailed ones. In the case of zooming out this process is the opposite: now the more detailed objects have to be 'switched' off, and the coarser objects will be made visible. Important is that all of this is handled in the client. Depending on the type of client this hiding/displaying could be based on event handling. When the GML data is visualized using SVG (scalable vector graphics) for example, the 'onzoom' event of the SVG DOM can be used.

When a new GetFeature request is necessary, the scenario is partly the same as in the previous case: first the right importance range for the new objects has to be calculated. The second step is then to request the new objects (only the ones that are not yet already in the client).

Finding out whether or not a new request is necessary implies that the client software should not only keep track of the spatial extent of the already received data, but also of the importance range(s) of these objects.

An issue for further research is how to calculate the right importance range for the 'new' set of objects to be displayed. The algorithm will have to be a function of the size of the map window (in pixels), the zooming factor for that particular zoom action, and some kind of optimal number of objects for that map size. To establish the optimal number of objects a straightforward rule of thumb could be "an optimal screen has a constant information density: so keep on adding objects with a lower importance until the specified number of objects is reached; for example 1000."

An alternative could be applying Töpfer's Radical Law: $n_f = n_a C \sqrt{(M_a/M_f)^x}$ where "n_f is the number of objects which can be shown at the derived scale, n_a is the number of objects shown on the source material, M_a is the scale denominator of the source map, M_f is the scale denominator of the derived map" [19]. The exponent 'x' depends on the symbol types (1 for point symbols, 2 for line symbols and 3 for area symbols) and 'C' is a constant depending on the nature of the data (an often used value is 1).

5.5.2 Extensions to OGC/ISO Standards?

Are extensions of the existing OGC WFS protocol necessary for the progressive transfer and refinement scenarios described in the previous paragraph? The first option is to use the existing GetFeature request and specify the importance range (imp_low, imp_high) as selection criteria in the Filter part of the request. And using the ogc:SortBy clause that is available since WFS version 1.1 the client can instruct the WFS service to return the objects in order of importance.

Still this is not an ideal solution. Somehow the Web service has to communicate to the client that it supports progressive transfer and refinement. The self-describing nature of the OGC WFS GetCapabilities response is an important part of creating interoperable service/client solutions. For the WFS protocol this means that somewhere in the GetCapabilities document it must be stated that this particular WFS server can send the objects sorted in order of importance. Another addition to the WFS Capabilities content is reporting the available importance range (imp_low – imp_high) of each feature type, comparable to the way the maximum spatial extent (in lat/long) of each feature type is given in the GetCapabilities response. For this reason (to supply solid service metadata that clearly state the capabilities of the service) it would be better to add a new request type to the WFS protocol, for example with the name GetFeatureByImportance. Just like there is, besides the Basic WFS, also a WFS-T (transactional WFS) with extra requests for editing via a Web service, we would then have a WFS-R (progressive refinement WFS) with an extra GetFeatureByImportance request and two parameters minImp and maxImp to specify the importance range of the features to be selected. When no value is specified for these parameters, all features are requested, but still in order of importance from high to low. And when minImp = maxImp exactly that importance level is requested, as a 'slice' of the data set (first query in Sect. 5.5.1). The HTTP Post request would then look like this (the 'D' in the ogc:SortBy is for 'descending'):

```
<wfs:GetFeatureByImportance service="WFS" version="1.1.0"
                            outputFormat="GML3" >
<wfs:Query typeName='gdmc:tgapface' minImp='50'
                            maxImp='150'>
  <ogc:Filter>
     <ogc:BBOX>
        <ogc:PropertyName>gdmc:geom</ogc:PropertyName>
        <gml:Box srsName="http://www.opengis.net/gml/srs/
                       epsg.xml#28992">
          <gml:coordinates>136931,416574 139382,418904
          </gml:coordinates>
        </gml:Box>
     </ogc:BBOX>
  </ogc:Filter>
  <ogc:SortBy>gdmc:imp_high D</ogc:SortBy>
</wfs:Query>
</wfs:GetFeatureByImportance>
```

The WFS will give a GML stream of vector data as output. To encode the components of the tGAP structure in GML, the standard GML 3 geometry and topology classes can be used without a problem, maybe with one exception: because of the constraint in the GML topology model that each gml:Edge should have 2 gml:directedNodes (see fragment below), also exactly two nodes per edge must be in the output GML. In the tGAP model an edge does not have start- and end-nodes: only the corresponding BLG-tree has start- and end-nodes.

```
<complexType name="EdgeType">
  <complexContent>
    <extension base="gml:AbstractTopoPrimitiveType">
      <sequence>
      <element ref="gml:directedNode" minOccurs="2"
                                 maxOccurs="2"/>
      <element ref="gml:directedFace" minOccurs="0"
                    maxOccurs="unbounded"/>
      <element ref="gml:curveProperty" minOccurs="0"/>
      </sequence>
      <attributeGroup
             ref="gml:AggregationAttributeGroup"/>
    </extension>
  </complexContent>
</complexType>
```

For the implementation of line simplification based on BLG-trees a new data type in GML would be useful. Client software then 'knows' how to react on receiving BLG-tree geometry and standards-based line simplification routines in the client will be enhanced.

For the same reason it would be useful to have standard attributes for the importance values of each feature (imp_low and imp_high), instead of having user-defined ones.

5.5.3 Implications for Server and Client Software

In order to accomplish the progressive refinement when the geo-data is retrieved from the WFS-R service an 'order by' expression (or functional equivalent) has to be included in the WFS query to the data source. Necessary requirement is therefore that the WFS service can retrieve the geo-objects from the data source in a sorted way. A 'Progressive Refinement' WFS-R service, that serves data in order of importance and in a certain importance range, does not need large extensions to WFS software. What is extra in the WFS layer is an follows. (1) adding 'order by' to the queries sent to the data source, and (2) adding importance selection to the filter conditions (either spatial or non-spatial) that the user already has specified in the request to the WFS.

A WFS-R client must be able to hide the coarser geo-objects when the more detailed geo-objects are received and visualized and vice versa (during zooming out). For polygons (with a solid color, not transparent) the painters algorithm takes care of hiding the previously received objects. For (partly) transparent polygons and for lines and points the visibility of the objects has to be manipulated by the client software, not only during zooming and panning, but also during progressive rendering in the initial request. In case of line simplification also the topology-to-polygon construction has to be handled in the client.

5.6 Conclusions

In this chapter we presented a generalization approach based on offline generalization that results in a variable scale, topological data structure (or rather a set of data structures): the tGAP structure. This approach has a number of advantages over multi-scale/multirepresentation databases: there is no geometry redundancy, data consistency is therefore easier to maintain under updates; the structure is very fine-grained, which makes smooth zooming possible; and the hierarchical (tree) structures enable progressive transfer.

This chapter did illustrate the functioning of a vario-scale structure in a Web service/client context. The tGAP structure is very well suited to a Web environment: the client requirements are relatively low (almost no geometric processing of the data at the client side) and progressive transfer of vector data is supported (allowing quick feedback to the user). That the tGAP structure can be used for progressive transfer is an important asset. In addition, because the faces and edges have a fine-grained 'importance lifetime,' smooth zooming (in small steps) is very well possible.

Crucial for the quality of the GAP-tree generalization is how to establish the appropriate importance ranges for the feature objects. An taxonomy of feature classes could be used, on which the compatibility functions between two different feature classes can be based (to find 'the most compatible neighbor'). More research is needed in this area to automatically obtain good generalization results for real-world data.

In this chapter the focus was on two generalization operations: area aggregation and line simplification, but the basic idea of the tGAP approach (assigning importance values to objects during 'offline' model generalization, which are then used during query and visualization for 'on-the-fly' refinement/generalization) could be extended to other generalization operations. This will also be a subject for future research.

Acknowledgments

This publication is the result of the research program 'Sustainable Urban Areas' (SUA) carried out by Delft University of Technology. Special thanks to Martijn Meijers for supplying test data and discussing first results of his MSc thesis project.

References

1. Ai T, Van Oosterom P (2002) GAP-tree Extensions Based on Skeletons. In: Richardson D, Van Oosterom P (eds) Advances in Spatial Data Handling, 10th International Symposium on Spatial Data Handling. Springer, Berlin
2. Bertolotto M, Egenhofer MJ (2001) Progressive Transmission of Vector Map Data over the World Wide Web. GeoInformatica 5(4):345–373
3. Burghardt D, Neun M, Weibel R (2005) Generalization Services on the Web – A Classification and an Initial Prototype Implementation. In: Proc. of Auto-Carto 2005, Las Vegas (USA)
4. Buttenfield BP (2002) Transmitting Vector Geospatial Data across the Internet. In: Egenhofer MJ, Mark DM (eds) Proc. of GIScience 2002, Berlin: Springer Verlag, Lecture Notes in Computer Science Vol 2478, 51–64
5. Cecconi A (2003) Integration of Cartographic Generalization and Multi-Scale Databases for Enhanced Web Mapping. PhD Thesis, University of Zurich, Zurich
6. Douglas DH, Peucker TK (1973) Algorithms for the Reduction of the Number of Points Required to Represent a Line or Its Caricature. The Canadian Cartographer 10(2): 112–122
7. Hampe M, Sester M, Harrie L (2004) Multiple Representation Databases to Support Visualisation on Mobile Devices. 10th ISPRS Congress – Commission IV, WG IV/2, Istanbul, Turkey.
8. Hildebrandt J, Owen M, Hollamby R (2000) CLUSTER RAPTOR: Dynamic Geospatial Imagery Visualisation using Backend Repositories. In: Proc. 5th International Command and Control Research and Technology Symposium (ICCRTS)

9. Jones CB, Abdelmoty AI, Lonergan ME, Van Der Poorten PM and Zhou S (2000) Multi-Scale Spatial Database Design for Online Generalisation. In: Proc. 9th International Symposium on Spatial Data Handling, sec. 7b, 34–44

10. Lazaridis I, Mehrotra S (2001) Progressive approximate aggregate queries with a multi-resolution tree structure. In: Proc. of 2001 ACM SIGMOD International Conference on Management of Data. Santa Barbara, California, United States

11. OGC (1998), The OpenGISÖ Guide, Introduction to Interoperable Geoprocessing, Part I of the Open Geodata Interoperability Specification. OpenGIS Consortium

12. OGC (2004), OpenGIS, Geography Markup Language (GML) Implementation Specification. Version 3.1.1, http://portal.opengeospatial.org/files/?artifact_id=4700

13. OGC (2005), OpenGIS, Web Feature Service Implementation Specification. Version 1.1, https://portal.opengeospatial.org/files/?artifact_id=8339

14. OGC (2006), OpenGIS, Web Map Service Implementation Specification. Version 1.3.0, http://portal.opengeospatial.org/files/?artifact_id=14416

15. Products Compliant to or Implementing OGC Specs or Interfaces, http://www.opengeospatial.org/resources/?page=products

16. Rosenbaum R, Schumann H (2004) Remote raster image browsing based on fast content reduction for mobile environments. In: Chambel T, Correia N, Jorge J, Pan Z (eds) Eurographics Multimedia Workshop

17. Samet H (1984) The Quadtree and Related Hierarchical Data Structures. ACM Computing Surveys archive 16(2):187–260

18. Sester M, C Brenner (2004) Continuous Generalization for Fast and Smooth Visualization on Small Displays. XXth ISPRS Congress - Commission IV, WG IV/3, Istanbul, Turkey

19. Töpfer F, Pillewizer W (1966) The Principles of Selection. Cartographic Journal 3:10–16

20. Van Oosterom P (1990) Reactive Data Structures for Geographic Information Systems. PhD thesis Department of Computer Science, Leiden University

21. Van Oosterom P (1992) A Storage Structure for a Multi-scale Database: The Reactive-tree. International Journal, Computers, Environment and Urban Systems 16(3):239–247

22. Van Oosterom P (1993) The GAP-tree, an approach to 'On-the-Fly' Map Generalization of an Area Partitioning. GISDATA Specialist Meeting on Generalization, Compienge, France, 15–19 December 1993. Chapter 9 in: Müller JC, Lagrange JP, Weibel R (eds) GIS and Generalization, Methodology and Practice, Taylor and Francis, London, 120–132

23. Van Oosterom P (2005) Variable-scale Topological Data Structures Suitable for Progressive Data Transfer: The GAP-face Tree and GAP-edge Forest. Cartography and Geographic Information Science 32(4):331–346

24. Vermeij M, Van Oosterom P, Quak W, Tijssen T (2003) Storing and using scale-less topological data efficiently in a client-server DBMS environment. In: Proc. 7th International Conference on GeoComputation, University of Southampton, Southampton, UK 8-10 September 2003

25. Zhou X, Prasher S, Sun S, Xu K (2004) Multiresolution Spatial Data-bases: Making Web-based Spatial Applications Faster. In: Yu JX, Lin X, Lu H, et al. (eds) Proc. 6th Asia Pacific Web Conference (APWebÆ04), 14-17 April, 2004, Hang-zhou, China, Lecture Notes in Computer Science 3007, 36–47

References containing URLs are validated as of October 1ˢᵗ, 2006.

Integration of Spatial Data Sources

Integration of Spatial Data Sources

6

Automated Geographical Information Fusion and Ontology Alignment

Matt Duckham[1] and Mike Worboys[2]

[1] University of Melbourne, Melbourne (Australia)
[2] University of Maine, Orono (USA)

6.1 Introduction

Geographical information fusion is the process of integrating geographical information from diverse sources to produce new information with added value, reliability, or usefulness (cf. [14, 67]). Geographical information fusion is an important function of interoperable and Web-based GIS. Increased reliance on distributed Web-based access to geographical information is correspondingly increasing the need to efficiently and rapidly fuse geographical information from multiple sources.

The overriding problem facing any geographical information fusion system is *semantic heterogeneity*, where the concepts and categories used in different geographical information sources have incompatible meanings. Most of today's geographical information fusion techniques are fundamentally dependent on human domain expertise. This chapter examines the foundations of *automated* geographical information fusion using *inductive inference*. Inductive inference concerns reasoning from specific cases to general rules. In the context of geographical information fusion, inductive inference can be used to infer semantic relationships between categories of geographical entities (general rules) from the spatial relationships between sets of specific entities. However, inductive inference is inherently unreliable, especially in the presence of uncertainty. Consequently, managing reliability is a key hurdle facing any automated fusion system based on inductive inference, especially in the domain of geographical information where uncertainty is endemic.

This chapter develops a model of automated geographical information fusion based on inductive inference. Central to this model are techniques by which unreliable inferences and data can be accommodated. The key contributions of this chapter are to

- define the foundations of automated geographical information fusion using inductive inference;
- explore some of the limitations of automated geographical information fusion, inherent in inductive inference;

- indicate initial techniques to adapt the automated fusion process to operate in the presence of imperfect and uncertain geographical information.

Following a brief motivational example (Sect. 6.1.1), Sect. 6.2 presents a review of the relevant literature. Section 6.3 then sets out the foundations of inductive inference for automated geographical information fusion. The limitations resulting from the unreliability of the inductive reasoning process are set out in Sect. 6.4, while Sect. 6.5 addresses the management of uncertainty in the input geographical data. Finally, Sect. 6.6 concludes the chapter with a look at future research.

6.1.1 Motivational Example

Data on the structural characteristics of buildings is often important to decision makers as part of an emergency response effort. It is not unusual for several different agencies to collect such data for the same geographical region and to make it available online. These agencies may use heterogeneous definitions or may even produce semistructured data without separate or fixed definitions (see Chap. 2). For example, the category "Reinforced concrete building" in spatial database A may not have the same meaning as the category "Non-wooden building" in spatial database B (this example is taken from a study of the 1995 Kobe Earthquake [64]). Current geographical information fusion techniques rely on the generation of a manual specification of the semantic relationships between different categories by a human domain expert. Such manual techniques can be slow, unreliable, and do not scale easily to Web-based information fusion scenarios.

However, if all the *instances* of buildings categorized as "Reinforced concrete building" in spatial database A are categorized as "Non-wooden building" in spatial database B, then this provides evidence that the *category* "Reinforced concrete building" is a subcategory of "Non-wooden building." Although this example is highly simplified, it does support the central intuition behind using inductive inference for geographical information fusion: that analysis of spatial relationships can be used to infer semantic relationships. It is important to note that this inference process does not necessarily require an understanding of the meaning of "Non-wooden building" or "Reinforced concrete building," and hence can be applied in the context of automated reasoning systems.

6.2 Background

The semantics of an information source may be described using an *ontology* (defined as "an explicit specification of a conceptualization" [32]). The task of fusing information compiled using different ontologies is a classical problem in information science (e.g. [69]), and continues to be a highly active research issue within many topics, including databases [40, 42, 57], interoperability [56, 65], the semantic Web [6, 18], medical information systems [28, 55], knowledge representation [10], data warehousing [68, 73], and, of course, geographical information fusion (Sect. 6.2.3).

The term "schema" is not identical to "ontology," but the two terms are often used near-interchangeably. A schema is a formally (or otherwise precisely) defined taxonomy. Thus, a schema is an ontology in the sense of [32]. However, the term "ontology" encompasses a broad spectrum of specification methods, from schemas at one extreme through to general logical systems that can be used to define and reason about sophisticated relationships and constraints between elements within a taxonomy [47]. From this point onward, the term "ontology" is preferred because this term covers both schemas and more sophisticated types of ontologies. However, it should be noted that the ontologies in this chapter are simply schemas.

A critical step in the fusion process is to fuse the ontologies for the different information sources. A variety of closely related terms are used in the literature to refer to aspects of this task, including

- integration and alignment;
- merging and matching;
- transformation and mapping.

To further confuse the issue, almost all of these terms may appear in the literature combined with any one of "ontology," "schema," or "semantic" (e.g. "ontology alignment," "schema matching," and "semantic integration"). The choice of which precise terms are adopted by particular researchers is often more a matter of preference and domain than a strict difference in definitions. However, a clear distinction is usually made between the process of identifying the relationships between corresponding elements in two heterogeneous ontologies (termed "alignment/matching/mapping") and the process of constructing a single combined ontology based on these identified relationships (termed "integration/merging/transformation") [45, 51]. For consistency, in this chapter we use the terms "(ontology) integration" and "(ontology) alignment" to distinguish these two concepts.

6.2.1 Ontology Integration and Mediators

The concept of a *mediator*, a software system that can assist humans in integrating heterogeneous information sources, was first explicitly described by Gio Wiederhold [69]. Based on his general vision, dozens of different mediation systems have been proposed and developed over the past decade (for a full survey of mediation systems see [66]). For example, TSIMMIS was one of the earliest mediator systems to be researched. The core idea behind TSIMMIS was to mark up information sources with standardized tags, which included labels describing the semantics of each data item [12, 30]. Rather than using unstructured tagging to describe information sources, subsequent mediators, such as SIMS [1], OBSERVER [48], InfoSleuth [3], and OntoSeek [34], use predefined domain ontologies as the basis for integration. More recently, a suite of Web-based languages and mediation technologies have emerged around the topic of the semantic Web (e.g. [31]). The primary focus of all these systems is efficient *integration* of information across multiple information sources with heterogeneous ontologies. Ontology alignment is a prerequisite for such systems to operate, but the question of how the alignment semantics are generated is not explicitly addressed by these systems. This chapter contains further information on mediator systems, in particular on the distinction between local-as-view (LAV) and global-as-view (GAV) approaches to mediation.

The theoretical foundations of most mediator systems are similarly focused more on ontology integration than ontology alignment (e.g. [33, 62]). Formal concept analysis (FCA [29]) is a widely used technique, dating back more than 20 years, for structuring and integrating ontologies based on *concept lattices* (a special type of ordering relation on categories). Although the integration of concept lattices has a precise formal definition and can be automated, human domain expertise is required to identify salient attributes and categories within a domain and their interrelationships. Another widely used technique for representing and reasoning about heterogeneous ontologies is description logics. Description logics are decidable, tractable fragments of first-order predicate calculus, and form the basis of many mediator systems and components, including SIMS, OBSERVER, and OWL (Web ontology language, one of the primary standards for the semantic Web). Description logics are especially useful in the context of ontology integration because they provide several important reasoning services, most notably a *subsumption* service which can classify the relationships between categories and derive a complete and consistent integrated ontology (see [11, 19] for more information on description logics and their reasoning services). However, while description logics can offer efficient formally founded reasoning about ontologies, defining the relationships and rules that connect elements of different ontologies remains a primarily human activity.

In summary, much of the existing research into mediators and ontology integration does not address the issue of ontology alignment directly. Instead, such research typically assumes that the semantic relationships between ontological elements will be established using user interaction, predefined mappings, preexisting top-level ontologies, or existing lexical correspondences [66]. Building these mappings is assumed to require an understanding of the underlying concepts, and so is at root a human activity.

6.2.2 Automating Ontology Alignment

Some researchers have turned their attention to creating semi-automated or fully automated ontology alignment systems (see [54] for an overview). Much of this research adopts an *intensional* approach: it aims to analyze the definitions (intensions) of the concepts and categories used in the input information sources. Intensional techniques usually analyze heterogeneous ontologies to identify lexical similarities (e.g. PROMPT [52], Active Atlas [61]), structural similarities (e.g. DIKE [53], ONION [50]), or some combination of these (e.g. CUPID [45], FCA-Merge [60]).

There are two main drawbacks of adopting a purely intensional approach to ontology alignment. First, how concepts are defined is not necessarily the same as how they are used. As an analogy, people who learn to speak language from a dictionary (definitions) often have very different speech patterns from native speakers, who also learn from example. Only by looking at *extensional* information (specific instances in data) is it possible to begin to determine how concepts are actually used. Second, extensional information forms a rich source of examples that can be used as the basis for automated pattern recognition techniques.

Recognizing the importance of instance-level information, an increasing number of researchers have turned to extensional approaches, including the following.

- The SemInt system clusters patterns in instance-level information, and uses these clusters to train a neural network to identify intensional relationships [43, 44].
- Doan and collaborators [17, 18] and the Autoplex system [5] use Bayesian machine learning techniques on instance-level information to identify intensional relationships.
- The Clio [49] and iMAP [16] systems search for filters that relate sets of instance-level information within a database. These filters are then used to infer intensional relationships.
- He and Chang make use of patterns of co-occurrence of related attributes for Web pages [36]. The positive correlation between related attributes, along with an expected negative correlation between synonyms, is used to automatically infer semantic mappings between attributes within a domain.

Fundamentally, all these extensional approaches apply different forms of inductive inference: they use the structure and patterns in instance-level information to infer semantic relationships. An inherent limitation of using inductive inference is that it is unreliable. In many of the extensional approaches outlined above unreliability is combated using probabilistic techniques (such as Bayesian probability). We return to the topic of reasoning reliability in Sect. 6.4.

6.2.3 Geographical Information Fusion

Research into geographical information fusion mirrors the more general approaches to information fusion cited above. Fonseca and coauthors have published a series of papers on the so-called *ontology-driven GIS* [25–27]. This work aims to augment conventional GIS with formal representations of geographical ontologies, leading to tools that enable improved ontology-based information integration. A wide variety of related work has addressed the issue of geographical information integration from a similar perspective (e.g. [2, 7, 15, 58, 63]). In common with the research presented in Sect. 6.2.1, such research focuses on the integration itself, but assumes the semantic relationships between different ontologies are already known.

A relatively small amount of work has begun to provide tools for geographical ontology alignment. Most of this work adopts an intensional approach. Kavouras, Kokla, and coauthors use FCA as the basis for their approach to geographical ontology alignment [37, 38, 41]. Manoah et al. applied the intensional machine learning techniques discussed in Sect. 6.2.2 to geographical data [46]. Duckham and Worboys have investigated using description logics [20] and a formal algebraic approach [71] to ontology alignment. Because of the diversity of geographical terms and concepts, this work is at best semi-automated, and still requires human domain experts at critical stages in the alignment process.

To our knowledge, the study by Duckham and Worboys [22] is the only research in the geographical domain that adopts an extensional approach to automating information fusion (although it is not the only work to acknowledge the importance of extensions in the representation of geographical knowledge, e.g. [8, 9]). Geographical information is a richly structured and voluminous source of instances upon which to base inductive reasoning processes, more so than many other types of information source. In this respect, it is well suited to extensional approaches to automated information fusion. However, the problems of unreliable inference introduced in Sect. 6.2.2 are exacerbated in the geographical domain because uncertainty is an endemic feature of geographical information. Applying an unreliable reasoning process to uncertain data could potentially generate information that is degraded to the point of being meaningless. Consequently, following a closer look at using induction as a basis for automated geographical information fusion in Sect. 6.3, we turn to the issues of unreliability in the reasoning process (Sect. 6.4) and reasoning under uncertainty (Sect. 6.5).

6.3 ROSETTA: Automated Extensional Geographical Information Fusion

At the core of an extensional approach to automated geographical information fusion is the process of inferring semantic relationships from spatial relationships. As already discussed, this process is an example of inductive inference: reasoning from specific cases to general rules. As an analogy, archaeologists were able to decipher the meaning of ancient Egyptian hieroglyphs following the discovery of the Rosetta Stone, a 2nd-century tablet that contained the same official decree in both Egyptian (hieroglyphs and text) and Greek (text). Before the discovery and subsequent analysis of the Rosetta Stone, all attempts to decipher hieroglyphs were unsuccessful and Egyptian hieroglyphics were considered to be merely primitive picture writing. Only by comparing examples (extensions) of the Greek text with Egyptian text and hieroglyphs on the Rosetta Stone were archaeologists able to correctly infer a "dictionary" (intensions) for translation between these different information sources. In a similar way, the extensional approach to geographical information fusion constructs a shared "dictionary" for translating between the ontologies of the different information sources, based on the relationship between the spatial extents of the categories used in those information sources. In the remainder of this chapter, we use the term "ROSETTA" to refer to the extensional approach to automating geographical information fusion.

To illustrate, Fig. 6.1 contains a much simplified example of a ROSETTA-based fusion. In Fig. 6.1, each data set comprises an extensional component (the mapped spatial data) and an intensional component (the ontology for that spatial data). On the left-hand side of Fig. 6.1, the intension for data set A contains the categories **Forest** and **Built-up area**, while the extension contains two regions, one of each category. Similarly, on the right-hand side of Fig. 6.1, data set B contains the intensions **Woodland** and **Urban** along with a map of the spatial extensions of the **Woodland**

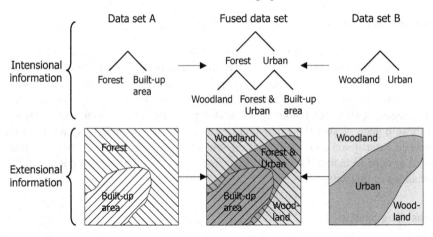

Fig. 6.1. Simplified example of inductive reasoning for automated geographical information fusion

and **Urban** categories. The geographical extents of data sets *A* and *B* are identical (i.e. the data sets cover the same area). Thus, Fig. 6.1 might represent the situation where two different environmental agencies have both mapped the same geographical region using different ontologies.

The fused data set is contained within the center of Fig. 6.1. Because all locations that are categorized as **Built-up area** in data set *A* are categorized as **Urban** in data set *B*, we have inferred in our integrated taxonomy that the category **Built-up area** is a subcategory of the category **Urban**. Similarly, because all locations that are categorized as **Woodland** in data set B are categorized as **Forest** in data set *A*, **Forest** subsumes **Woodland** in the integrated ontology. A new category **Forest & Urban** has been created to represent those regions that are categorized as **Forest** in data set *A* and **Urban** in data set *B*. In other words, although there exists no subsumption relationship between **Forest** and **Urban**, we have inferred that these categories overlap, on the grounds that their extensions overlap. Note that, although highly simplified, the process illustrated by Fig. 6.1 is more than a simple overlay. The data sets have been fused, in the sense that we have gained (a small amount of) new information about the relationships between the categories represented in each of the input data sets.

6.3.1 Computational Approaches

Although the discussion above provides an informal description of a ROSETTA system, recent work by the authors does provide a formal basis for such a system [22]. In this section we provide a brief overview and synthesis of some of the central ideas and results of this work.

Extensional Form

A key concept in the development of a ROSETTA system is to consider categories in geographical ontologies in their extensional form. The extensional form of a category is the set of all instances of that category. For example, one way to describe what is meant by the category "Car" is to refer to the set of all objects that we call cars. Subcategories, such as "Tan-colored Chevrolet Lumina," will contain only a subset of those objects. Using the extensional form of a category makes explicit the link between extensional and intensional information, enabling an automated computational system to manipulate categories without any requirement to understand the semantics of that category.

Reasoning System

An initially attractive route to realizing a ROSETTA system, such as described informally above, is to formalize the rules required for the inductive inference process, and then implement those rules within an automated reasoning system. We might start by representing a taxonomy as a partially ordered set $(C, \leq)$, where C is a set of categories and $\leq$ is the ordering (subsumption relationships) on those categories. Now, a *geographical data set* can be represented as a set S that is a partition of a region of space, a *taxonomy* $(C, \leq)$, and a *function* $e : C \rightarrow 2^S$ that defines which spatial regions are labeled with which categories (2^S is the power set of S). Thus, e associates each category in the taxonomy with a unique set of elements from the partition of space S. We call e an extension function because it provides the extensional form of each category within the context of its data set.

To illustrate, for data set A in Fig. 6.1 the taxonomy $(C_A, \leq_A)$ is represented by hierarchy of categories; the partition of space S_A is represented by the map itself, comprised of jointly exhaustive and pairwise disjoint regions; and the extension function e_A is represented by the labels on both the taxonomy and the map (i.e. for each category we can identify on the map the set of locations that are labeled as that category).

From this basis, it is possible to start to define simple first-order logical rules that embody our inductive inference process. For two data sets $G_1 = \langle S_1, (C_1, \leq_1), e_1 \rangle$ and $G_2 = \langle S_2, (C_2, \leq_2), e_2 \rangle$ we wish to construct the fused data set $G_f = \langle S_f, (C_f, \leq_f), e_f \rangle$. We might specify as a first rule:

$$\text{for all } x \in C_1 \text{ and } y \in C_2 \begin{cases} \text{if } e_1(x) \subseteq e_2(y) & \text{then } x \leq_f y \\ \text{if } e_2(y) \subseteq e_1(x) & \text{then } y \leq_f x \end{cases}$$

In other words, where the spatial extent of a category a contains the spatial extent of a category b, we infer that a is a subcategory of b in our fused taxonomy. Similarly, we could formulate further rules dealing with more of the possibilities for spatial relationships between the extensional forms of two categories, such as the following:

$$\text{for all } x \in C_1 \text{ and } y \in C_2 \text{ if } e_1(x) \cap e_2(y) = \emptyset \text{ then } x \nleq_f y \text{ and } y \nleq_f x$$

In plain language, the rule above states that if the extensions of two categories x and y are disjoint then we infer that the categories themselves are incomparable. A further obvious rule, suggested by the category **Forest & Urban** in Fig. 6.1, is to create a new category corresponding to two overlapping category extents as follows:

$$\text{for all } x \in C_1 \text{ and } y \in C_2 \text{ if } e_1(x) \cap e_2(y) \neq \emptyset$$
$$\text{and } e_1(x) \not\subseteq e_2(y) \text{ and } e_2(x) \not\subseteq e_1(y)$$
$$\text{then } x \cap y \in C_f \text{ and } x \cap y \leq_f x \text{ and } x \cap y \leq_f y$$

The rule above creates a new category $x \cap y$ in the fused taxonomy that lies at the intersection of categories x and y. For two data sets G_1 and G_2, the conclusions from such rules form an ordering relation that relates categories in the two source taxonomies. Together with those source-ordering relations, this enables the derivation of a new fused partial order in G_f that defines the subsumption relationships between categories within the different taxonomies of G_1 and G_2.

Once formalized, these rules can be implemented within an automated reasoning system. Indeed, early versions of our ROSETTA system adopted this approach, using the RACER description logic engine [35] for automated reasoning. An advantage of using description logics for this purpose is that any inconsistencies between the chosen rules can be automatically detected using the consistency and satisfiability services provided by any description logic. Having generated the ontology alignment, the spatial data itself can then be automatically fused based on the standard geographical information integration techniques (i.e. overlay the two spatial data sets, and assign to each fused region the category in the fused partial order that lies at the intersection of the two source categories for the fused region).

Algebraic System

The reasoning system approach described above provides an important step on the road to practical automated geographical information fusion systems. However, it has at least two important shortcomings.

First, a partial order is a rather too general structure for describing a geographical ontology. For each pair of input categories we need to be able to identify a unique category in our fused taxonomy that corresponds to the fusion of those input categories. Using partial orders, it may not be possible to guarantee that such a unique fused category exists, since a pair of elements in a partial order may have multiple incomparable least upper and greatest lower bounds. A more appropriate structure is a lattice, which as we have already seen is commonly used in formal approaches to ontological information [29, 37, 60]. A lattice is a special type of partial order, where all subsets of elements have a unique least upper bound and a unique greatest lower bound in the lattice. The simplified taxonomies in Fig. 6.1 and subsequent figures can be represented as lattices.[1]

---

[1] Strictly, the taxonomies in the figures in this chapter are shown as join semi-lattices, but any finite join semi-lattice can be trivially transformed into a lattice with the addition of a bottom element.

Second, developing ad hoc fusion rules, such as illustrated above, may not always lead to an associative and commutative fusion system. Thus, we could add further rules to our reasoning system that would result in different fusion products, depending on what order we input data into the system. This is clearly undesirable. To be well formed we would expect a fusion process to produce a unique fusion product for a set of inputs irrespective of the order in which they are fused. As a parallel, GIS would be considerably less useful if the overlay operator were defined in such a way that the order in which source data sets were overlaid affected the output results of the overlay operation.

An important result of [22] is to formalize geographical information fusion in such a way that

1. the taxonomy associated with a data set can be represented as a lattice;
2. the fusion process is represented as an associative and commutative binary operator;
3. the fusion process is closed, in the sense that the fusion product is itself a valid geographical data set that can be used in subsequent fusion operations.

Formally, [22] shows that the geographical data sets (represented as a partition of space, a lattice, and an extension function) combined with the fusion operator form a fusion algebra with the properties of a commutative semigroup (closed, associative, commutative). The reader is referred to [22] for more detail on this topic; the remainder of this chapter turns to issues of reliability and uncertainty rather than formalization of fusion systems.

6.4 Reliability

The ROSETTA system outlined above is simple, effective, and has a clear theoretical basis. However, in developing practical automated geographical information fusion systems, there are two main issues that must be addressed: unreliability and uncertainty in the fusion process. In this section, we first examine the issue of the unreliability of inductive inference.

6.4.1 Deductive Validity

An inherent limitation of the extensional approach to geographical information fusion is that inductive inference is not deductively valid. In general, an inference is said to be deductively valid if, given that all the premises are true, the conclusion is also true. Using inductive inference, it is entirely possible to formulate deductively invalid inferences. For example, given the premise that all the birds I have ever seen can fly, I might inductively infer the conclusion that all birds can fly. Clearly, this conclusion is not necessarily valid, even though the premise may be. A similar problem can occur with a ROSETTA system. In the example in Fig. 6.1, it might be that if we had used data sets with greater spatial extents, we would have discovered a region of built-up area in data set *A* that overlapped a region of Woodland in data set *B*. In

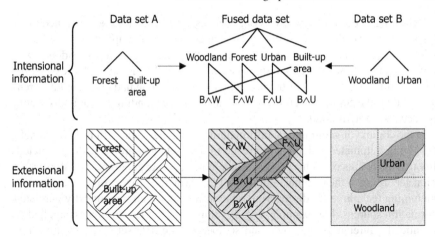

Fig. 6.2. Unrepresentative spatial extents (*dotted line*) may lead to invalid inferences

this case, the inductive inference procedure would have different premises, leading to a different fused data set and ontology. Figure 6.2 illustrates this situation. Note also that the resulting data set no longer contains new information about the semantic relationships between the different categories. In this case the "fusion" has degraded to a simple overlay. We return to this issue later on in this chapter (Sect. 6.5.4).

The primary guard against deductive invalidity is to ensure that the data sets to be fused are large enough to contain a representative range of the possible spatial relationships between the different categories represented in the data sets. Thus, a feature of ROSETTA systems is that they are "data-hungry," in the sense that we expect the fusion process to become more reliable the more data we can feed into the process. Small fragments of data sets will tend to yield integrated ontologies that embody chance, rather than real semantic, relationships.

By way of analogy, when the Rosetta Stone was discovered, almost half the text on the artifact was damaged in some way (even missing in the case of hieroglyphs). More extensive damage would have further reduced the availability of corresponding words upon which to base lexicographic inferences. With fewer examples of correspondences between the different languages, any process of deciphering would be more likely to lead to incorrect inferences.

6.4.2 Semantic and Spatial Extents

An underlying assumption of the extensional approach to geographical information fusion is that the thematic domains for the input data sets are semantically related. In our example in Fig. 6.1, both input data sets concerned land cover. Similarly, in earlier examples, we considered the fusion of data sets that concerned the structural characteristics of buildings. Using a spatial metaphor, we can say that for information fusion to take place we expect the semantic extents of two information sources to overlap.

Returning to our analogy, it was only because the Rosetta Stone contained three copies of the same decree in different languages that the attempt to derive a meaningful Egyptian–Greek dictionary was successful. The direct correspondence was known about because it is explicitly stated in the Greek version of the text. If, instead, the different versions of the text on the Rosetta Stone had contained different decrees, then the Stone's usefulness as an aid to understanding hieroglyphs would have been severely limited.

In the context of geographical information fusion, there may still be some benefit to applying automated inductive inference to data sets that are topically unrelated. Although the results of such a process would not constitute geographical information fusion according to the original definition of the term, the process may be useful as a data mining technique for discovering relationships between semantically unrelated information sources. For example, Fig. 6.3 illustrates the fusion process applied to semantically unrelated land cover and socioeconomic data sets. The relationships generated between categories in the input data sets are not subsumption relationships (it would not be true to say that Woodland is a subcategory of Low income), but might provide useful summarizations of the semantic relationships embedded in the data set.

It may also be important to consider the spatial extents of the information sources. In the simple automated information fusion systems discussed in this chapter, the inference process is driven by direct spatial coincidence. Thus, only those locations that are represented in both information sources to be fused provide premises for the inductive inference process. However, current research is also investigating the possibility of using other types of spatial relationships, such as proximity or topology, to drive inductive inferences about spatial data that is not necessarily coincident.

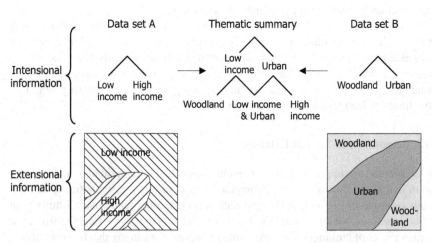

Fig. 6.3. "Fusion" of semantically unrelated data sets

6.5 Uncertainty

Geographical information is inherently imperfect, leading to uncertainty about the real features represented in a geographical data set. Imperfection is often represented and quantified using spatial data quality elements and standards (Chap. 7). However, there are many different spatial data quality elements that have been proposed in standards and the research literature. Three fundamental types of imperfection are commonly identified in the literature: *inaccuracy*, *imprecision*, and *vagueness* [21, 70, 72]. In this section we look at the effects of each of these types of imperfection in turn, followed by an overview of ongoing research into ways to regulate uncertainty in a ROSETTA system.

6.5.1 Inaccuracy

Inaccuracy in geographical information concerns a lack of correspondence between information and the actual state of affairs in the physical world. In a ROSETTA system, inaccuracy degrades the reliability of the inductive inference process, potentially leading to semantic relationships being inferred between categories that are, in reality, unrelated. Conversely, inaccuracy may lead to a failure to identify semantic relationships between categories that are, in reality, related. For example, suppose that in our land cover data set B part of the Urban region has been misclassified as Woodland such that it overlaps the Built-up area in data set A. In turn, this might lead to the incorrect inference that Woodland and Built-up area are semantically overlapping (Fig. 6.4). Note that the inaccuracy has again produced a fused ontology that is not particularly informative, in the sense that we have gained no new information about the relationships between the categories in the input data sets (we could have achieved the same results using a simple overlay).

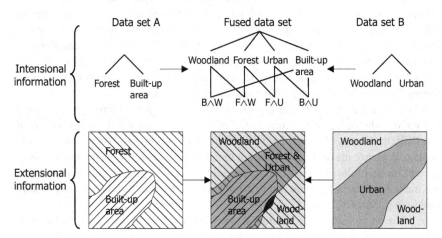

Fig. 6.4. Inaccuracy in input data sets (*black region indicates sliver polygon*)

We can imagine what might have happened if some of the words on the Rosetta Stone had been incorrectly drafted or inscribed. It is possible that such inaccuracies would lead to incorrect lexicographic inferences, especially in the case of systematic inaccuracies. To guard against inaccuracy, it is important to ensure the extensions used in the inference process are large enough such that examples of incorrect correspondences due to random inaccuracies will be greatly under-represented when compared with examples of correct correspondences. In terms of a ROSETTA system, the situation is a little more complex. However, in principle, the possibility of random inaccuracies is another reason why the ROSETTA systems are fundamentally data hungry: the more examples used in the inference process, the more likely it is that these examples will provide a basis for valid inferences.

In addition to spatial inaccuracy, inaccuracies may occasionally occur within the taxonomy itself (e.g. where one category is incorrectly labeled or incorrectly positioned within the taxonomy). Since the taxonomy is central to the fusion process, it is difficult to see how the automated fusion process described here (or indeed any of the fusion systems encountered in this chapter) could hope to effectively combat such inaccuracies.

6.5.2 Imprecision

Imprecision, a lack of detail in information, is another intrinsic feature of geographical information. Imprecision leads to *granularity*: the existence of "clumps" or "grains" in the data. The granularity at which geographical phenomena are represented strongly influences what features are observed. Like inaccuracy, heterogeneous levels of granularity degrade the reliability of the inductive inference process. For example, imagine that land cover data set A has been collected at a coarser level of spatial granularity than data set B. Then it will be likely that the detailed features found in data set B will simply not be represented in data set A (such as small pockets of Woodland within the predominately Urban area that are represented in data set B, but have no correspondent pockets in data set A). As a result, a naive inductive inference process may again incorrectly infer that Woodland and Built-up area are semantically overlapping, as in Fig. 6.5 (similar to the effects of inaccuracy in Fig. 6.4). As for inaccuracy, the fusion product in Fig. 6.5 is not particularly informative, as it is essentially a simple overlay of the data.

It is difficult to say how the efforts to decipher Egyptian hieroglyphics would have fared if the different versions of the official decree on the Rosetta Stone contained different levels of detail about the official declarations. The structure of natural language does not make it easy to automatically infer relationships between texts at different levels of detail. However, the spatial structure of geographical information does make inferences between information sources at different levels of detail more feasible (e.g. Sects. 6.5.4 and [20]).

In addition to spatial imprecision, it may be important to also consider the possibility of heterogeneity in taxonomic granularity. In this case, semantic differences that are distinguished apart in the taxonomy for one data set may not be distinguished in the taxonomy for a different data set. For example, the category Woodland is at

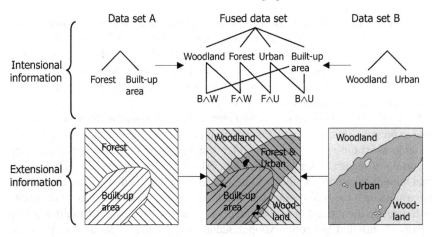

Fig. 6.5. Granularity in input data sets (*black regions in fused spatial data show fine grained "pockets" of Woodland*)

a coarser level of semantic granularity than the category Broadleaved native woodland. In general, the inductive inference process is able to operate satisfactorily in the presence of taxonomic imprecision: after all granularity is an integral feature of the hierarchical structure of taxonomies themselves.

Nevertheless, geographical information sources are especially interesting in this respect as they often exhibit *contravariant* granularity, where an information source is at a relatively fine spatial granularity but relatively coarse taxonomic granularity when compared with another information source. This situation may occur as a result of the economies of scale for spatial data capture. The high cost of performing large-scale spatially detailed data capture tends to ensure that such data is collected in a general purpose form (taxonomically coarse granularity), so as to maximize its utility to the widest possible range of potential uses. Conversely, limited resources mean that spatial data collected for specific application domains (taxonomically fine granularity) tends to be at a spatially coarse granularity. An example of data sets at contravariant granularities is the topographic data collected by the UK national mapping agency, Ordnance Survey, when compared with the CORINE land cover data set for the UK. Ordnance Survey topographic data is at a much higher spatial granularity than the CORINE data set, being derived from ground survey rather than satellite imagery. Conversely, the CORINE data set is at a much higher taxonomic granularity than Ordnance Survey topographic data, providing more detailed information about the actual land cover categories present at a particular location [20].

6.5.3 Vagueness

Vagueness concerns the existence of borderline cases in information. For example, the category "mountain" is vague, because for any particular mountain we expect there to exist locations which are definitely on the mountain, locations that are definitely not on the mountain, and locations for which is it indeterminate whether or not they are on the mountain. Unlike imprecision and inaccuracy, which may occur independently in both extensional and intensional aspects of the data, vagueness is directly associated with the intensional aspects of the data. In other words, we regard vagueness as a type of imperfection in definition, rather than imperfection in observation (i.e. we adopt an epistemic view of vagueness, leaving to one side for the moment debates about ontic vagueness [39]).

Although vagueness is an intensional phenomenon, vagueness can have an extensional expression in spatial data sets, which typically impose precise spatial boundaries around spatial regions. If, as is often the case in spatial data, the underlying categories are vague (such as the categories "Mountain" or "Forest" [4, 24]) then the actual boundaries imposed will be somewhat arbitrary. The effect of such boundary arbitrariness on a ROSETTA system will be similar to those resulting from inaccuracy: it will degrade the reliability of the inductive inference process, potentially leading to errors of omission and commission in identifying semantic relationships between categories represented in the source data sets.

In order to tackle vagueness, it is first necessary to provide an explicit representation of the existence of vagueness. Typically, this is done by replacing the crisp boundaries for regions used in conventional spatial data with a representation of regions with broad boundaries, such as fuzzy sets [23], rough sets [21], two-stage sets [59], or egg-yolk representations [13]. For example, Fig. 6.6 shows a hypothetical fusion of data sets A and B, containing broad boundaries between the regions Built-up area and Forest in data set A, and Urban and Woodland in data set B.

The question of exactly how such a fusion operator should be constructed is the topic of current research (hence, unlike previous figures, Fig. 6.6 is a hypothetical fusion product). The structure of the data in Fig. 6.6 is incompatible with the formal structures discussed so far. Either the extensions in Fig. 6.6 contain regions that have no corresponding intensions in the taxonomy (i.e. the unlabeled broad boundaries are themselves separate regions); or from another perspective the extensions do not form

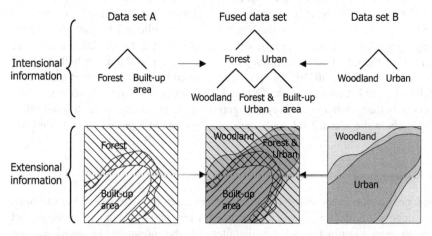

Fig. 6.6. Fusion of information sources including regions with broad boundaries

a partition of space (i.e. the broad boundaries constitute an overlap between two or more neighboring regions). Thus, the formal mechanisms currently being developed for fusing data-containing regions with broad boundaries are generalizations of those formalizations already discussed.

Whatever the formal structures used, the goal is to infer crisp semantic relationships between vague categories based on indeterminate spatial extents. For example, we may be certain that a "Copse" is a sub-category of "Woodland," even if both categories are vague. In the case of Fig. 6.6, we might devise new inference rules like those in Sect. 6.3 that only consider the *core* of the extent of each category (those parts of space that are classified as definitely belonging to the category). Conversely, a weaker inference system could be developed by allowing semantic relationships to be inferred where the core of one category is contained within the entirety of another category.

6.5.4 Computation with Uncertain Data

From the discussion above, we can begin to suggest simple mechanisms for incorporating inaccuracy and imprecision into the automated information fusion process (vagueness is the topic of current research). One such mechanism for incorporating inaccuracy and granularity arises from noting that sliver polygons (resulting from inaccuracy) or regions of fine-grained detail (resulting from fine granularity) are expected to make up a relatively small proportion of the entire regions being fused. For example, if the overlap between two regions is smaller than 5% of the total area of either regions, this might constitute evidence that the overlap arises from inaccuracy in the input regions. Similarly, if the overlap between region A and region B is less than 5% of the total area of region A and more than, say, 95% of the total area of region B, this might constitute evidence that the overlap arises from heterogeneous granularity in the data sets (i.e. region B is at a finer granularity than region A).

Consequently, setting thresholds for the proportion of overlap between two extensions of a category provides a basis for detecting spatial relationships that can be attributed to inaccuracy or heterogeneous granularity. Spatial relationships that are attributed to inaccuracy or imprecision then can be omitted from premises for the inductive inference process. The effect of using such an approach is illustrated for our example ROSETTA system in Fig. 6.7, based on Fig. 6.4. Here, the small sliver overlap between the extents of **Built-up area** and **Woodland** comprises less than 5% of the total area of these extents. This overlap is omitted from the inductive inference process, leading to a fused taxonomy as for Fig. 6.1. However, in the fused data set, the omitted region (black region) then becomes *unclassifiable* (has no category associated with it).

The thresholds can be set arbitrarily, or by a human user. As the thresholds increase, more overlaps are omitted from the inference process, usually leading to more direct subsumption relationships in the fused taxonomy (cf. the taxonomies in Figs. 6.4 and 6.7). Fused taxonomies containing more direct subsumption relationships are generally more desirable because they provide more *new* information about the relationships between categories in the source taxonomies (the fused taxonomies

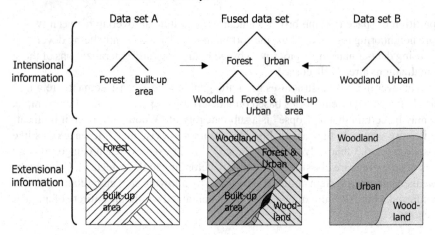

Fig. 6.7. Sliver polygon, resulting from inaccuracy as in Fig. 6.4, is eliminated from inductive inference process using overlap thresholds

in Figs. 6.2, 6.4, and 6.5 are degenerate cases that contain no new information that could not have been derived from a simple overlay of the two data sets). Thus, in setting such thresholds, there is a balance to be struck between the quality of extensional and intensional information in the fused data set. Tolerating higher levels of inaccuracy or imprecision generally leads to more useful intensional information, but at the same time lower quality extensional information with more unclassifiable regions. Conversely, tolerating lower levels of inaccuracy or imprecision leads to less useful intensional information, but higher quality extensional information with fewer unclassifiable regions. Current research is investigating techniques for automatically setting the thresholds in such a way as to maximize some overall measure of the usefulness of the fused intensions (e.g. measures of the information content of the fused taxonomy) or quality of the fused extensions (e.g. measures of the area of unclassifiable regions).

6.6 Conclusions

This chapter has provided the conceptual basis for an extensional approach to automated geographical information fusion. The key innovation in this approach is to infer semantic relationships between those data sets based on their spatial relationships. This process is an example of inductive inference, reasoning from specific cases to general rules. The main obstacles to using inductive inference for automated geographical information fusion are the unreliability of inductive inference and imperfection in both extensional and intensional information. However, this chapter argues that these obstacles are surmountable, and indicates some of the ways they may be overcome.

The approach holds considerable promise for application to Web-based environments. The increasing availability of geographical information from Web-based sources is only of limited use unless it is accompanied by concomitant ability to combine those information sources in a meaningful way. Non-expert users cannot be expected to do this unaided, so automation is an essential step in extending the usability of Web-based GIS into a range of new applications and domains.

However, there are several research issues to be addressed before practical automated geographical information fusion systems become a reality, including the following.

- *Inclusion of Human Expert Domain Knowledge*: Although the ROSETTA approach aims to enable fully automated information fusion, it is also important to allow the inclusion of partial human expert domain knowledge where it already exists, and integrate this knowledge with automatically inferred knowledge. Some initial techniques for dealing with this issue are presented in [22].
- *Integration with Existing Mediator Architectures*: The extensive work on existing mediator architectures and ontology-based GIS (cf. Chap. 7) is complementary to the goals of a ROSETTA system. Future work aims to integrate both in an "intelligent geomediator architecture," which provides the integration capabilities of a mediator with the alignment capabilities of a ROSETTA system.
- *Regions with Broad Boundaries*: A high-priority goal of current research is to extend the existing formal ROSETTA systems with the ability to operate with vague categories, where the extents of those categories have broad boundaries.
- *Automated Thresholding*: In addition to developing new techniques for dealing with imperfection, current research is investigating developing automated thresholds for reasoning in the presence of inaccuracy and imperfection, as discussed in Sect. 6.5.4.
- *Further Spatial Relationships*: The inferences discussed in this chapter all concern containment or overlap between extensions of categories. However, given the rich variety of spatial relationships embedded within spatial data, it is expected that many more types of spatial relationships might be useful as a basis for inductive inference, including topological and metric relationships (cf. Chap. 8).
- *Spatially Varying Alignment*: The approach presented in this chapter aims to infer alignments that are non-spatial, in that they hold for all locations in space. Developing ROSETTA systems that can infer spatially varying ontology alignments (i.e. semantic relationships that hold only in specific regions of geographical space) will potentially provide much greater flexibility in defining future fusion systems.

Acknowledgments

Matt Duckham is supported by the Australian Research Council under ARC Discovery Grant DP0662906, entitled "Automatic fusion of geoinformation: The intelligent geomediator architecture (iGMA)." Collaboration between Matt Duckham and Mike Worboys has been partially supported by funding from the Australian Academy of

Science (AAS) and a University of Melbourne ICR (international collaborative research) grant. Mike Worboys' work is supported by the National Science Foundation under NSF grant numbers IIS-0429644 and BCS-0327615. A preliminary version of this chapter was presented at the GISPLANET 2005 conference, Lisbon, Portugal.

References

1. Arens Y, Knoblock C, Shen W-M (1996) Query Reformulation for Dynamic Information Integration. Journal of Intelligent Information Systems, 6(2/3):99–130
2. Baru C, Gupta A, Ludäscher B, Marciano R, Papakonstantinou Y, Velikhov P, Chu V (1999) XML-based Information Mediation with MIX. In: Proc. of ACM SIGMOD Int. Conference on Management of Data. Philadelphia, Pennsylvania, USA, 597–599
3. Bayardo R, Bohrer W, Brice R, Cichocki A, Fowler J, Helal A, Kashyap V, Ksiezyk T, Martin G, Nodine M, Rashid M, Rusinkiewicz M, Shea R, Unnikrishnan C, Unruh A, Woelk D (1997) InfoSleuth: Agent-based Semantic Integration of Information in Open and Dynamic Environments. In: Proc. of the ACM SIGMOD Int. Conf. on Management of Data. Tucson, Arizona, USA, 195–206
4. Bennett B (2001) What is a Forest? On the Vagueness of Certain Geographic Concepts. Topoi, 20:189–201
5. Berlin J, Motro A (2001) Autoplex: Automated Discovery of Contents for Virtual Databases. In: Proc. of the 9th Int. Conf. On Cooperative Information Systems. Trento, Italy, Lecture Notes in Computer Science, Springer, Berlin Heidelberg New York, 108–122
6. Berners-Lee T, Hendler J, Lassila O (2001) The Semantic Web. Scientific American, 279(5):34–43
7. Boucelma O, Garinet J-Y, Lacroix Z (2003) The VirGIS WFS-based Spatial Mediation System. In: Proc. of the 12th Int Conf. on Information and Knowledge Management. New Orleans, Louisiana, USA, 370–374
8. Brodaric B, Gahegan M (2001) Learning Geoscience Categories *In Situ*: Implications for Geographic Knowledge Representation. In: Proc. of the 9th Int. Symp. on Advances in GIS. Atlanta, Georgia, USA, 130–135
9. Brodaric B, Gahegan M (2002) Distinguishing Instances and Evidence of Geographical Concepts for Geospatial Database Design. In: Proc. of the 2nd Int. Conf. on Geographic Information Science. Boulder, Colorado, USA, 22–37
10. Calvanese D, De Giacomo G, Lenzerini M, Nardi D, Rosati R (1998) Description Logic Framework for Information Integration. In: Proc. of the 6th Int. Conf. on Principles of Knowledge Representation and Reasoning. Trento, Italy, 2–13
11. Calvanese D, De Giacomo G, Nardi D, Lenzerini M (2001) Reasoning in Expressive Description Logics. In: Robinson A, Voronkov A (eds) Handbook of Automated Reasoning, 2:1581–1634
12. Chawathe S, Garcia-Molina H, Hammer J, Ireland K, Papakonstantinou Y, Ullman U, Widom J (1994) The TSIMMIS Project: Integration of Heterogeneous Information Sources. In: Proc. of the 10th Anniversary Meeting of the Information Processing Society of Japan. Tokyo, Japan, 7–18
13. Cohn A, Gotts N (1996) The "egg-yolk" Representation of Regions with Indeterminate Boundaries. In: Burrough P, Frank A (eds) Proc. of the GISDATA Specialist Meeting on Geographical Objects with Undetermined Boundaries. 171–188

14. Dasarathy B (2001) Information Fusion–What, Where, Why, When, and How? Information Fusion, 2(2):75–76
15. Devogele T, Parent C, Spaccapietra S (1998) On Spatial Database Integration. Int. Journal of Geographical Information Science, 4(1):335–352
16. Dhamankar R, Lee Y, Doan A, Halevy A, Domingos P (2004) iMAP: Discovering Complex Mappings Between Database Schemas. In: Weikum G, König A, Deßloch S (eds) Proc. of 2006 ACM SIGMOD Int. Conf. on Management of Data. Paris, France, 383–394
17. Doan A, Domingos P, Levy A (2000) Learning Source Description for Data Integration. In: Proc. of the ACM SIGMOD Workshop on The Web and Databases (Informal Proceedings). Dallas, Texas, USA, 81–86
18. Doan A, Madhavan J, Domingos P, Halevy A (2002) Learning to Map Between Ontologies on the Semantic Web. In: Proc. of the 11th Int. World Wide Web Conf. 662–673
19. Dononi F, Lenzerini M, Nardi D, Schaerf A (1996) Reasoning in Description Logics. In: Brewka G (ed) Principles of Knowledge Representation and Reasoning. CSLI Publications, 193–238
20. Duckham M, Lingham J, Mason K, Worboys M (2006) Qualitative Reasoning about Consistency in Geographic Information. Information Sciences, 176(6):601–627
21. Duckham M, Mason K, Stell J, Worboys M (2001) A Formal Approach to Imperfection in Geographic Information. Computers, Environment and Urban Systems, 25:89–103
22. Duckham M, Worboys M (2005) An Algebraic Approach to Automated Geospatial Information Fusion. Int. Journal of Geographic Information Science, 19(5):537–557
23. Fisher P (1996) Boolean and Fuzzy Regions. In: Burrough P, Frank A (eds) Geographic Objects with Indeterminate Boundaries. Taylor and Francis, 87–94
24. Fisher P, Wood J (1998) What is a Mountain? Or the Englishman Who Went Up a Boolean Geographical Concept but Realised It Was Fuzzy. Geography, 83(3):247–256
25. Fonseca F, Davis C, Camara G (2003) Bridging Ontologies and Conceptual Schemas in Geographic Information Integration. Geoinformatica, 7(4):355–378
26. Fonseca F, Egenhofer M (1999) Ontology-driven Geographic Information Systems. In: Medeiros CB (ed) Proc. of the 7th Symp. on Advances in GIS. Kansas City, USA, 14–19
27. Fonseca F, Egenhofer M, Agouris P, Câmara G (2002) Using Ontologies for Integrated Geographic Information Systems. Transactions in GIS, 6(3):231–257
28. Gangemi A, Pisanelli D, Steve G (1998) Ontology Integration: Experiences with Medical Terminologies. In: Guarino N (ed), Proc. of the 1st Int. Conf. on Formal Ontology in Information Systems. Trento, Italy, IOS Press, 163–178
29. Ganter B, Wille R (1999) Formal Concept Analysis. Springer, Berlin Heidelberg New York
30. Garcia-Molina H, Papakonstantinou Y, Quass D, Rajaman A, Sagir Y, Ullman J, Vassalos V, Widom J (1997) The TSIMMIS Approach to Mediation: Data Models and Languages. Journal of Intelligent Information Systems, 8(2):117–132
31. Gómez-Pérez A, Corcho O (2002) Ontology Languages for the Semantic Web. IEEE Intelligent Systems, 17(1):54–60
32. Gruber T (1993) A Translation Approach to Portable Ontology Specifications. Knowledge Acquisition, 5(2):199–220
33. Guarino N (1998) Formal Ontology and Information Systems. In: Guarino N (ed), Proc. of the 1st Int. Conf. on Formal Ontology in Information Systems. Trento, Italy, IOS Press, 3–15
34. Guarino N, Masolo C, Vetere G (1999) Ontoseek: Content-based Access to the Web. IEEE Intelligent Systems, 14(3):70–80

35. Haarslev V, Möller R (2001). Description of the RACER System and Its Applications. In: Goble C, Möller R, Patel-Schneider P (eds) Proc. of the Int. Work. in Description Logics. Stanford, USA

36. He B, Chang K C-C (2006) Automatic Complex Schema Matching across Web Query Interfaces: A Correlation Mining Approach. ACM Transactions on Database Systems, 31(1): 346–395

37. Kavouras M, Kokla M (2002) A Method for the Formalization and Integration of Geographical Categorizations. Int. Journal of Geographical Information Science, 16(5): 439–453

38. Kavouras M, Kokla M, Tomai E (2005) Comparing Categories Among Geographic Ontologies. Computers & Geosciences, 31(2):145–154

39. Keefe R, Smith P (1996) In: Keefe R, Smith P (eds) Vagueness: A Reader. MIT Press, Cambridge, MA

40. Kim W, Sea J (1992) Classifying Schematic and Data Heterogeneity in Multidatabase Systems. IEEE Computer, 24(12):12–18

41. Kokla M, Kavouras M (2001) Fusion of Top-Level and Geographic Domain Ontologies based on Context Formation and Complementarity. Int. Journal of Geographical Information Science, 15(7):679–687

42. Lakshmanan L, Sadri F, Subramanian I (1993) On the Logical Foundations of Schema Integration and Evolution in Heterogeneous Database Systems. In: Ceri S, Tanaka K, Tsur S (eds) Proc. 3rd Int. Conf. on Deductive and Object-Oriented Databases, Lecture Notes in Computer Science, Springer, Berlin Heidelberg New York, 81–100

43. Li W-S, Clifton C (1994) Semantic Integration in Heterogeneous Databases Using Neural Networks. In: Bocca JB, Jarke M, Zaniolo C (eds) Proc. of the 20th Int. Conf. on Very Large Data Bases. Santiago de Chile, Chile, Morgan Kaufmann, 1–12

44. Li W-S, Clifton C (2000) SEMINT: A Tool for Identifying Attribute Correspondences in Heterogeneous Databases Using Neural Networks. Data and Knowledge Engineering, 33(1):49–84

45. Madhavan J, Bernstein P, Rahm E (2001) Generic Schema Matching with Cupid. In: Apers P, Atzeni P, Ceri S, Paraboschi S, Ramamohanarao K, Snodgrass R (eds) Proc. of the 20th Int. Conf. on Very Large Data Bases. Roma, Italy, 49–58

46. Manoah S, Boucelma O, Lassoued Y (2004) Schema Matching in GIS. In: Bussler C, Fensel D (eds) Proc. of the 11th Int. Conf. on Artificial Intelligence: Methodology, Systems, and Applications. Varna, Bulgaria, Lecture Notes in Computer Science, Springer, Berlin Heidelberg New York, 500–509

47. McGuinness D (2003) Ontologies for Information Fusion. In: Proc. of the 6th Int. Conf. of Information Fusion. Carins, Queensland, Australia, 650–657

48. Mena E, Illarramendi A, Kashyap V, Sheth AP (2000) OBSERVER: An Approach for Query Processing in Global Information Systems based on Interoperation Across Pre-Existing Ontologies. Distributed Parallel Databases, 8(2):223–271

49. Miller R, Haas L, Hernandez M (2000) Schema Mapping as Query Discovery. In: Proc. of the 26th Int. Conf.on Very Large Data Bases. Cairo, Egypt. Morgan Kaufmann, 77–88

50. Mitra P, Wiederhold G, Kersten M (2000) A Graph-Oriented Model for Articulation of Ontology Interdependencies. In: Zaniolo C, Lockemann PC, Scholl MH, Grust T (eds) Proc. of the 7th Conf. on Extending Database Technology. Lecture Notes in Computer Science, Springer, Berlin Heidelberg New York, 86–100

51. Noy N, Musen M (1999) An Algorithm for Merging and Aligning Ontologies: Automation and Tool Support. In: Proc. of the AAAI99 Work. on Ontology Management. Orlando, Florida, USA, AAAI Press, 17–27

52. Noy N, Musen M (2003) The PROMPT Suite: Interactive Tools for Ontology Merging and Mapping. Int. Journal of Human-Computer Studies, 59(6):983–1024
53. Palopoli L, Rosaci D, Terracina G, Ursino D (2005) A Graph-based Approach for Extracting Terminological Properties from Information Sources with Heterogeneous Formats. Knowledge and Information Systems, 8(4):462–497
54. Rahm E, Bernstein P (2001) A Survey of Approaches to Automatic Schema Matching. The VLDB Journal, 10(4):334–350
55. Rosse C, Mejino J (2003) A Reference Ontology for Biomedical Informatics: the Foundational Model of Anatomy. Journal of Biomedical Informatics, 36(4):478–500
56. Sheth A (1999) Interoperability and Spatial Information Theory. In: Goodchild M, Egenhofer M, Fegeas R, Kottman C (eds) Interoperating Geographic Information Systems, Kluwer, Dordrecht, Netherlands, 5–29
57. Sheth A, Kashyap V (1993) So Far (Schematically) Yet so Near (Semantically). In: Hsiao D, Neuhold E, Sacks-Davis R (eds) Proc.of the IFIP DS-5 Conf. on Semantics of Interoperable Database Systems. Lorne, Australia, 283–312
58. Spaccapietra S, Parent C, Dupont Y (1992) Model Independent Assertions for Integration of Heterogeneous Schemas. The VLDB Journal, 1(1):81–126
59. Stell J, Worboys M (1997) The Algebraic Structure of Sets of Regions. In: Hirtle S, Frank A (eds) Proc. of the Int. Conf. on Spatial Information Theory. Laurel Highlands, Pennsylvania, USA, Lecture Notes in Computer Science, Springer, Berlin Heidelberg New York, 163–174
60. Stumme G, Maedche A (2001) FCA-Merge: Bottom-up Merging of Ontologies. In: Proc. of the 17th Int. Conf. on Artificial Intelligence. Seattle, Washington, USA, 225–230
61. Tejada S, Knoblock C, Minton S (2001) Learning Object Identification Rules for Information Integration. Information Systems, 26(8):607–633
62. Tzitzikas Y, Spyratos N, Constantopoulos P (2001) Mediators over Ontology-based Information Sources. In: Proc of the 2nd Int. Conf. on Web Information Systems Engineering. Kyoto, Japan, 31–40
63. Uitermark H, Oosterom P, Mars N, Molenaar M (1999) Ontology-based Geographic Data Set Integration. In: Böhlen M, Jensen C, Scholl M (eds) Proc. of the Int. Work. on Spatio-Temporal Database Management. Edinburgh, Scotland, Lecture Notes in Computer Science, Springer, Berlin Heidelberg New York, 60–79
64. Umemura K, Murao O, Yamazaki F (2000) Development of GIS-based Building Damage Database for the 1995 Kobe Earthquake. In: Proc. of the 21st Asian Conf. on Remote Sensing. 389–394
65. Vckovski A (1998) Interoperable and Distributed Processing in GIS. Taylor & Francis, London
66. Wache H, Vögele T, Visser U, Stuckenschmidt H, Schuster G, Neumann H, Hübner S (2001) Ontology-based Integration of Information–A Survey of Existing Approaches. In: Stuckenschmidt H (ed) Proc. of the IJCAI-01 Work. on Ontologies and Information Sharing. Seattle, Washington, USA, 108–117
67. Wald L (1999) Definitions and Terms of Reference in Data Fusion. In: Baltsavias E, Csatho B, Hahn M, Koch B, Sieber A, Wald L, Wang D (eds) Int. Archives of Photogrammetry and Remote Sensing. 2–6
68. Widom J (1995) Research Problems in Data Warehousing. In: Proc. of the 4th Int. Conf. on Information and Knowledge Management. Baltimore, Maryland, USA
69. Wiederhold G (1992) Mediators in the Architecture of Future Information Systems. IEEE Computer, 25(3):38–49
70. Worboys MF, Clementini E (2001) Integration of Imperfect Spatial Information. Journal of Visual Languages and Computing, 12:61–80

71. Worboys MF, Duckham M (2002) Integrating Spatio-Thematic Information. In: Egenhofer M, Mark D (eds) Geographic Information Science. Lecture Notes in Computer Science, Springer, Berlin Heidelberg New York, 346–361
72. Worboys MF, Duckham M (2004) GIS: A Computing Perspective, 2nd edition. CRC Press.
73. Zhou G, Hull R, King R, Franchitti J-C (1995) Data Integration and Warehousing Using H2O. IEEE Data Engineering Bulletin, 18(2):29–40

7

A Quality-enabled Spatial Integration System

Omar Boucelma, Mehdi Essid, and Yassine Lassoued

Université Paul Cézanne, Marseille (France)

7.1 Introduction

The proliferation of spatial data on the Internet is beginning to allow a much larger audience to share data currently available in various geographical information systems (GIS). As spatial data increases in importance, many public and private organizations need to disseminate and have access to the latest data at a minimum (right) cost and as fast as possible. In order to move to a real Web-based spatial data system, we need to provide flexible and powerful GIS data integration solutions. Indeed, GIS are highly heterogeneous: not only do they differ in their data representation, but they also offer radically different query languages. The main problems resulting from data integration are the data modeling (how to integrate different source schemas) and their querying (how to answer correctly to the queries posed on the global schema). The first issue addressed in this chapter is related to geographical data *integration*.

The second important issue we address in this chapter is spatial data *quality*. Indeed, data quality descriptions are crucial for the development of GIS, since they play a central role in many fields: environmental hazards, risk prevention, and so on. Note that quality is an issue for general purpose information systems (IS). As an example, the IS research community has addressed several aspects such as data cleaning [15], quality estimation [28], or quality assessment [33], to cite a few. The GIS community has been very proactive in metadata quality modeling: several quality models have been suggested such as ISO/TC 211 [22], FGDC [13], IGN [11], or more recently the ISO 19115 quality model [21].

In this chapter, we will describe how we tackled both issues and will present our solution, that is a quality-enabled geographical integration system. The chapter is organized as follows. We first discuss some related work in Sect. 7.2 then, in Sect. 7.3, we present a motivating example that will be used all along this chapter. In Sect. 7.4 we discuss the schema mapping process, while in Sect. 7.5 the query rewriting process is detailed. Section 7.6 discusses the design of a quality-enabled geographical mediator, and the processing of quality queries is described in Sect. 7.7. Finally we conclude in Sect. 7.8.

7.2 Related Work

In this section, we describe some related work that will help the reader to understand our approaches to both spatial data integration and geographical data quality management.

7.2.1 Data Integration

The DataBase (DB) community has extensively studied and developed data integration approaches and systems leading to, among others, a virtual approach to data integration called *mediation* [40]. Several systems and prototypes have been developed: examples of such systems are TSIMMIS [16], PICSEL [18], Information Manifold [23], AGORA [26], Styx [1]. A mediation system provides to the user a uniform interface of the different data sources via a common data model. Schema integration issues are the sources heterogeneity, global schema modeling and definition, the definition and the management of the mapping rules that express the correspondences between the global schema and the data source ones, the source semantics and the schema evolution.

There are two main approaches to data mediation: in the *global as view* (GAV) approach, the global schema is defined as a set of views over local schemas, while in the *local as view* (LAV) one, local sources are defined as a set of views over a given global schema, pertaining sometimes to some domain ontology. Pros and cons of these two approaches are well-known: *query rewriting* is straightforward in GAV while adding a new data source is made easy in LAV. Alternative approaches, such as *global local as view* (GLAV) [14] or *both as view* (BAV) [6], have been proposed. GLAV combines the expressive power of both LAV and GAV, allowing flexible schema definitions; BAV is based on the use of reversible schema transformation sequences.

The fundamental question, when attempting to interoperate several data sources, is twofold: on the one end, the identification of different objects having a semantic link, and coming from different data sources, on the other end, the resolution of structural (schematic) differences between objects having a semantic link. Schematic conflicts between data may arise when equivalent concepts are represented differently in local data sources. Those conflicts can be associated with concept names, their data types (e.g. a building can be a polygon in a data source and a point in another), the unit (perimeter can be expressed in meters in a data source and in kilometers in another one), the attributes (some attributes may be absent in some data sources). Another kind of schematic conflict is the difference in the representation of an attribute. For example, one can represent an address with a single attribute and another can represent it with a tuple (Number, Street, City, Zip code).

In addition to the query language, the power of an integration system is based on how schema mapping is performed and how efficient is the query rewriting algorithm. Most of the existing mediation systems use views to express correspondences between the real data source schemas and the integrated one. For example, in Information Manifold, the real data sources are expressed as a set of relational views over the global schema, and query rewriting is performed with the *Bucket* algorithm [25]. Another algorithm, *MiniCon* [34], extended *Bucket* in using the input/output common variables between the query subgoals to reduce the bucket size. *MiniCon* reduces the query rewriting time and returns some solutions ignored by *Bucket*. The *Styx* algorithm is an ontology-based algorithm which uses the parent–child dependencies of query variables for query decomposition.

As for the GIS community, essential work has focused on interoperability aspects [10, 19]. Most of the approaches propose to enrich the data models in order to conform to a "unified model", and the creation of the open geospatial consortium (OGC) [32] is the most visible output of this trend. Other research work has focused on schema integration, for example [12], object fusion [3], or ontology-based GIS integration: Chapter 6 provides a clear description of ontology usage in the field of data integration.

7.2.2 Data Quality

Data quality descriptions are crucial for the development of an organized business with geographical data. Potential users of data must understand beforehand the quality of the data they intend to acquire. One of the issues is to assign a meaning to the geographical data in a context—a standard problem now lumped together with other similar problems in metadata description. The standardized descriptions, however—in an equally long tradition—describe the quality of data by giving details of the data collection process, so-called lineage data (e.g. Dublin Core). For some aspects of spatial data quality, quantified measures are often used: for instance, the ISO 19113-5 standards [20, 21] propose about 15 (mostly quantifiable) quality elements and sub-elements. The issue is exacerbated when it relates to the manipulation of multiple heterogeneous GIS, each of them providing (or not) some quality criteria and information.

There is a growing awareness of the data quality problem in both DB and GIS research communities. Work in DB projects have been interested in several aspects of quality-based data integration techniques [27], [29], [17]. Within the GIS community, several geographical data quality models have been suggested by organizations such as FGDC [13], IGN [11] and converge to an ISO/TC 211 model [22].

Classical integration approaches assume that data stored at different sources have the same high quality and correctness. Actually, data sources are of different qualities but sufficient for the applications they were initially dedicated to. However, they may lead to low-quality integrated data in the context of a heterogeneous application. Furthermore, the "high quality" definition is subjective.

7.3 Motivating Example

The example is drawn from a real geographical data integration problem being studied in the REV!GIS project [35].

7.3.1 Data Sources and Schemas

The example used in this chapter consists of three local data sources referred to as BDTQ-21ee201 (TQe), BDTQ-31hh202 (TQh) (Bases de Données Topographiques du Québec) [2], and NTDB-21e05 (NT) the National Topographic DataBase [8]. TQe and TQh have the same schema which consists of 25 classes covering geographical entities at the scale of 1/20 000. NT schema refers to 196 classes covering geographical entities at a scale of 1/50 000. As illustrated in Fig. 7.1, TQe and NT do overlap.

For the sake of simplicity, the example is limited to one global class *Road*, and two local classes *BDTQ.vCommL* and *NTDB.roadL* which, respectively, represent linear objects in the transportation network, and linear roads. Figure 7.2 illustrates the classes (schemas) with their relationships. Figure 7.3 details the meaning of attributes.

GIS providers may supply some data quality information as illustrated in Fig. 7.4.

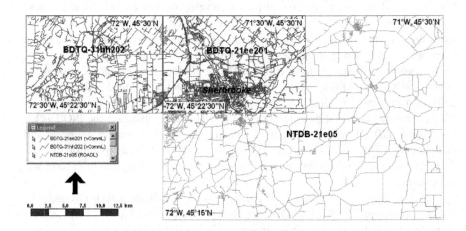

Fig. 7.1. Sources' spatial covers

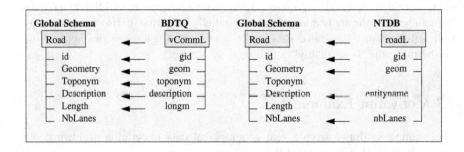

Fig. 7.2. Global and local schema mappings

Attribute	Meaning
gid	Object identifier
geom	Object geometry: point, line, polygon, etc.
toponym	Object toponyme
description, entityname	Describe the kind of a road: "Railway," "Street," "Bridge," "Unpaved street," etc.
longm	Length of a road or a bridge
nbLanes	Number of lanes of a road or a bridge

Fig. 7.3. Attribute definitions

Quality parameter	BDTQ-21ee201 & BDTQ-31hh202	NTDB-21e05
Scale	1/20 000	1/50 000
Spatial cover	21ee201 : [(-720, 45022′30″),(-71030′ , 45030′)] 31hh202 : [(-72030′ 45022′30″),(-720, 45030′)]	[(-720, 45015′),(-710, 45030′)]
Provider	Ministry of Natural Resources - Quebec	Topo. Info. Center - Sherbrooke
Validation date, date of validity	1995	1996
Geometric accuracy	4m	10m
Nomenclature source	Quebec Toponym Commission	
Roads classification source	Transportation Ministry	
Frequency of updates	5 years	
deficit, surplus, classification error	1%	5%

Fig. 7.4. Quality metadata

Note 7.1. We are using OGC's abstract feature model with its GML encoding [30] for DB modeling and schema specification. A GML data source provides two main elements: an XML schema description (XSD) and a GML document. The XSD tree describes the structure of the GML document that represents a source schema. A non-terminal node corresponds to a class (a complex-type element) and a leaf corresponds to an attribute (a simple-type element or a geometric property).

7.3.2 Schema Correspondences

The problem is to seamlessly access the various data sources described above. Our solution consists in a mediation system that supplies a *mediation schema* or *global schema*. Figure 7.2 depicts the relationships between the *mediation* class *Road* and the local classes *vCommL* and *roadL*. These relationships represent semantic links and are expressed by a set of schema mapping rules. Mappings identify portions of a source schema and the global schema that correspond (totally or partially). Note that correspondences between class elements are not usually trivial (e.g. one-to-one). They may need restrictions (in case of partial mappings between classes, such as between *vCommL* and *Road*, for example) or conversion functions (in case of attribute mappings with heterogeneous representations such as for *description* and *Description*, for example).

7.4 Schema Mapping

7.4.1 Dealing with Geographical Heterogeneity

One of the main issues faced when integrating geographical data is schema heterogeneity. As a matter of fact, each geographical data source is an abstraction of the real world according to a particular point of view and purpose. A data source may represent regular topographic maps (e.g. ordnance survey) or a statistical survey (national Census), or even a satellite image (weather forecast or land-cover). For each dataset, the underlying motivation leads to a particular representation of geographical objects. As a result, depending on the source point view, its scale, its producer and its final user, the data source has its own way of classifying geographical objects. For example, NTDB (as well as the global schema) distinguishes bridges and roads, whereas BDTQ groups all transportation network objects in two classes according to their geometries. Often different classifications have different levels of granularity, and this situation leads to partial correspondences between objects. We can refer, for instance, to the correspondence between class roadL from NTDB and class vCommL from BDTQ. Note that the latter uses attribute description to distinguish real object functions, that is to make a more precise classification of the objects; possible values for this attribute are *Railway, Street, Highway, Bridge, Unpaved street, Cross-street, Road under construction.*

As mentioned above, class Road has a partial match with BDTQ class vCommL. This match is valid under a constraint on attribute description in vCommL. Hence, we are not dealing with simple one-to-one mappings, and we need to express partial correspondences between classes using some constraints, as illustrated below:

$$\text{Road} \xleftarrow{\sigma_{vr}} \text{vCommL}$$

which means that Road corresponds to vCommL under the restriction σ_{vr}, where σ_{vr} is defined by the constraint

$$\text{description} \in \{\text{"Street," "Highway,"} \cdots \}.$$

7.4.2 More on Mappings

Data integration is a process by which several local schemas are integrated to form a single virtual (global) schema. As seen in Sect. 7.1, data integration approaches are known either as GAV, LAV, GLAV, or BAV to cite a few. Whatever approach we adopt, we need to provide a framework for schema transformations, that is a language for schema mappings. We adopted a GLAV-like approach, with simple *mapping rules* that allow the specification of one-to-one schema transformations under some constraints.

A mapping is composed of a right term, a left term, and a restriction (in the case of a class mapping). The left term is a global schema construct, while the right one consists of one or more paths of the source schema construct (possibly an empty path). Such a mapping means that the global schema element of the left term corresponds to the local schema path(s) under the given restriction. A *mapping rule* is identified to a part of the global schema where each element is expressed as depending on one or more paths of the local schemas using a mapping.

For example, rule R_1 in Fig. 7.5 describes the mapping of global class Road in NTDB. Similarly, rule R_2 illustrates the mapping of class Road in BDTQ. More generally, for each local source, we describe the mapping for each class of the global

```
<catalog>                          </rule>
<mapping sourcePath="http://...">  <mapping sourcePath="http://...">
<sourceName>NTDB-21e05</sourceName><sourceName>BDTQ-21ee201</sourceName>
<rule label="R1" >                 <rule label="R2" >
<classMapp>                        <classMapp>
<targetTag>Road</targetTag>        <targetTag>Road</targetTag>
<sourceTag>roadL</sourceTag>       <sourceTag>vCommL</sourceTag>
<attributeMapp>                    <condition> <OR>
<targetTag>id</targetTag>          <GeneralComp> =
<sourceTag>gid</sourceTag>         <sourceTag>Sescription</sourceTag>
</attributeMapp>                   <Literal>Street</Literal>
<attributeMapp>                    </GeneralComp>
<targetTag>Description</targetTag> <GeneralComp> =
<sourceTag>entityname</sourceTag>> <sourceTag>Description</sourceTag>
</attributeMapp>                   <Literal>Highway</Literal>
<attributeMapp>                    </GeneralComp>
<targetTag>NbLanes</targetTag>     </OR> </condition>
<sourceTag>nbLanes</sourceTag>     <attributeMapp> ...
</attributeMapp>                   <!-- gid -> id, Geom. -> geom
<attributeMapp>                        description -> Description
<targetTag>Geometry</targetTag>        toponym -> Toponym,
<sourceTag>geom</sourceTag>            longm -> Length -->
</attributeMapp>                   </rule> </mapping>
</classMapp>                       </catalog>
```

Fig. 7.5. Mapping rules $R_1(roadL, Road)$ and $R_2(vCommL, Road)$

schema. We are using <targetTag> (resp. <sourceTag>) conditions tags to denote local schema constructs (rep. global ones); this syntax is inspired from XQueryX [38]. We clearly see that class Road corresponds to vCommL under the restriction that attribute description is "Street" or "Highway." In the same way, attribute Description corresponds to description, Toponym to toponym, and so on. Let us note also that, for a given local source, not all global attributes have a correspondent in the local schema. For example, in the BDTQ mapping, attribute NbLanes has no correspondent.

7.5 Query Rewriting

In this section, we first give an overview of the VirGIS [5] integration system and the query languages being used. Then, we describe the query processing internals.

7.5.1 VirGIS Overview

The VirGIS is a geographical mediation system that complies with OGC's recommendations with respect to some of its core software components, that is, GML [30] and WFS [31]. In addition, it provides a spatial query language GQuery [9] for querying purposes and expressing complex schema mappings. VirGIS is in charge of processing (geographical) user queries which are expressed either in GQuery or in XML/KVP (keyword-value-pair), the WFS language. Sources are published on the VirGIS portal via WFS servers. An *admin interface* allows the addition, modification, or deletion of sources located by their WFS addresses. Local schemas and capabilities are extracted using WFS queries (*GetCapabilities* and *DescribeFeature-Type*). The *global schema* is specified as an XSD [37]. To understand how the system works, let us describe the path followed by a query as illustrated in Fig. 7.6.

When submitted to the system, a (global) user query is rewritten into subqueries expressed in terms of source schemas. First of all, the *decomposition* module decomposes the user query into a set of elementary subqueries (ESQs) expressed in terms of the global schema. Each such (elementary) subquery aims to extract information about a single feature. The decomposition module computes a global execution plan (GEP) expressed as a join over ESQs.

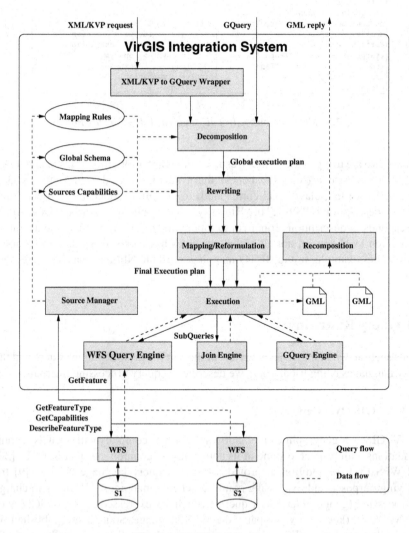

Fig. 7.6. VirGIS architecture

Next, subqueries of the GEP are processed by the *Rewriting* module in order to be expressed in terms of source schemas. This is performed by this module in using the mapping rules; global to local bindings are also computed at this stage. Hence, we obtain a local execution plan for each ESQ of the global execution plan. Each local execution plan is a combination of queries over source schemas using joins, unions, and other (possibly spatial) operations. The *Mapping/Reformulation* module reformulates these queries such that the results are compliant with the global schema. This leads to a final execution plan which may be composed of three kinds of queries:

1. HTTP queries are WFS queries in charge of extracting a feature from a local source. These are processed by the WFS Query Engine.
2. GQuery queries are (mainly) used to express spatial operations, and are processed by the GQuery engine.
3. JOIN queries play a central role in the execution plan. Since we are handling a large amount of data, and because existing XQuery implementations do not supply efficient join modules, we implemented a specific Join engine.

The final execution plan is then processed by the *Execution* module, which is in charge of dispatching the queries to the appropriate engine. This module uses a cache repository to save temporary result of the different queries. When the execution is finished, the *Recomposition* module eliminates the duplicated answers and returns a GML file that can be either processed (if needed) or simply (generally) displayed by a GML viewer.

7.5.2 Query Language

In the VirGIS system, we are using GQuery, a spatial query language based on XQuery [39], that allows users to perform both attribute and spatial queries against GML documents. Spatial semantics is extracted and interpreted from XML tags without affecting XQuery syntax, that is a spatial query looks like a regular XQuery expression.

A GQuery sentence is composed of expressions, each of them being built-in or user-defined functions. An expression has a value, or generates an error. The result of an expression can be the input of a new one. A value is an ordered sequence of items. An item is a node or an atomic value [36]. There is no distinction between an item and a sequence containing one value. There are seven types of nodes: document, element, attribute, text, comment, processing-instruction, and namespace nodes. Writing a query consists of combining simple expressions (like atomic values), XPath expressions [36], FLOWER expressions (For-Let-Where-Return), test expressions (if-then-return-else-return), or (pre-or user-defined) functions. Non-spatial operators are arithmetic operators $(+,-,\times,/,mod)$, operators over sequences (concatenation, union, difference), comparison operators (between atomic values, nodes, and sequences), and Boolean operators.

Spatial operators are applied to sequences. A sample of operators are described below:

- operators that return numeric values:
  area, length : *sequence* = (*node*) → *numeric value* where *node* is a GML data node.
- operators that return GML values:
  convexhull, centroid : *sequence* = (*node*) → *node*
- operators that return Boolean values:
  equal, within, touches : *sequence* = (*node, node*) → *boolean*

Each result of a GQuery expression is part of the data model. The input and output of every query or subexpression within a query is an instance of the data model. GQuery is closed under this query data model. When computing *area(node)*, if *node* is a Polygon, the function returns a numeric value, otherwise it raises an error. In both cases, results are instances of the data model. Spatial operators can be nested.

7.5.3 Obtaining a Global Execution Plan

The GEP is obtained in decomposing the user query into a set of ESQs expressed in terms of the global schema. Actually, data sources are queried through a WFS interface. Although providing easy access to any geographical repository, WFS is unable to handle complex queries and does not provide support for data integration. Thus execution plans are a combination (join, union, and operation) of elementary queries, each of them being posed against a single feature.

As an example, let us consider query of Fig. 7.7 (denoted Q_0 in the sequel) which consists in extracting all two-lane roads crossing bridges and having a length up to 1000 meters.

As we can see, this query involves two features (bridge and road) and three conditions. The condition (cross(x/*geometry*,y/geometry)) is a complex one because it involves two attributes from two different features. Consequently, Q_0 is split into three subqueries as illustrated in Fig. 7.8. The first subquery consists in extracting all two-lane roads having length up to 1000 metres. The second one consists in extracting all the bridges and the last one consists in joining the first query and the second one with the condition cross(x/*geometry*,y/geometry) = true.

The execution plan is represented as an XML document, where nodes describe queries: a priority attribute indicates the query execution order. For example, `query0` and `query1` are executed first, followed by `query2`, and so on.

```
for $x in document(bridge), $y in document(road)
where cross( $x/geometry, $y/geometry) = true
and $x/length > 1000 and $x/NbLanes = 2 return $x
```

Fig. 7.7. User-query example

```
<Queries>
<Query id="query0" priority="0">
for $x in document(road) where $x/length > 1000 and $x/NbLanes = 2
return $x
</Query>
<Query id="query1" priority="0">
for $x in document(bridge)
return $x
</Query>
<Query id="query2" priority="1">
for $x in document(query0), $y in document(query1) where
cross($x/geometry,$y/geometry) = true return $x
</Query>
</Queries>
```

Fig. 7.8. Global execution plan for Q_0

```
Input       : User Query Q0
Output      : Global Execution Plan
Algorithm   :
  ExecutionPlan EP
  SimpleQueries SQ = ExtractSimpleQueries(Q0)
  QueryCpt = 0
  For each Q  in SQ Do
      Q.setId("query"+ QueryCpt)
      QueryCpt++
      EP.AddQuery(Q)
  End For
  RemoveSimpleConditions(Q0)
  ReplaceFeatureByQueryID(Q0)
  EP.AddQuery(Q0)
```

Fig. 7.9. Decomposition algorithm

The decomposition algorithm, illustrated in Fig. 7.9, relies mainly on three functions which are described below:

- ExtractSimpleQueries: extracts simple queries as well as their simple conditions (queries consisting in extracting features) from the user query.
- RemoveSimpleConditions: removes all the simple conditions from the user query because they have already been treated in the simple queries.
- ReplaceFeatureByQueryID: replaces the features by the id of their corresponding simple query.

7.5.4 Obtaining a Final Execution Plan

Assuming that the GEP contains $N + 1$ subqueries, we rewrite each of the first N subqueries in terms of the local schemas, the last one being used to compute the final result. To this purpose, we adapted the approach described in [1]. For each query, we start in computing its *binding*. A binding of a basic query is the set of mappings that support (totally or partially) the feature of the query. When a source supports all the attributes of the query (totally), we say that this query has a *full-binding* with this source. The binding of the query allows query rewriting into a set of subqueries expressed in terms of the local schemas.

Let us consider, for instance, a basic query Q_e and its binding $\mathcal{B} = \{M_i, i \in 1 \ldots n\}$ where n is the number of sources that support the feature queried by Q_e and M_i, $i \in 1 \ldots n$, are their respective mappings. To extract the maximum of information from the local sources, first we rewrite Q_e with a given mapping of the binding, then we try to look for the unsupported attributes (if they exist) using the other mappings of the binding. To do this we use the notion of *prefix query* and *suffix query*. Given a mapping M_i of a source S_i, a prefix query Q_p is a subquery of Q_e that consists in extracting only the attributes supported by S_i. Note that M_i provides a full-binding for query Q_p. The suffix query Q_s is the subquery consisting in extracting the remaining attributes (plus the key's attributes).

Given these definitions, for each mapping $M_i \in \mathcal{B}$, Q_e is processed as follows:

1. compute Q_p and Q_s of the query Q_e and the mapping M_i;
2. let $\mathcal{B}_s = \{M_j, M_j \in \mathcal{B}, j \neq i\}$;
3. let $\mathcal{B}_i = \{M_i\}$;
4. the execution plan of query Q_e using the binding $\mathcal{B}_i$ is the join between Q_p and the execution plan of Q_s using the binding $\mathcal{B}_s$.

Since queries that are responsible for extracting feature information are executed by the WFS servers, they are reformulated, using the mapping rules, into queries that are expressed in terms of local schemas (each feature is replaced by the corresponding feature, each attribute is replaced by the corresponding attribute, and the conditions of the rule (cf. Sect. 7.4) are added to the set of conditions of the query). This reformulation generates a problem because the answer to this query (the answer of the WFS) is written in terms of the local schema. To alleviate this problem, an additional GQuery query is added to perform the inverse transformation. Queries `query03` and `query04` illustrated in Fig. 7.11 are examples of such queries.

Let us now discuss constraints of the prefix and the suffix queries. It is easy to see that the prefix query contains only the subset of conditions over attributes of the prefix query. However, the suffix query contains the subset of conditions over attributes of the suffix query extended by the set of conditions of the prefix query. This extended set is used to refine the prefix query in adding more conditions. This allows us to minimize the number of extracted tuples, hence minimizing the execution time. Finally, a condition that is not supported by a data source is added to the GQuery expression, which is responsible for the transformation of the result: this is the case where the source contains the attributes on which the condition is made but the capabilities of the source cannot support it.

Figure 7.10 illustrates the rewriting algorithm which, given an elementary query Q_e and its binding $\mathcal{B}$, computes the execution plan expressed in terms of the local sources. As we may notice, $\mathcal{B}_s$ is defined as $\{M_j \in \mathcal{B}, j > i\}$ and not as $\{M_j \in \mathcal{B}, j \neq i\}$ as mentioned in the beginning of this section. We made this change in order to eliminate duplicated answers. In fact, the subset of mappings $\{M_k \in \mathcal{B}, k < i\}$ has already been treated in the previous step; hence it is useless to compute it again.

To compute the final execution plan, we rewrite each elementary query of the GEP. As an example, let us process the first query of the global execution plan of

```
Input      : Elementary query Q_e, Binding  B
Output     : Execution Plan expressed in term of local schemas
Algorithm  : RW(Q_e, B)
  ExecutionPlan EP
  If (fullbinding (Q_e, M_1))
      tmpEP = Q_e
  Else
      Q_p = getPrefixQuery(M_1)
      Q_s = getSuffixQuery(M_1)
      B_s = {M_j ∈ B, j > 1}
      tmpEP = (Q_p ⋈ RW(Q_s, B_s))
  End if
  EP = EP ∪ tmpEP
```

Fig. 7.10. Rewriting algorithm

```
<Queries>
<Query id="query01" priority="0"
 Type="HTTP" Adress="http://...">
<GetFeature>
<Query typeName="roadL">
<PropertyName>gid</PropertyName>
<PropertyName>entityname</PropertyName>
<PropertyName>nbLanes</PropertyName>
<PropertyName>geom</PropertyName>
<Filter>
<PropertyIsEqualTo>
<PropertyName>nbLanes</PropertyName>
<Literal>2</Literal>
</PropertyIsEqualTo>
</Filter>
</Query>
</GetFeature>
</Query>
<Query id="query02" priority="0"
 Type="HTTP" Adress="http://...">
<GetFeature">
<Query typeName="vCommL">
<PropertyName>gid</PropertyName>
<PropertyName>toponym</PropertyName>
<PropertyName>longm</PropertyName>
<Filter>
<PropertyIsGreaterThan>
<PropertyName>longm</PropertyName>
<Literal>1000</Literal>
</PropertyIsGreaterThan>
</Filter>
</Query>
</GetFeature>
</Query>
<Query id="query03"
 priority="1" Type="GQUERY">
```

```
for $x in document(QUERY01)//roadL
return ( <featureMember>
<Road>
<id>$x/gid</id>
<Desc>$x/entityname</Desc>
<NbLanes>$x/nbLanes</NbLanes>
<Geometry>$x/geom</Geometry>
</Road> </featureMember>)
</Query>
<Query id="query04"
 priority="1" Type="GQUERY">
for $x in document(QUERY02)//vCommL
return (<featureMember>
<Road>
<id>$x/gid</id>
<Toponym>$x/toponym</Toponym>
<Length>$x/longm</Length>
</Road>
</featureMember>)
</Query>
<Query ID="query0"
 Priorite="2" Type="JOIN">
for $x in document(query03)//Road
 , $y in document(QUERY04)//Road
where $x/id = $y/id return (
<featureMember>
<Road>
<id>$x/id</id>
<Desc>$x/Description</Desc>
<NbLanes>$x/NbLanes</NbLanes>
<Geometry>$x/Geometry</Geometry>
<Toponym>$y/toponym</Toponym>
<Length>$y/length</Length>
</Road>
</featureMember>)
</Query>
```

Fig. 7.11. Final execution plan

Q_0. This is done in extracting all Roads having two lanes and a length greater than 1000 meters. The schemas illustrated in Fig. 7.2 shows that attributes Toponym and Length have no correspondent in class roadL (NTDB). Hence, the query is divided in two subqueries: a prefix query that extracts NTDB attributes ID, Description, Lanes, and Geometry that satisfy the condition NbLanes = 2, and a suffix query which extracts BDTQ attributes ID (the key), Toponym, and Length satisfying the condition

Length > 1000. A join is then performed between the results of the prefix and the suffix queries in order to build the final result. Figure 7.11 illustrates the execution plan of query query0 of the GEP.

7.6 VirGIS/Q: A Quality Geographical Mediator

In this section, we describe how to handle quality parameters in answering a mediated query that may involve many data sources with different qualities.

7.6.1 Problem Sketch

Consider a user who is looking for highways of a given region b that satisfy a given quality condition. The user may express a SQL like query Q_0:

Q_0: **Select** *all information about road objects of type Highway* **such that** *data cover region b and have been validated during the last 10 years (e.g. 1995–2005), with a deficit rate of road objects that is less than 1%, a geometric accuracy of at most 5 meters and a high semantic accuracy of the operational state of the roads.*

First of all, query processing may be different depending on the choice of the spatial cover b. For example, if we respectively choose region B, B', or B'' of Fig. 7.12, the number of data sources that will participate in query processing is respectively two, three, or zero. The case of B'' is trivial since the query has no answer. Examples with B and B' are more complicated for the reasons explained below.

Case of Region B: Both TQh and NT are eligible over the whole surface B. Both sources satisfy the condition for the validation date. The deficit rate of the first source (1%) is sufficient; that of the second is higher (i.e. worse) than required (5% > 1%); however, we cannot reject the data source because when considered with the first

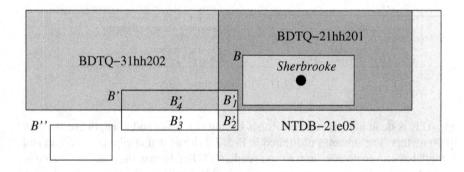

Fig. 7.12. Two sources participate with B as spatial cover, three sources with B', and none with B''

one it will improve (decrease) the total deficit rate. According to the quality conditions, the geometric property must be extracted from TQe, information about the operational state must be extracted from NT, while rest of the information can be extracted from any source.

Note that in modifying the quality conditions we may also modify the origins of attributes. For example, requiring a validation date between 1996 and 2004 instead of requiring a geometric accuracy of at most 5 m implies that the geometric property should be extracted from NT instead of TQe.

Suppose now that the deficit rate required by the user is 0.5% (instead of 1%). In this case, even though both data sources do not satisfy the condition with the deficit rate, they are still interesting and, if used together, they may lead to a good result. In fact, the probability that an object is absent from both data sources is $1\% \times 5\%$, that is 0.05%, which is less than 0.5%.

Case of Region B': In region B', TQh and NT cover different zones. Zones B'_1, B'_2, B'_3, and B'_4 do not involve the same data sources. However, each of them is homogeneous and can be treated in the same way we did above for B and B''.

7.6.2 Requirement Analysis

The goal of our (quality) mediation system is to allow a community of users to share a set of heterogeneous, autonomous geographical data sources of various qualities. A user poses a query (against the mediated schema) and asks for geographical objects of a particular kind, with a specific quality (conditions over the spatial cover, the geometric accuracy, the date of validity, the deficit and surplus rates, etc.). The system must explore the different data sources with their information quality in order to retrieve interesting data from appropriate sources and integrate them to generate, if possible, a result with an acceptable quality.

In order to achieve the quality mediation goal, we need to take into account many parameters:

1. Geometry: because geographical objects have geometric properties (punctual, linear, polygonal, etc.).
2. Spatiality: since data are linked to the ground and cover different zones, depending on the query zone the data sources used by the system may be different (see Sect. 7.6.1, for example).
3. Heterogeneity in data representation, different classifications, and different schemas.
4. Data Quality: its definition (model) and the requirements it imposes on the schema mapping language, the query model, and the query rewriting mechanism.

7.6.3 Solution Sketch

The mediation system uses a set of *mapping rules* that describe correspondences between global and local sources. A mapping rule associates pairs of corresponding elements (from the global and a source schema) and quality information. These rules allow the mediator (i) to select the appropriate sources and data for the user's query and (ii) to rewrite the user's query into queries over the data sources' schemas. Mapping rules are expressed in a language that supports restrictions and conversion functions.

In order to facilitate query rewriting and to ensure system extensibility (addition and deletion of data sources), a mapping rule describes a part (subtree) of the global schema according to only one data source schema. Combining information of various data sources is performed automatically by means of a *global key* mechanism. The idea behind this mechanism is to relate each attribute of a class to a key on which it depends. Finally, we use quality aggregation operations in order to estimate the quality of integrated data sets.

7.6.4 Quality Mediator

In order to define the quality mediator, we extended Lenzerini's definition for data integration systems [24], in adding a quality model, and a global key mechanism.

Definition 7.1 (Quality Mediation System). *A quality mediation system is defined as a tuple* $I = (G, S, \mathcal{D}, \mathcal{Q}, key)$, *where G is a global schema,* $S = \{S_i, i \in [1, \ldots, n]\}$ *is a set of source schemas,* $\mathcal{D} = \{M_i, i \in [1, \ldots, n]\}$ *is the set of their respective descriptions, that is for each* $i \in [1, \ldots, n]$, M_i *is the set of mapping rules between* S_i *and G. The term* $\mathcal{Q}$ *denotes a quality model and key denotes a function that associates to each class of the global schema a global key.*

Definition 7.2 (Global Key). *Given a class C of a global schema G, a global key is an attribute or a set of attributes of class C allowing identification of class instances (i.e. on which the values of the remaining attributes depend), whatever the (local) data source they are extracted from.*

The key mechanism we are using here is a simplified version of *XML keys* defined in [7] for XML documents and is similar to the one described in [1]. A global key plays the role of an object identifier. We suppose that each class C has a global key. As an example, the key attribute of class *Road* is its geometric property. Note that the *id* attribute cannot be used as a global key because it allows one to identify only objects of the same data source; thus it is called a local key [1].

7.6.5 Quality Model

A *quality model* allows the representation of various data quality parameters. It is mainly composed of quality criteria called *quality parameters* or *quality elements*. Several geographical quality models have been proposed by standard or public institutions such as FGDC [13], the ISO/TC 211 model [22], or the IGN [11] that we adopted as our quality reference model.

Quality Parameters

The IGN quality model we have adopted is composed of the following parameters:

$$Q_0 = B_b : \text{ Spatial cover} \qquad Q_1 = P_1 : \text{ Lineage}$$
$$Q_2 = P_2 : \text{ Currentness} \qquad Q_3 = P_3 : \text{ Geometric accuracy}$$
$$Q_4 = P_4 : \text{ Semantic accuracy} \qquad Q_5 = C_1 : \text{ Completeness}$$
$$Q_6 = C_2 : \text{ Consistency}$$

Each of the parameters Q_i, $i \in [0..6]$, is evaluated through a tuple of sub-parameters $(Q_i^1, \cdots, Q_i^{k_i})$. For the sake of simplicity, we only consider the IGN sub-parameters listed in Table 7.1. Because Q_5 and Q_6 have a different behavior, they are denoted respectively by C_1 and C_2. In fact, as explained later, these are percentages, which can be improved or degraded by combining data sets.

The quality model is attached to the global schema, each parameter applies to one or more types in the global schema. Column 2 of Table 7.1 indicates to which type of elements each parameter (or sub-parameter) is applicable. Each quality sub-parameter, when applied to a global schema element, is evaluated using a value of a given type. The last column of Table 7.1 represents the type of each sub-parameter.

Quality Assertions

Definition 7.3 (Quality Assertion). *A quality assertion κ associates to each quality sub-parameter Q_i^j, $i \in [0..6]$, $j \in [1..k_i]$, and an element E of the global schema (such as Q_i^j is applicable to E) a quality value $\kappa(Q_i^j, E)$. For ordered quality sub-parameters, κ returns an interval (a rectangle in the case of the spatial cover), whereas for others it returns a finite set of values.*

Table 7.1. Quality parameters and sub-parameters

parameter	applicable to	sub-parameter	significance	type
Q_0	geographical classes	$Q_0^1 = B_b$	bounding box	*boundingbox*
$Q_1 = P_1$	geometric properties	$Q_1^1 = P_1^1$	scale	*float*
	classes, attributes	$Q_1^2 = P_1^2$	provider	*string*
		$Q_1^3 = P_1^3$	source	*string*
		$Q_1^4 = P_1^4$	source organization	*string*
$Q_2 = P_2$	classes, attributes	$Q_2^1 = P_2^1$	validation date	*date*
		$Q_2^2 = P_2^2$	date of validity	*date*
		$Q_2^3 = P_2^3$	expiry date	*date*
		$Q_2^4 = P_2^4$	frequency of updates	*integer*
$Q_3 = P_3$	geometric properties	$Q_3^1 = P_3^1$	position accuracy	*float*
		$Q_3^2 = P_3^2$	shape accuracy	*float*
$Q_4 = P_4$	attributes	$Q_4^1 = P_4^1$	semantic accuracy	*enumeration/float*
$Q_5 = C_1$	classes, attributes	$Q_5^1 = C_1^1$	deficit rate	*percentage*
		$Q_5^2 = C_1^2$	surplus rate	*percentage*
$Q_6 = C_2$	classes	$Q_6^1 = C_2^1$	accordance rate	*percentage*
	attributes	$Q_6^2 = C_2^2$	violation rate	*percentage*

For example, if κ_{vr} denotes the quality assertion that describes the element *vCommL* of data source TQe that corresponds to class *Road* of the global schema, then we have

$$\kappa_{vr}(Q_0^1, Road) = [-72, -71.5] \times [45.375, 45.5],$$
$$\kappa_{vr}(Q_1^2, Road) = \{\text{'Ministry of Natural Resources of Quebec'}\},$$
$$\ldots,$$
$$\kappa_{vr}(Q_6^2, Road) = [0, 1\%].$$

The advantage of such representation is that it allows the expression of implicit preference relationships without having to define them. For example, to express that the deficit rate of a class E is better than 1% according to a quality assertion κ, we use the expression:

$$\kappa(C_1^1, E) = [0, 1\%].$$

Quality Comparison

In order to answer users' queries, we need to know if the quality of some dataset satisfies a given (quality) condition. As we use quality assertions to represent quality information as well as quality conditions, we need to make these assertions comparable.

Consider two quality assertions κ and γ. If the parameter Q_0 is applicable to a given element E, then we say that κ *satisfies* γ for the element E according to the sub-parameter Q_0 if the following condition is satisfied:

$$\kappa(Q_0, E) \cap \gamma(Q_0, E) \neq \emptyset.$$

In other words, the spatial cover specified by assertion κ satisfies that of γ if both of them overlap.

An example of quality comparison is illustrated in Fig. 7.13. Each parallelepiped in this figure represents the quality value for a given assertion, but related to the same element. Only the spatial cover and the accordance rate are represented here and they form a three-dimensional space. A satisfying quality value for that represented by P corresponds to the parallelepiped P'. On the contrary, P'' and P''' do not satisfy P.

Source Description

A source description specifies the mappings between the global schema and the source schema. Source descriptions are used by the mediator during query rewriting. In order to ensure expressive and generic correspondences between schemas, we use *mapping functions* (such as restrictions, concatenation, conversion functions, geometrical, or topological functions, specific administrator-defined functions, etc.). In the definition of a mapping below, we introduce a quality assertion, while in Sect. 7.4, we used constraints.

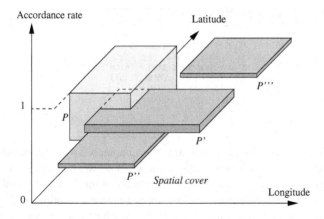

Fig. 7.13. A 3D illustration of the quality satisfaction problem between four quality values P, P', P'' and P'''

Definition 7.4 (Mapping). *A mapping is a couple* $(E \longleftarrow e, \kappa)$, *where E is an element of the global schema, e is an element or a tuple of elements of the source schema (identified by their absolute or relative paths), and κ is a quality assertion describing E.*

E is called the *target* of the mapping, e is its *source*, and κ is its quality assertion. The quality assertion κ of the mapping describes the quality of element E as being corresponding to e. For example, mapping between classes *Road* and *vCommL* of TQe can be expressed as follows:

$$(Road \longleftarrow vCommL, \kappa_{vr}),$$

where κ_{vr} is the quality assertion defined in Sect. 7.6.5.

7.7 Processing Quality-mediated Queries

7.7.1 Quality Query Language

The quality query language is based on GQuery [9], extended by a syntactic sugar clause SUCH THAT to express quality constraints. A query expression is defined as a block FOR-LET-WHERE-SUCH THAT followed by a RETURN clause. We assume that clause WHERE contains conditions with only an attribute, an operator (=, <, ≤, ≠, etc.), and a constant; it does contain no joins and no attribute comparison. For example, query Q_0, defined in Sect. 7.6.1, can be written as follows:

```
FOR $x in document("Region/Road")
WHERE $x/Type = "Highway"
SUCH THAT
    Bb(Road) = [-71.95, -71.65]*[45.38, 45.44]
    AND P.2.2(Road) = [1994, 2004]
    AND P.3.1(Road/lineStringProperty) = [0,5m]
    AND P.4.1(Road/Operational) = [High,VHigh]
    AND C.1.1(Road) = [0, 1%]
RETURN
<Sherbrooke>
    <Road> $x </Road>
</Sherbrooke>
```

The quality query language described in this chapter has some limitations: it supposes that users have a good knowledge of the schema/data they are going to query. Unfortunately this is not the case, due to the quality of the data and possibly multiscale representation. To overcome this limitation, a possible solution could consist in introducing some query relaxation mechanism, by which approximate answers are returned to the user. A preliminary work toward this direction is described in [4], while the foundation of this work are highlighted in Chap. 8.

7.7.2 Rewriting Quality Queries

Query rewriting is done in four steps: (1) decomposition of a query into ESQs, this leads to a GEP, (2) correspondence discovery and subquery reformulation over local sources, (3) space partitioning which leads to sectoral execution plans (SEP), and (4) construction of the final execution plan (FEP). In the next subsections, we shortly describe these steps.

Decomposing Quality Queries

A query is decomposed into ESQs, each of them returns an attribute or a key together with the key it depends on. In doing so, we can rewrite the initial query as the join of these EQ, hence building the GEP. For example, query Q_0 can be decomposed into 5 EQ ($q_0 \cdots q_4$) returning respectively attributes *Toponym, Description, Classification, Status,* and *NbLanes* together with the geometric property *Geom*.

Note 7.2. We are using here additional attributes of global class Road, namely *Classification* and *Status*, which map differently with attributes *classification* or *status* of local classes vCommL and roadL as illustrated in Fig. 7.14. Possible values for *classification* are 0 for unknown, 1 for highway, 2 for main, 3 for secondary, and so on. Possible values for *status* are 0 for unknown, 1 for operational, and so on.

Extraction of Correspondences

The purpose of this phase is to explore source descriptions in order to identify relevant sources and interesting mappings for each ESQ. The result is a *source evaluation* for each subquery and data source, showing how to express the subquery over the source schema and specifying the quality of the corresponding result.

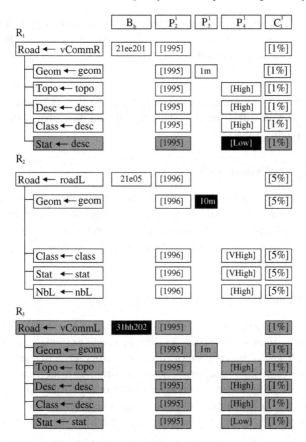

Fig. 7.14. Mapping rules for element *Road*

Given q (an EQ) and a source description M of S, the evaluation of q over S is the set of mappings $E \longleftarrow e$, where E traverses the whole set of elements required by q. The quality assertion associated with this evaluation is calculated as the union of the quality assertions of the mappings composing it.

Example 7.1. Consider query Q_0 and its subqueries q_i, $i \in [0..4]$, and suppose that the required coverage is rectangle B of Fig. 7.12. Consider the descriptions M_1, M_2, and M_3 of TQe, NT, and TQh respectively. Let $\mathcal{R}_1$, $\mathcal{R}_2$, and $\mathcal{R}_3$ be the mapping rules for element *Road* respectively with elements *vCommL* of TQe, *roadL* of NT, and *vCommL* of TQh, as illustrated in Fig. 7.14, where nodes of the left-side trees represent mappings.

Consider query q_2 which extracts attribute *Classification* together with key *Geom* of class *Road*. Only rules $\mathcal{R}_1$ and $\mathcal{R}_2$ allow a full evaluation of q_2. In fact, according to $\mathcal{R}_1$, each element of q_2 has a correspondent with satisfactory quality. According to $\mathcal{R}_2$, each element of q_2 has a correspondent. Although the geometrical accuracy of property *Geom* is not satisfied, the evaluation is taken into account because *Geom* is a global key. Element *Road* has a mapping in $\mathcal{R}_3$, but its quality is not that good because the spatial covers do not overlap. Consequently, we can reject this mapping. Since it is not useful anymore to use this rule for the rest of the elementary queries, we can prune it.

In the same way, we proceed with rest of the elementary queries without taking into account the pruned branches of the mapping rules. Figure 7.14 depicts, respectively in white and gray, the branches that are kept and those that are pruned: black cells show the quality values that are violated.

Elementary Query Reformulation

Once the ESQs evaluations are performed, it is possible to reformulate them in terms of the source schemas. This is done in replacing subquery elements by their correspondents while adding the restrictions resulting from the mappings. For each subquery evaluation, there corresponds a reformulated query having the same quality assertion as the evaluation.

Sectoral Execution Plans

Each ESQ of a query Q may involve a set of different data sources that play a certain role in different regions of the space, see, for instance, example of rectangle B' of Fig. 7.12. This leads to the creation of different sectors, each of them generating a different way for evaluating subqueries. Sector partitioning is motivated by two main reasons: (i) it avoids useless computations (while trying to join information of an area to those of a disjoined area) and (ii) it allows one to estimate the quality of the final execution plan. For each region, we compute a SEP, together with its quality assertion.

Final Execution Plan

The FEP of a query Q corresponds to the selection of an nSEP for each sector. The goal of the query rewriting process is to obtain a FEP with a *good quality*, that is each of its SEPs has a good quality. Our goal is to choose, among the reformulated queries, those that will participate in the FEP in a way that guarantees a satisfying quality. Queries are ranked according to their qualities, in order to improve the quality of the result, as well as the process speed. Making such a choice is difficult because quality parameters are often contradictory. For example, improving the deficit rate means degrading the surplus rate and the logical consistency. Thus, the problem is to obtain a compromise leading to a rather complete result set without affecting its surplus rate or its logical consistency.

7.7.3 Query Execution

Once the FEP is generated, it is ready to be executed. Each SEP is considered separately and its reformulated queries are extended with a spatial restriction over the sector. In order to limit the number of transactions, we gather for each sector the reformulated subqueries according to their sources. The decomposition of the result will be carried out at the mediator level. If a data source does not support a space operation, the latter is executed at the mediator level. Once the subqueries are executed, we obtain their output data. These data are fused thereafter to obtain a result by sector.

7.8 Conclusions

Traditional mediation systems do not take into account quality information in the data integration process. Most of the time they assume that data are of equal quality by means of some data-cleaning mechanism for instance. In this chapter, we described a quality-mediation approach that we developed in the context of a GIS integration project. Besides the well-known semantic heterogeneities, we also had to address a very difficult problem, that is assessing data quality of the mediated geographical system. One of the problems is the complexity of query rewriting. Incorporating quality information in the rewriting algorithm increases the complexity. For instance, some quality criteria are contradictory: improving the deficit rate implies degradation of the surplus or the violation rates. The trade-off is to deliver a rather complete set of answers without affecting its consistency. We believe our approach has three main advantages. First, it allows dynamic combination of mapping rules accordingly with the quality requirements, leading to quality rating of the result before data extraction. Second, it takes into account other metadata such as the spatial cover in the query rewriting process, leading to a significant reduction of useless operations (join of information related to different areas). Finally, it ensures system extensibility: adding or removing a data source does not significantly affect the global schema, and may result in some new mapping rules to be inserted in the system.

Acknowledgments

We would like to acknowledge the French RNTL research framework for the support of the VirGIS project.

References

1. Amann B, Beeri C, Fundulaki I, Scholl M (2002) Querying XML Sources Using an Ontology-based Mediator. In: Proc. of the 10th Int. Conf. on Cooperative Information Systems. Irvine, California, USA, 429–448
2. Base de Données Topographiques du Québec (2001) felix.geog.mcgill.ca/heeslib/bdtq_20000.html
3. Beeri C, Kanza Y, Safra E, Sagiv Y (2004) Object Fusion in Geographic Information Systems. In: Proc. of the 30th Int. Conf. on Very Large Data Base. Toronto, Canada, 816–827

4. Belussi A, Boucelma O, Catania B, Lassoued Y, Podestà P (2006) Towards Similarity-based Topological Query Languages. In: Proc. of the EDBT Work. on Query Language and Query Processing. Munich, Germany

5. Boucelma O, Essid M, Lacroix Z, Vinel J, Garinet JY, Bétari A (2004) VirGIS: Mediation for Geographical Information Systems. In: Proc. of the 20th Int. Conf. on Data Engineering. Boston, MA, USA, 855–856

6. Boyd M, Kittivoravitkul S, Lazanitis C (2004) AutoMed: A BAV Data Integration System for Heterogeneous Data Sources. In: Proc. of 16th Int. Conf. on the Advanced Information Systems Engineering. Riga, Latvia, 82–97

7. Buneman P, Davidson S, Fan W, Hara C, Tan WC (2001) Keys for XML. In: Proc. of the 10th Int. Conf. on World Wide Web. 201–210

8. Centre for Topographic Information (2006) www.cits.rncan.gc.ca

9. Colonna FM, Boucelma O (2003) Querying GML Data. In: 1ère Conférence en Sciences et Techniques de l'Information et de la Communication. Rabat, Morocco

10. Cranston C, Brabec F, Hjaltason G, Nebert D, Samet H (1999) Adding an Interoperable Server Interface to a Spatial Database: Implementation Experiences with OpenMap. In: Vckovski A, Brassel KE, Schek HJ (eds) Proc. of the 2nd Int. Conf. on Interoperating Geographic Information Systems. Zurich, Switzerland, Lecture Notes in Computer Science, Springer, Berlin Heidelberg New York, 115–128

11. David B, Fasquel P (1997) Qualiti d'une Base de Donnies Giographique: Concepts et Terminologie. LIV 67, IGN, Direction Technique, Service de La Recherche. Saint-Mandi, Cedex

12. Devogele T, Parent C, Spaccapietra S (1998) On Spatial Database Integration. Int. Journal of Geographical Information Science, 12(4):335–352

13. Federal Geographic Data Committee (1998) Content Standard for Digital Geospatial Metadata. Technical Report FGDC-STD-001-1998, Federal Geographic Data Committee, www.fgdc.gov/metadata/csdgm/

14. Friedman M, Levy A, Millstein T (1999) Navigational Plans for Data Integration. In: Proc. of the 16th Nat. Conf. on Artificial Intelligence. Orlando, Florida, USA, 67–73

15. Galhardas H, Florescu D, Shasha D, Simon E (2000) AJAX: An Extensible Data Cleaning Tool. In: Chen W, Naughton J, Bernstein PA (eds) Proc. of 2000 ACM SIGMOD Int. Conf. on Management of Data. Dallas, Texas, USA, 590–590

16. Garcia-Molina H, Papakonstantinou Y, Quass D, Rajaman A, Sagir Y, Ullman J, Vassalos V, Widom J (1997) The TSIMMIS Approach to Mediation: Data Models and Languages. Journal of Intelligent Information Systems, 8(2):117–132

17. Gert M, Schmit I (1998) Data Integration Techniques based on Data Quality Aspects. In: Schmitt I, Türker C, Hildebrandt E, Höding M (eds), Proc. of the 3rd Work. Föderierte Datenbanken, 1–19

18. Goasdoue F, Lattes V, Rousset MC (2000) The Use of CARIN Language and Algorithms for Information Integration: The PICSEL System. Int. Journal of Cooperative Information Systems, 9:383–401

19. Goodchild MF (1999) Interoperating Geographic Information System

20. ISO (International Standards Organization) (2006) 19113: Geographic Information Quality Principles. www.iso.org

21. ISO (International Standards Organization) (2006) 19115: Geographic Information Metadata. www.iso.org

22. ISO/TC 211 (International Standards Organization/Technical Committee) (2006) Geographic Information/Geomatics. www.isotc211.org

23. Kirk T, Levy AY, Sagiv Y, Srivastava D (1995) The Information Manifold. In: Knoblock C, Levy A (eds) Information Gathering from Heterogeneous, Distributed Environments. Stanford University, Stanford, California, USA
24. Lenzerini M (2002) Data Integration: A Theoretical Perspective. In: Proc. of the 20th ACM Symp. on Principles of Database Systems. Madison, Wisconsin, USA, 233–246
25. Levy A, Rajaraman A, Ordille J (1996) Querying Heterogeneous Information Sources Using Source Descriptions. In: Proc. of the 22th Int. Conf. on Very Large Data Bases. Bombay, India, 251–262
26. Manolescu I, Florescu D, Kossmann D (2001) Answering XML Queries over Heterogeneous Data Sources. In: Proc. of the 27th Int. Conf. on Very Large Data Bases. Rome, Italy, 241–250
27. Mihaila G, Raschid L, Vidal ME (2000) Using Quality of Data Metadata for Source Selection and Ranking. In: Proc. of the ACM SIGMOD Work. on The Web and Databases. Dallas, Texas, USA, 93–98
28. Motro A, Rakov I (1996) Estimating the Quality of Data in Relational Databases. In: Wang RY (ed), Proc. of the 1st Int. Conf. on Information Quality, Cambridge, MA, USA, 94–106
29. Naumann F, Leser U, Freytag J (1999) Quality-driven Integration of Heterogeneous Information Systems. In: Proc. of the 25th Int. Conf. on Very Large Data Base. Edinburgh, Scotland, 447–458
30. OGC (OpenGeoSpatial Consortium) (2001) Geography Markup Language (GML) 2.0. http://portal.opengeospatial.org/files/?artifact_id=1034
31. OGC (OpenGeoSpatial Consortium) (2002) Web Feature Server Implementation Specification
32. OGC (OpenGeoSpatial Consortium) (2006) Initiatives - Interoperability Program www.opengeospatial.org/initiatives
33. Pipino L, Lee Y, Wang, R (2002) Data Quality Assessment. Communications of the ACM, 45:211–218
34. Pottinger R, Levy A (2000) A Scalable Algorithm for Answering Queries Using Views. The VLDB Journal, 484–495
35. REV!GIS: Revision of the Uncertain Geographic Information (2000-2004). www.lsis.org/REVIGIS/Full/index.html
36. W3C (World Wide Web Consortium) W3C Specifications. www.w3c.org
37. W3C (World Wide Web Consortium) (2004) XML Schema Part 1: Structures. www.w3.org/TR/2001/REC-xmlschema-1-20010502
38. W3C (World Wide Web Consortium) (2006) XML Syntax for XQuery 1.0 (XQueryX). www.w3c.org/TR/2003/WD-xqueryx-20031219/
39. W3C (World Wide Web Consortium) (2006) XQuery 1.0: An XML Query Language.
40. Wiederhold G (1992) Mediators in the Architecture of Future Information Systems. IEEE Computer, 38–49

References containing URLs are validated as of October 1st, 2006.

8

Using Qualitative Information in Query Processing over Multiresolution Maps

Paola Podestà[1], Barbara Catania[1], and Alberto Belussi[2]

[1] University of Genoa, Genoa (Italy)
[2] University of Verona, Verona (Italy)

8.1 Introduction

Recently, the availability of huge amounts of spatial data representing geographical information has significantly increased. This is essentially due to both the increasing number of different devices collecting such data (i.e. remote sensing systems, environmental monitoring devices and, in general, all devices linked to location-aware technologies) and to the amazing development of distributed computing infrastructures (e.g. the Web) as platforms to share and access any type of information.

Geographical data usually correspond to data sets collected and integrated from different sources (private or public institutions), produced by different processes (e.g. social, ecological, economical) over a geographical area, at different times (e.g. every 10 years), possibly using different devices. Such geographical data sets are sets of geographical features spatially described as geometric objects embedded in a reference space and/or spatial relationships existing among them. Information concerning spatial relationships can be derived from geometric data but it can also be directly provided and related to features for which no geometric information is available. In this chapter, using a quite common terminology in geographical database systems, we call such data sets *maps*.

In a distributed environment, depending on the application context, it is often possible to find different, that is, *multiresolution*, representations of the same or overlapping maps. Multiresolution may have different meanings in different contexts. It may correspond to a different pixel size in raster data sets, to the number of points used to visualize a line in vector data sets, or to the different dimensions assigned to the same geographical feature in distinct maps. For example, a road can be represented as a region for an ecological process whereas it can be represented as a line for a traffic analysis process. In this chapter, we consider multiresolution maps according to this last meaning.

Managing multiresolution data sets in a distributed environment is an interesting but complex problem that can be addressed under two different points of view. From a system point of view, multiresolution may lead to query processing and integration problems, as the same concept can be represented in different ways and at different resolutions, leading to accuracy, precision, and consistency issues. On the other side, from a user point of view, multiresolution may result in a gap between the available data and the user's knowledge of such data during query specification, reducing user satisfaction in using a given application.

A possible approach to deal with such heterogeneity would be that of extending spatial data integration solutions to cope with specific data quality parameters. As we have seen in Chap. 7, this may lead to the definition of specific quality-aware query languages. However, when the user does not know exactly the schema of the data set to be queried, as often happens in distributed environments, this solution may not work. Another approach consists in tackling the problem from the top, by extending query specification and processing to directly exploit multiresolution. For this purpose, in Chap. 4, an overview of methods for the progressive transmission of multiresolution spatial data over the Web has been presented. In this chapter, we consider a complementary problem concerning how query processing techniques, in a distributed environment, have to be revised in order to use information concerning multiresolution maps, thus improving the quality of query processing results. The proposed techniques focus on three main GIS application contexts:

1. *Similarity-based Processing.* Due to the presence of multiresolution data sets over distributed architectures, the user may not know exactly the spatial domain she wants to query in terms of properties, available features, and geometric feature types. This gap may impact the quality of the results obtained by a query execution, reducing user satisfaction in using a given application as the result obtained may not exactly correspond to user needs.

2. *GIS Mediation Architectures.* Mediation systems provide users with a uniform access to a multitude of data sources via a common model, without duplicating such data. The user poses their query against a virtual global schema and the query is in turn rewritten into queries against the real local sources, taking into account differences in the data models and query languages. Results obtained from the various sources are then merged and returned to the user (see Chap. 6 for a review of geographical information fusion techniques). In the context of GIS data, VirGIS is a mediation system based on open geospatial consortium (OGC) standards that addresses the issue of integrating GIS data and tools [3, 4, 20] (see Chap. 7). In general, mediator systems, including VirGIS, take into account differences concerning feature representation in local sources. Differences in data sources may depend on how each single data source models spatial objects in terms of their descriptive attributes (length of a river, population in a town), their geometric type (region, line, point), and their spatial relations. Unfortunately, mediators usually do not consider the impact of spatial relationship information on query rewriting. The problem here is that, since not all spatial relationships are defined for any pair of geometric types, depending on object types in each local source schema, different spatial predicates should be considered for execution at the local level in order to return results consistent with the global request.

3. *Consistency Checking.* In order to compare or use together multiresolution maps in processing activities, tools for consistency analysis of data sets obtained from different sources are required in order to establish whether maps may lead to contradictory results if a user performs some kind of analysis over them.

In all the contexts presented above, the specification of equality-based queries and consistency checking methods—by which the user specifies in an exact way the constraints that data to be retrieved or to be checked must satisfy—may not be the right choice as multiresolution is not effectively used during such processing. In order to exploit multiresolution, a possible approach would be that of introducing some mechanism of query relaxation, by which the specific characteristics of multiresolution maps are taken into account and, as a consequence, approximated answers are returned to the user. This approach may generate some false hits but at the same time makes query answers more satisfactory from the user point of view.

In this chapter, we tackle this problem by discussing some qualitative techniques for query relaxation over geographical maps. Here "qualitative" means that we rely on the use of spatial relationships. More precisely, we restrict ourselves to consider topological and cardinal spatial relationships, due to their importance in real applications. Under this assumption, relaxation can be defined by introducing some distance functions for topological and cardinal directional relationships. Such functions can then be used in similarity-based query processing, to relax spatial predicates specified in the user query, and in mediator query processing to detect the closest predicates to the global one, in each local source. We notice that spatial relations are a good choice for also checking map consistency. Indeed, even if both geometry and properties concerning spatial relationships are available in a multiresolution map, it seems reasonable to discard geometry. Indeed, geometric consistency would be reduced to an equality test between two geometric map representations and similarity will require a sort of object extension measure in order to compare the geometric changes between two objects; in both cases, after a change of resolution (i.e. dimension), these properties cannot be preserved. On the other hand, map representations based on spatial relationships seem more suitable for checking consistency. Indeed, information about spatial relationships is more abstract than the geometric one and describes properties that are preserved after object dimension changes.

We remark that, despite a lot of work already exists concerning multiresolution modeling and spatial relations (see Sect. 8.2), only a few approaches directly exploit spatial relations combined with multiresolution in query processing. In particular, existing distance functions for spatial relationships are defined over pairs of objects with fixed dimension; thus they cannot be directly applied to multiresolution maps.

The aim of this chapter is therefore to: (i) provide a model for topological and cardinal directional relationships in multiresolution spatial data sets, suitable for defining relaxed query processing techniques; (ii) extend existing distance functions for topological and cardinal directional relationships; (iii) use distance functions for checking consistency of multiresolution spatial data sets; (iv) present examples about how the previous concepts can be used for spatial query processing in a distributed environment.

The remainder of this chapter is organized as follows. In Sect. 8.2 we briefly survey related work on topological and cardinal relationships. The reference spatial model and motivating scenarios are discussed in Sect. 8.3. Distance functions for topological and cardinal directional relationships are then presented in Sects. 8.4 and 8.5, respectively. In Sect. 8.6, the defined functions are used to provide consistency checking methods. Examples of the application of the proposed approaches are then presented in Sect. 8.7. Finally, Sect. 8.8 presents some conclusions and outlines future research directions.

8.2 Related Work

As remarked in Sect. 8.1, the techniques we propose for checking map similarity and consistency rely on the definition of distance functions for topological and cardinal directional relationships. Therefore, in the following we survey the most relevant existing approaches for modeling and analyzing such spatial relationships.

8.2.1 Topological Relationships

Topological relations capture the essential spatial relationships between the objects belonging to a map; they consider connectivity and adjacency of objects and are invariant under continuous transformations of space such as translation, rotation, and scaling.

The most relevant papers concerning the representation of topological properties are [6, 9, 10]. According to the model presented in [9], known as *4-intersection model*, spatial objects are modeled as point-sets describing their interior and their boundary, as shown in Fig. 8.1(a). Only regions are considered and topological relations are defined as 2×2 matrices, containing the value empty ($\oslash$ or 0) or nonempty ($\neg\oslash$ or 1) for each intersection of the point-sets of the objects involved (see Fig. 8.1(b)). As an example, Fig. 8.1(c) presents the 4-intersection model configuration for the topological relation *Overlap* between two regions.

The 4-intersection model has subsequently been extended by considering also object exterior [10], leading to the definition of the *9-intersection model*. In the 9-intersection model, each spatial object A is represented by 3 point-sets: its interior $A°$, its exterior A^-, and its boundary ∂A (see Fig. 8.1(d)). The definition of binary topological relations between two spatial objects A and B is then based on the 9 intersections between each pair of object components. Thus, a topological relation can be represented as a 3×3 matrix, called *9-intersection matrix* (see in Fig. 8.1(e)). As an example, Fig. 8.1(f) presents the 9-intersection model configuration for the topological relation *Overlap* between two regions.

By considering the value empty or non-empty for each intersection, one can categorize in a complete way all the relationships between regions, lines, and points embedded in $\mathbb{R}^2$. In [6], the 9-intersection model has been extended by considering the dimension of the intersection. As the number of the resulting relationships is quite high, a partition of extended 9-intersection matrices has been proposed in [6],

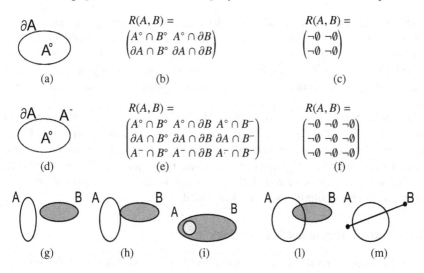

Fig. 8.1. Models for topological relationships: (**a**) object representation in the 4-intersection model; (**b**) the 4-intersection matrix; (**c**) 4-intersection matrix for *Overlap* between regions; (**d**) object representation in the 9-intersection model; (**e**) the 9-intersection matrix; (**f**) 9-intersection matrix for *Overlap* between regions; (**g**) *A Disjoint B*; (**h**) *A Touch B*; (**i**) *A In B*; (**l**) *A Overlap B*; (**m**) *B Cross A*

grouping together similar matrices and assigning a name to each group. The result is the definition of the following set of binary, jointly exhaustive, and pairwise disjoint topological relationships: *REL* = {*Disjoint, Touch, In, Overlap, Cross*}, shown in Fig. 8.1(g)–(m) for regions. Such relationships are now at the basis of query processors of well-known spatial and geographical database systems.

It is important to remark that not all relationships can be defined for any pair of dimensions; for example, relation *Overlap* (see Fig. 8.1(l)) is defined only between pairs of regions or pairs of lines.

Multiresolution has been mainly considered in the definition of ad hoc operators to generalize/specialize map objects while maintaining map consistency. In [8], topological relationships are considered and a similarity measure between maps, defined as a deviation from consistency, has also been provided. In [15], a formal model based on planar abstract cell complexes, for representing multiresolution maps and for their consistent generalization has been proposed. The consistency test is defined at the combinatorial level by means of homeomorphisms.

In [7], a method for checking similarity between topological relations over regions, defined according to the 9-intersection model, is presented. Similarity is determined by comparing two 9-intersection matrices and computing the number of different values of each intersection (*topology distance*).

Using this function, a partial order over topological relations has been defined and used to evaluate how two relations are far from each other. In [5], topology distance is used to evaluate similarity of spatial scenes, by taking into account also direction and distance relations. In [11], topology distance is used to define a model (snapshot model) to compare two different topological relations between lines and regions. In all the papers cited above, similarity is computed only between pairs of objects with the same dimension. Multiple object representations are not considered at all.

Two different consistency issues are addressed in [18, 19]. In [18], consistency among networks, defined as sets of lines (homogeneous networks) and sets of lines and regions (heterogeneous networks), is investigated. In [19], the problem of checking consistency under aggregation operations, that merge together disconnected parts of the same region, is considered. Both approaches do not consider changes in object dimension and no similarity measure is provided.

8.2.2 Cardinal Directional Relationships

Cardinal directional relations provide a way to determine what is the relative position of a target object with respect to a reference object by considering some cardinal directions.

The idea behind a qualitative representation of cardinal relations is to map quantitative directional information (i.e. the degrees of an angle) into a set of symbols. More precisely, directional relationships are binary functions that map two object points $(P1, P2)$ in the plane (representing respectively the reference object used to define directions and the target object whose direction with respect to the reference object has to be detected) onto a symbolic direction d. The number of direction symbols available depends on the model of cardinal directions used. The basic models for representing cardinal directions divide the space around the reference object into cone-shaped (or triangular) areas or into half planes, as shown in Fig. 8.2(a)–(c). When the reference object is not a point, the most used model, called D_9 model, is based on the space decomposition presented in Fig. 8.2(d) and approximates the reference object with its minimum bounding box (MBB). The space around the reference object is divided into distinct areas, called *tiles*, using the infinite extensions of the sides

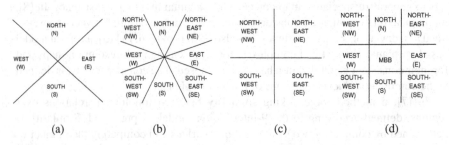

(a)   (b)   (c)   (d)

Fig. 8.2. Different space division: (**a**) four cones; (**b**) eight cones; (**c**) four half-planes; (**d**) nine tiles

of its box. The tiles result in the usual set of directions {*N, NE, E, MBB, SE, S, SW, W, NW*}. We notice that all tiles, excluding *MBB*, are unbounded. Their union coincides with $\mathbb{R}^2$.

Based on the D_9 model, cardinal relations between two regions can be represented as 3×3 matrices containing the value empty or non-empty for each intersection between the target object *A* and the tiles generated by a reference object *B* (see Fig. 8.3(a)). In [12, 13], this basic model has been extended to deal with regions without holes, lines, and points. In this case, a cardinal directional relation is represented again using a 3×3 matrix, but now each cell contains a 9-cells vector, called *neighbor code* (see Fig. 8.3(b)). A neighbor code records information concerning

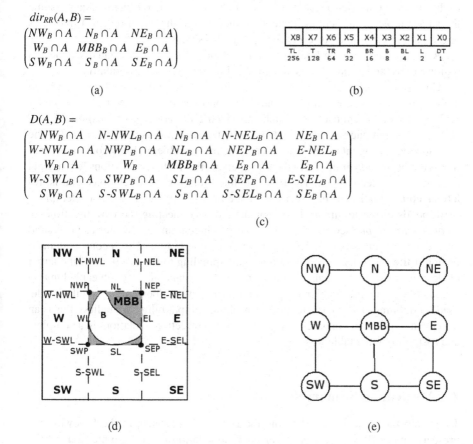

$dir_{RR}(A, B) =$
$$\begin{pmatrix} NW_B \cap A & N_B \cap A & NE_B \cap A \\ W_B \cap A & MBB_B \cap A & E_B \cap A \\ SW_B \cap A & S_B \cap A & SE_B \cap A \end{pmatrix}$$

(a)

X8	X7	X6	X5	X4	X3	X2	X1	X0
TL	T	TR	R	BR	B	BL	L	DT
256	128	64	32	16	8	4	2	1

(b)

$D(A, B) =$
$$\begin{pmatrix} NW_B \cap A & N\text{-}NWL_B \cap A & N_B \cap A & N\text{-}NEL_B \cap A & NE_B \cap A \\ W\text{-}NWL_B \cap A & NWP_B \cap A & NL_B \cap A & NEP_B \cap A & E\text{-}NEL_B \\ W_B \cap A & W_B & MBB_B \cap A & E_B \cap A & E_B \cap A \\ W\text{-}SWL_B \cap A & SWP_B \cap A & SL_B \cap A & SEP_B \cap A & E\text{-}SEL_B \cap A \\ SW_B \cap A & S\text{-}SWL_B \cap A & S_B \cap A & S\text{-}SEL_B \cap A & SE_B \cap A \end{pmatrix}$$

(c)

(d)

(e)

Fig. 8.3. Models for cardinal directional relationships: (**a**) the 3×3 directional matrix; (**b**) the neighbor code; (**c**) the 5×5 directional matrix; (**d**) space subdivision corresponding to the directional matrix; (**e**) the cardinal directional conceptual graph

the intersections between the boundaries of a particular tile and the target object A, using nine bits (x_0-x_8). Bit 0 (x_0) records the value of the intersection between A and the direction tile the vector refers to, say DT, and bits x_1-x_8 record the values of the intersections between A and the Left (L), Bottom-Left (BL), Bottom (B), Bottom-Right (BR), Right (R), Top-Right (TR), Top (T), and Top-Left (TL) boundaries of DT, respectively. Each neighbor code corresponds to a binary number between 0 and 256. Thus, each matrix can be seen as a 3×3 matrix of integer numbers. Different matrix configurations correspond to different cardinal relations. However, by considering only connected objects, not all possible configurations represent a correct cardinal directional relation.

The information contained in neighbor codes can also be represented by using a 5×5 matrix, called *directional matrix* (see Fig. 8.3(c)). In such a matrix, a row and a column exist for each tile interior and each boundary between two tiles, according to the space subdivision presented in Fig. 8.2(d). Each matrix element can assume the value empty or non-empty, depending on whether the target object A intersects or does not intersect the corresponding portion of space.

A formal model for cardinal directional relations for connected and disconnected regions, lines, and points is provided in [16, 17], based on the 5×5 matrix.

Consistency of cardinal directional relations has been discussed in [12, 13], in the context of specialization and generalization operations, by considering binary 3×3 matrices as reference model and introducing the concept of *compatibility*. A matrix D_1 is compatible with a matrix D_2 if, for each non-zero element in D_2, the corresponding element in D_1 is non-zero. The main problem of this consistency notion is that it is not symmetric. In the same papers, two distance functions for cardinal relations have been defined. The first is defined for single-tile relations; that is, relations corresponding to intersections of the target object with a single tile, and it corresponds to the minimum length of the paths connecting the two directions in a *conceptual graph* (see Fig. 8.3(e)). Such graph contains a node for each tile and one edge between pairs of tiles sharing at least one border. The second is defined for multi-tile relations; that is, relations corresponding to intersections of the target object with multiple tiles, and it considers the percentage of target object belonging to each tile. The main problem of this approach is that it does not rely on the model used for defining cardinal relations and checking compatibility. We believe this is an important requirement in order to support cardinal directional relations in a complete and easily implementable way.

8.3 Motivating Scenarios

To explain the basic idea underlying the proposed approach, in the following we present a more detailed example for each application context pointed out in the introduction: (i) similarity-based processing in a distributed environment; (ii) query processing in GIS mediation architectures; (iii) consistency checking for spatial data sets. Before presenting such scenarios, we introduce the reference spatial model used in the rest of this chapter.

8.3.1 The Reference Spatial Data Model

We define a *map schema* as a set of feature types, object classes representing real-world entities (e.g. lakes, rivers, etc.). Each feature type has some descriptive attributes, including a feature identifier and a *spatial attribute*, having a given dimension. We assume that values for the spatial attribute are modeled according to the OGC *simple feature* geometric model [14], restricted to consider only *connected objects*. In such a model, the geometry of an object can be of type point (dimension 0, also denoted by *P*), line (dimension 1, also denoted by *L*), or polygon – more generally called region (dimension 2, also denoted by *R*). A point describes a single location in the coordinate space; a line represents a linear interpolation of an ordered sequence of points; a polygon is defined as an ordered sequence of closed lines defining the exterior and interior boundaries (holes) of an area. In the following, we denote with *REG*, *LINE*, and *POINT* the set of all the connected regions (possibly with holes), lines, and points in $\mathbb{R}^2$, respectively. We denote with *OBJ* the set $REG \cup LINE \cup POINT$.

Given a set of map schemas $MS_1, ..., MS_n$ and a set of feature types $ft_1, ..., ft_m$, we assume that a feature type may belong to one or more map schemas, possibly with different dimensions. For example, the feature type *river* may belong to MS_1 with dimension 1 (i.e. rivers are represented as lines in MS_1) and to MS_2 with dimension 2 (i.e. rivers are represented as regions in MS_2). Thus, each feature type is assigned multiple dimensions, one for each map schema in which it appears.

An instance M_i of a map schema MS_i is called *map* and is a set of features, instances of the feature types belonging to the map schema. Given a set of map instances $M_1, ..., M_n$, instances of the map schemas $MS_1, ..., MS_n$, respectively, a feature f_j over a feature type ft_i may belong to one or more maps, associated with possibly different geometries and dimensions, according to the map schemas. When this happens, we say that $M_1, ..., M_n$ are *multiresolution maps*. We assume that the feature identifier is used to determine whether a feature belongs to a map.

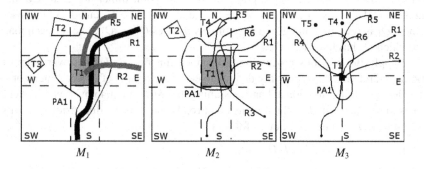

Fig. 8.4. Sketch of the content of the map examples

In the following, we consider three distinct maps M_1, M_2, and M_3, sketched in Fig. 8.4. They represent roads (identified by R_i), towns (identified by T_i), and pollution areas (identified by PA_i). Roads, towns, and pollution areas have dimensions (2,2,2) in map M_1, (1,2,2), in map M_2, and (1,0,2) in map M_3.

8.3.2 Similarity-based Processing Scenario

Suppose a user wants to query some spatial data available on the Web, without having a detailed knowledge about such data. When the users specify the query, they may not know the resolution of the underlying database, therefore they may not be able to specify the query in an exact way as spatial predicates could change when changing object dimensions. As a consequence, the quality of the result obtained may be reduced because interesting pairs may not be returned.

As an example, suppose the user wants to know which roads enter town T_1 (see Fig. 8.4), this query can be specified as follows:

$$Q_1 = \{r | r \text{ is a road}, r \text{ Overlap } T_1\}.$$

If roads and towns are represented as regions, as in map M_1, predicate *Overlap* is defined and can be executed (see Sect. 8.4 for details). However, if roads are represented as lines and towns as regions, as in map M_2, predicate *Overlap* is not defined. In this context, a similarity-based approach could be very useful. The user could specify the query by: (i) assuming data have the maximal dimension, that is all feature types have dimension 2 (in order to made available to the user the largest set of topological predicates); (ii) providing a threshold value. Such value can be used to increase the quality of the generated result, for example, to return more information, even if not necessarily significant, to the user.

In general, suppose the user wants to execute query Q_1 up to an error ϵ_t. Actually, this error depends on the user's application and needs. The basic idea is to rewrite each topological relation θ in Q_1 into one or more topological relations with distance at most ϵ_t from θ, say $\theta_1, \ldots, \theta_n$, by considering the possible dimension change of the features involved in θ in the map where the query has to be executed. As a consequence, Q_1 is rewritten into a sequence of new queries $Q_1^1, \ldots, Q_1^m$, one for each map over which Q_1 has to be executed.

To apply the previous processing, a distance function d_t defined over pairs of topological relationships, possibly defined over different pairs of object dimensions, is needed. With this function at hand, assuming that M_i is the map over which query Q_1 has to be executed, d_3 is the dimension of roads and d_4 is the dimension of towns in M_i, query Q_1 can be rewritten as follows:

$$Q_1^i = \{r | r \text{ is a road}, \exists \theta'(d_t(Overlap, (R, R), \theta', (d_3, d_4)) \leq \epsilon_t \wedge r \, \theta' \, T_1)\}.$$

In the previous query specification, $d_t(Overlap, (R, R), \theta', (d_3, d_4))$ denotes the distance between relation *Overlap* defined over dimensions (R, R) and relation θ' defined over dimensions (d_3, d_4). In order to apply the previous processing, in Sects. 8.4 and 8.5 we present two distance functions, one for topological relations (d_t) and one for cardinal relations (d_c), taking into account feature dimensions.

8.3.3 Query Processing in GIS Mediation Architectures

The second scenario deals with mediation systems. The basic architecture of a mediation system is based on two main components: the mediator and the wrappers. We focus our attention on the mediator that rewrites the user's query into local queries, afterwards translated by the wrapper in the data source query language (see Chap. 7).

In general, mediators do not consider, in query rewriting, the possible change of spatial relations caused by multiresolution. As an example, assume that the maps in Fig. 8.4 represent three local sources to be integrated and that, at the global level, features are represented with the maximal dimension by which they appear in the local sources. In our example, this means that at the global level, roads, towns, and pollution areas are represented as regions. Actually, in more general cases, the choice of feature dimensions depends on the user's application and needs; moreover, it is possible to create specific interfaces that may impose their own feature representation to the users.

Suppose now that the user, at the global level, wants to know the roads that enter town T_1 from North to NorthEast. As we will see in Sect. 8.5, this query can be specified as follows:

$$Q_2 = \{r | r \text{ is a road}, r \text{ Overlap } T_1 \text{ and } r \text{ MBB:N:NE } T_1\}.$$

As predicates *Overlap* and *MBB:N:NE* may not be defined over some local sources, such query may not return any result from some of them. For each local source, a more reasonable approach would be that of rewriting such predicates into a set of predicates which are consistent with the given ones and are defined over the given local source.

In order to apply this processing, a notion of consistency among topological and cardinal directional relations has to be defined. Assume that given a topological (cardinal directional) relation θ defined over dimensions d_1 and d_2, the set of topological (cardinal directional) relationships defined over dimensions d_3 and d_4 which are consistent with respect to θ is denoted by $r_t((d_1, d_2), \theta, (d_3, d_4))$ $(r_c((d_1, d_2), \theta, (d_3, d_4)))$. With those sets at hand, if d_3 is the dimension of roads and d_4 is the dimension of towns in a local source M_i, query Q_2 can be rewritten for execution against M_i as follows:

$$Q_2^i = \{r | r \text{ is a road}, \exists \theta' \in r_t((R, R), Overlap, (d_3, d_4))(\exists \theta'' \in r_c((R, R),$$
$$MBB:N:NE, (d_3, d_4))(r \, \theta' \, T_1 \wedge r \, \theta'' \, T_1))\}.$$

In Sect. 8.6, we present two distinct consistency notions that can be used to apply the processing described above.

8.3.4 Consistency Checking

The evaluation of consistency among a set of maps is an important issue today since often GIS applications of different nature have to integrate and compare spatial information coming from different sources, leading to maps with different resolutions. In order to use such data sets in a multiresolution GIS, the problem arises of determining their consistency.

Consistency checking can be performed by using various consistency measures and, in case of negative result, pairs of features which more contribute to the inconsistency situation should be identified and ordered according to the distance value. These pairs could be considered as candidate errors and the list of all inconsistent objects can be used to generate, by difference, some consistent views of the input maps, to be consistently used inside GIS applications.

For example, we may be interested in determining whether maps M_1, M_2, and M_3 in Fig. 8.4 are consistent. The main problem in checking consistency is due to the fact that the same object can be represented in distinct maps with different dimensions (consider, for example, objects T_4 and R_5 in maps M_1, M_2, and M_3). In order to evaluate consistency between pairs of maps, in Sect. 8.6 we define two distinct notions of consistency based on the proposed topological and cardinal direction distances.

8.4 Topological Relations

In the following, we first present a formal model for topological relations; then, we introduce a distance function for comparing two topological relationships, possibly defined over multiresolution objects.

8.4.1 The Model

In defining a distance function for topological relationships, we consider the following set of topological relationships: $TREL = \{Disjoint, Touch, In, Contain, Equal, Cross, Overlap, Cover, CoveredBy\}$. Each topological relation corresponds to a set of 9-intersection matrices (see Sect. 8.2.1), according to the model presented in [6]. We notice that relations *Cover* and *CoveredBy* are here defined as refinements of relations *Contain* and *In* and are not considered in [6]. The semantics of the topological predicates of *TREL* is provided in Table 8.1.

As we can see from Table 8.1, not all relationships can be defined for any pair of dimensions. Therefore, given two dimensions $d_1, d_2 \in \{R, L, P\}$, in the following we denote with $TREL(d_1, d_2)$ the set of topological relationships that can be defined for d_1 and d_2. Moreover, we denote with $I_9(\theta, d_1, d_2)$ the set of 9-intersection matrices defining predicate $\theta \in TREL$ between two objects of dimension d_1 and d_2.

8.4.2 Distance Function for Topological Relationships

In order to define a distance function for topological relationships in *TREL*, as each topological relationship in *TREL* corresponds to a set of 9-intersection matrices, we use a two-step approach: first, a distance function between two 9-intersection matrices is defined then such function is used for computing the final result.

Table 8.1. Definition of the reference set of topological relationships

name	definition	object type
disjoint (d)	$f_1 \cap f_2 = \emptyset$	all
touch (t)	$(f_1^\circ \cap f_2^\circ = \emptyset) \wedge (f_1 \cap f_2) \neq \emptyset$	R/R, R/L, R/P, L/L, L/P
in (i)	$(f_1 \cap f_2 = f_1) \wedge (f_1^\circ \cap f_2^\circ) \neq \emptyset$	R/R, L/L, L/R, P/R, P/L
contain (c)	$(f_1 \cap f_2 = f_2) \wedge (f_1^\circ \cap f_2^\circ) \neq \emptyset$	R/R, R/L, R/P, L/L, L/P
equal (e)	$f_1 = f_2$	R/R, L/L, P/P
cross (r)	$dim(f_1^\circ \cap f_2^\circ) =$	L/R
	$(max(dim(f_1^\circ), dim(f_2^\circ)) - 1) \wedge$	
	$(f_1 \cap f_2) \neq f_1 \wedge (f_1 \cap f_2) \neq f_2$	L/L
overlap (o)	$dim(f_1^\circ) = dim(f_2^\circ) = dim(f_1^\circ \cap f_2^\circ) \wedge$	R/R
	$(f_1 \cap f_2) \neq f_1 \wedge (f_1 \cap f_2) \neq f_2$	L/L
cover (v)	$(f_2 \cap f_1) = f_2 \wedge (f_2^\circ \cap f_1^\circ) \neq \emptyset \wedge$	R/R, R/L, L/L
	$(f_1 - f_1^\circ) \cap (f_2 - f_2^\circ) \neq \emptyset$	
coveredby (vb)	$(f_1 \cap f_2 = f_1) \wedge (f_1^\circ \cap f_2^\circ) \neq \emptyset \wedge$	R/R, L/L, L/R
	$(f_1 - f_1^\circ) \cap (f_2 - f_2^\circ) \neq \emptyset$	

In defining the distance between two 9-intersection matrices ψ_1 and ψ_2 (denoted by $d_9(\psi_1, \psi_2)$), we adopt the approach proposed in [7] and we define it as the fraction between the number of different cells in the two matrices and the total number of cells (9). Two cells are considered different if one corresponds to a non-empty intersection (whatever is its dimension) and the other to an empty intersection.[3]

Based on this distance, given two relationships ψ_1 and ψ_2 in *TREL*, their distance can now be computed as the minimum distance between any 9-intersection matrix defining ψ_1 and any 9-intersection matrix defining ψ_2. We have chosen the minimum and not the average or other functions as we are interested in maximizing similarity among topological relations after dimension changes.

Definition 8.1 (Topology Distance). *Let* $\theta_1 \in TREL(d_1, d_2)$ *and* $\theta_2 \in TREL(d_3, d_4)$. *The topology distance between* θ_1 *and* θ_2 *is defined as follows:*

$$d_t(\theta_1, (d_1, d_2), \theta_2, (d_3, d_4)) =$$

$$min\{d_9(\psi_1, \psi_2) | \psi_1 \in I_9(\theta_1, d_1, d_2), \psi_2 \in I_9(\theta_2, d_3, d_4)\}.$$

Example 8.1. Suppose we want to compute $d_t(Contain, (R, L), In, (L, R))$. Based on Table 8.1 and on the results presented in [2], it is possible to show that *Contain* over (R, L) corresponds to the following two 9-intersection matrices:

$$Contain_1 = \begin{pmatrix} \neg\emptyset & \neg\emptyset & \neg\emptyset \\ \emptyset & \emptyset & \neg\emptyset \\ \emptyset & \emptyset & \neg\emptyset \end{pmatrix} Contain_2 = \begin{pmatrix} \neg\emptyset & \neg\emptyset & \neg\emptyset \\ \neg\emptyset & \emptyset & \neg\emptyset \\ \emptyset & \emptyset & \neg\emptyset \end{pmatrix}$$

---

[3] Note that the dimension of the intersection is not taken into account when computing the distance.

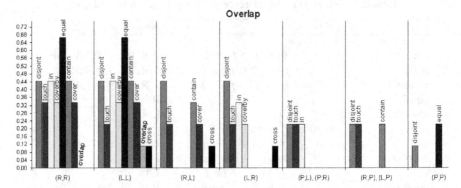

Fig. 8.5. Distance values for the *Overlap* topological relationship defined over pairs of regions

On the other hand, *In* over (L, R) corresponds to the following two 9-intersection matrices:

$$In_1 = \begin{pmatrix} \neg\emptyset & \emptyset & \emptyset \\ \neg\emptyset & \emptyset & \emptyset \\ \neg\emptyset & \neg\emptyset & \neg\emptyset \end{pmatrix} In_2 = \begin{pmatrix} \neg\emptyset & \neg\emptyset & \emptyset \\ \neg\emptyset & \emptyset & \emptyset \\ \neg\emptyset & \neg\emptyset & \neg\emptyset \end{pmatrix}$$

In order to compute their distance, we first need to compute all the distances between a matrix for *Contain* and a matrix for *In*, that is, $d_9(Contain_i, In_j)$, $i, j = 1, 2$, and then take the minimum value. For example, $d_9(Contain_1, In_1) = 6/9$, since 6 positions out of 9 are different. Similarly, $d_9(Contain_1, In_2) = 5/9$, $d_9(Contain_2, In_1) = 5/9$, $d_9(Contain_2, In_2) = 4/9$. Thus, $d_t(Contain, (R, L), In, (L, R)) = 4/9$.

All values for $d_t(\theta_1, (d_1, d_2), \theta_2, (d_3, d_4))$ can be found in [1]. Figure 8.5 just presents distances $d_t(Overlap, (R, R), \theta_2, (d_3, d_4))$, for $d_3, d_4 \in \{R, L, P\}$.

8.5 Cardinal Directional Relations

In the following, we first present a formal model for cardinal directional relations, then we introduce a distance function for comparing two cardinal directional relationships, possibly defined over multiresolution objects.

8.5.1 The Model

In defining a distance function for cardinal directional relationships, we rely on the 5×5 direction matrices model proposed in [12, 13] and then formalized in [16, 17] for connected and disconnected regions (see Sect. 8.2 for additional details). Here, we extend this model to deal with regions, lines, and points for reference and target objects. For the sake of simplicity, we however consider only connected objects, consistently with the type of objects considered for topological relations.

Depending on the dimension and the shape of the reference object, a different number of tiles is generated. Figure 8.6 presents tiles generated when the reference object is a region (*REG*), or a sideway line (*S_LINE*), a vertical line (*V_LINE*), a horizontal line (*H_LINE*), or a point (*POINT*).

As the 5×5 matrix model leads to a huge number of different configurations, following what has been done in [12, 13, 16, 17], we group cardinal relations into sets. The group of relationships we consider in this chapter corresponds to the following set of basic cardinal relationships *BCR*={*NorthWest (NW), North (N), NorthEast (NE), West (W), MinimumBoundingBox (MBB), East (E), SouthEast (SE), South (S), SoutWest (SW)*}, possibly combined together.[4] To make cardinal directional relations pairwise disjoint, the boundary between two tiles is assumed to define just one cardinal directional relation among {*N,W,E,S*} (the most relevant one). Additionally, we assume that *MBB* contains the boundary of the corresponding tile. The overall set of cardinal directional relationships can be defined as follows.

Definition 8.2 (Cardinal Directional Relationships). *The set of cardinal directional relationships, called CREL, is composed of any expression θ of the form $R_1 : \dots : R_k$, $1 \le k \le 9$, such that:*

(i) $R_i \in BCR$, $i = 1, \dots, k$.
(ii) $R_i \ne R_j$, $i \ne j$, $1 \le i, j \le k$.
(iii) R_i and R_{i+1} $i = 1, \dots, k - 1$ share at least one line boundary.

If $k = 1$, the cardinal directional relation is called single-tile, *otherwise it is called* multi-tile. *The set $\{R_1, \dots, R_k\}$ is called* support *of θ and it is denoted by $S(\theta)$.*

From the previous definition it follows that a multi-tile relation is obtained by listing a set of single-tile relations (item (i)). Item (ii) avoids the duplication of the

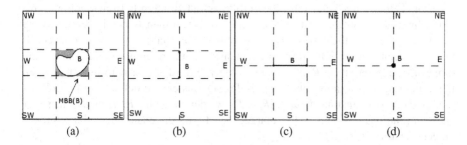

Fig. 8.6. Different grids according to different reference object dimension: (a) region; (b) vertical line; (c) horizontal line; (d) point

---

[4] We remark that other groups of cardinal directional relationships can be defined. For example, a relation for each portion of space considered in the 5×5 matrix can also be defined. We believe, however, that from a user point of view, it is more convenient to cope with a small and well-known set of relationships. This choice is also motivated by the fact that, as far as we know, no commercial system directly supports cardinal directional relations.

same single-tile relation inside a multi-tile one. Finally, item (iii) provides a normal form for cardinal directional relations, specifying that two single-tile relations are adjacent in the list if and only if their corresponding tiles are adjacent in the plane. This is a reasonable assumption as we consider only connected objects.

The semantics of the basic cardinal directional relations can be defined by considering the intersection of the tiles with the target object. For this purpose, it is useful to introduce the concepts of greatest lower bound and lowest upper bound of an object on a given axis. Given an object A, the *greatest lower bound* of the projection of A on the x-axis (respectively y-axis) is denoted by $inf_x(A)$ (respectively $inf_y(A)$). The *lowest upper bound* of the projection of A on the x-axis (respectively y-axis) is denoted by $sup_x(A)$ (respectively $sup_y(A)$). The *minimum bounding box* of A, denoted by $mbb(A)$, is the box formed by the straight lines $x = inf_x(A)$, $x = sup_x(A)$, $y = inf_y(A)$, and $y = sup_y(A)$. Table 8.2 presents the semantics of the relations of the set BCR. Such semantics, which guarantees that basic cardinal directional relations are mutually exclusive, has been obtained by extending that presented in [16, 17] to deal with reference and target objects with possibly different dimensions.[5]

The semantics of the relations of $CREL$ can be defined considering the semantics of the basic cardinal directional relations as follows.

Definition 8.3. *Let A, $B \in OBJ$, let $\theta \in CREL$, $\theta = R_1 : \ldots : R_k$. $A \theta B$ is satisfied if and only if there exist $A_1, \ldots, A_k \in OBJ$, $A_i \neq A_j$, $i \neq j$, $1 \leq i, j \leq k$, such that $A = A_1 \cup, \ldots, \cup A_k$ and $A_1 R_1 B, A_2 R_2 B, \ldots, A_k R_k B$ are true.*

We remark that the previous definition does not take into account what percentage of object intersects a given tile. Rather, it assumes that when an object intersects more than one tile, it is uniformly distributed among them. We also notice that the semantics of single-tile relations coincides with the semantics of basic cardinal directional relations.

From Definition 8.3 and Table 8.2, it follows that, differently from topological relationships, all single-tile relations are defined for any dimension of the reference and target object, even if the corresponding definition may change. As a consequence, all multi-tile relations are defined for any dimension of the reference object and for any dimension, different from P, of the target object. Indeed, it is easy to show that when the target object is a point, only single-tile relations can be defined. In the following, given two dimensions $d_1, d_2 \in \{R, L, P\}$, we denote with $CREL(d_1, d_2)$ the set of cardinal relationships defined for a target object with dimension d_1 and a reference object with dimension d_2.

8.5.2 Distance Function for Cardinal Directional Relations

In defining the distance function for cardinal directional relations, we first notice that the approach used for topological relationships is not sufficient. To show this, we first compute the distance between two 5×5 matrices, similarly to what we have

---

[5] Notice that this is not true in [16, 17].

Table 8.2. Definition of the semantics of the basic cardinal directional relations *BCR*. *A* is the target object and *B* is the reference object

name	definition	type of the refer. obj. *B*
NW	$inf_y(A) > sup_y(B), inf_x(B) > sup_x(A)$	$REG \cup S\_LINE$
	$inf_y(A) > y_B, inf_x(B) > sup_x(A)$	H_LINE
	$inf_y(A) > sup_y(B), x_B > sup_x(A)$	V_LINE
	$inf_y(A) > y_B, x_B > sup_x(A)$	$POINT$
N	$inf_x(B) \leq inf_x(A), sup_x(A) \leq sup_x(B), sup_y(B) < inf_y(A)$	$REG \cup S\_LINE$
	$y_B < inf_y(A), inf_x(B) \leq inf_x(A), sup_x(A) \leq sup_x(B)$	H_LINE
	$inf_x(A) \leq x_B \leq sup_x(A), inf_y(A) < sup_y(B)$	V_LINE
	$inf_x(A) \leq x_B \leq sup_x(A), y_B < inf_y(A)$	$POINT$
NE	$inf_y(A) > sup_y(B), sup_x(B) < inf_x(A)$	$REG \cup S\_LINE$
	$inf_y(A) > y_B, sup_x(B) < inf_x(A)$	H_LINE
	$inf_y(A) > sup_y(B), x_B < inf_x(A)$	V_LINE
	$inf_y(A) > y_B, x_B < inf_x(A)$	$POINT$
W	$inf_y(B) \leq inf_y(A), sup_y(A) \leq sup_y(B), inf_x(B) > sup_x(A)$	$REG \cup S\_LINE$
	$sup_x(A) < inf_x(B), inf_y(A) \leq y_B \leq sup_y(A)$	H_LINE
	$sup_x(A) < x_B, inf_y(B) \leq inf_y(A), sup_y(A) \leq sup_y(B)$	V_LINE
	$inf_y(A) \leq y_B \leq sup_y(A), x_B > sup_x(A)$	$POINT$
MBB	$inf_x(B) \leq inf_x(A), sup_x(A) \leq sup_x(B), inf_y(B) \leq inf_y(A), sup_y(A) \leq sup_y(B)$	$REG \cup S\_LINE$
	$inf_x(B) \leq inf_x(A), sup_x(A) \leq sup_x(B), inf_y(A) \leq y_B \leq sup_y(A)$	H_LINE
	$inf_y(B) \leq inf_y(A), sup_y(A) \leq sup_y(B), inf_x(A) \leq x_B \leq sup_x(A)$	V_LINE
	$(x_B, y_B) \in A$	$POINT$
E	$inf_y(B) \leq inf_y(A), sup_y(A) \leq sup_y(B), sup_x(B) < inf_x(A)$	$REG \cup S\_LINE$
	$inf_y(A) \leq y_B \leq sup_y(A), sup_x(B) < inf_x(A)$	H_LINE
	$inf_y(B) \leq inf_y(A), sup_y(A) \leq sup_y(B), x_B < inf_x(A)$	V_LINE
	$inf_y(A) \leq y_B \leq sup_y(A), x_B < inf_x(A)$	$POINT$
SW	$sup_y(A) < inf_y(B), inf_x(B) > sup_x(A)$	$REG \cup S\_LINE$
	$sup_y(A) < y_B, inf_x(B) > sup_x(A)$	H_LINE
	$sup_y(A) < inf_y(B), x_B > sup_x(A)$	V_LINE
	$sup_y(A) < y_B, x_B > sup_x(A)$	$POINT$
S	$inf_x(B) \leq inf_x(A), sup_x(A) \leq sup_x(B), inf_y(B) > sup_y(A)$	$REG \cup S\_LINE$
	$sup_y(A) < y_B, inf_x(B) \leq inf_x(A), sup_x(A) \leq sup_x(B)$	H_LINE
	$inf_x(A) \leq x_B \leq sup_x(A), sup_y(A) < inf_y(B)$	V_LINE
	$inf_x(A) \leq x_B \leq sup_x(A), sup_y(A) < y_B$	$POINT$
SE	$sup_y(A) < inf_y(B), sup_x(B) < inf_x(A)$	$REG \cup S\_LINE$
	$sup_y(A) < y_B, sup_x(B) < inf_x(A)$	H_LINE
	$sup_y(A) < inf_y(B), x_B < inf_x(A)$	V_LINE
	$sup_y(A) < y_B, x_B < inf_x(A)$	$POINT$

done for topological relations. Such a distance, denoted by d_{25}, is defined as the fraction between the number of different cells in the two matrices and the total number of cells (25). Two cells are considered different if one corresponds to a non-empty intersection and the other to an empty intersection. Based on this distance, given two cardinal relationships $\theta_1 \in CREL(d_1, d_2)$ and $\theta_2 \in CREL(d_3, d_4)$, their matrix-based distance, denoted by $d_m(\theta_1, (d_1, d_2), \theta_2, (d_3, d_4))$, can now be computed as the minimum distance between any 5×5 matrix defining θ_1 and any 5×5 matrix defining θ_2. In [2], the following result has been obtained stating that the matrix-based distance just depends on the number of different single-tile relations in θ_1 and θ_2.

Proposition 8.1. Let $d_1, d_2, d_3, d_4 \in \{R, L, P\}$, $\theta_1 \in CREL(d_1, d_2)$, $\theta_2 \in CREL(d_3, d_4)$. Then, $d_m(\theta_1, (d_1, d_2), \theta_2, (d_3, d_4)) = |(S(\theta_1) \cup S(\theta_2)) - (S(\theta_1) \cap S(\theta_2))|$.

Based on the previous results, we can show that function d_m is not a good distance for cardinal relations. To this end, consider the following example.

Example 8.2. Consider maps M_4, M_5, M_6 shown in Fig. 8.7. The target object A is a region in M_4 and becomes a line in map M_5 and in map M_6, while the reference object B is a region in all the maps. Intuitively, map M_4 seems more similar to map M_5 than to map M_6 as target objects in M_4 and M_5 are spatially closer than target objects in M_4 and M_6. However, the matrix-based distance between M_4 and M_5 coincides with the matrix-based distance between M_4 and M_6. Indeed, $d_m(NW{:}N, (2,2), NE{:}E, (1,2)) = d_m(NW{:}N, (2,2), SE{:}S, (1,2)) = 4$. Thus, function d_m is not able to discriminate between the two situations that are different.

The problem in the previous example is that function d_m does not take into account the distance between the two configurations in the plane. To consider this aspect, given two cardinal directional relations θ_1 and θ_2, it seems more reasonable to define a distance function which takes into account the distance in the plane among the single-tile relations contained in the support of θ_1 and θ_2. In order to define such a distance, we rely on the *cardinal directional conceptual graph*, shown in Fig. 8.3(e). The conceptual graph contains a node for each single-tile and an edge between any pair of adjacent tiles. Given two single-tile relations θ_1^s and $\theta_2^s \in CREL$, that is, given two nodes of the conceptual graph, we denote with $Path_{Min}(\theta_1^s, \theta_2^s)$ the length of the shortest path between θ_1^s and θ_2^s in the conceptual graph. The path-based distance between two single or multi-tile relations, denoted by d_{path}, is now defined by computing the average $Path_{Min}$ distance between each single tile of one relation and each single tile of the other relation, which is not contained in the first one. Function d_{path} can be defined as follows.

Definition 8.4 (Path Distance). *Let $\theta_1 \in CREL(d_1, d_2)$, $\theta_2 \in CREL(d_3, d_4)$, with $d_1, d_2, d_3, d_4 \in \{R, L, P\}$. Let $\theta_1 = R_1 : \ldots : R_k$ and $\theta_2 = S_1 : \ldots : S_h$. The path distance between θ_1 and θ_2, denoted by $d_{path}(\theta_1, (d_1, d_2), \theta_2, (d_3, d_4))$, is defined as follows:*

(i) If $S(\theta_1) = S(\theta_2)$, then 0.

(ii) If $S(\theta_1) \subset S(\theta_2)$, then $\dfrac{1}{|S(\theta_1)| \times |S(\theta_2) - S(\theta_1)|} \displaystyle\sum_{\substack{\theta_i^s \in S(\theta_1) \\ \theta_j^s \in S(\theta_2) - S(\theta_1)}} Path_{Min}(\theta_i^s, \theta_j^s).$

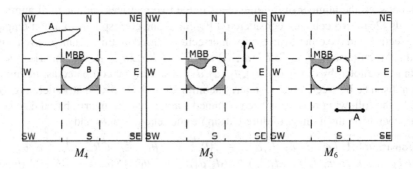

Fig. 8.7. Comparing different cardinal relations

(iii) If $S(\theta_2) \subset S(\theta_1)$, then $\frac{1}{|S(\theta_2)| \times |(S(\theta_1)-S(\theta_2)|} \sum_{\substack{\theta_i^s \in S(\theta_2) \\ \theta_j^s \in S(\theta_1)-S(\theta_2)}} Path_{Min}(\theta_i^s, \theta_j^s).$

(iv) In all the other cases

$$\frac{1}{X+Y} \left(\sum_{\substack{\theta_i^s \in S(\theta_2) \\ \theta_j^s \in S(\theta_1)-S(\theta_2)}} Path_{Min}(\theta_i^s, \theta_j^s) + \sum_{\substack{\theta_i^s \in S(\theta_1) \\ \theta_j^s \in S(\theta_2)-S(\theta_1)}} Path_{Min}(\theta_i^s, \theta_j^s) \right)$$

where $X = |S(\theta_1)| \times |S(\theta_2) - S(\theta_1)|$ and $Y = |S(\theta_2)| \times |S(\theta_1) - S(\theta_2)|$.

It is easy to show that function d_{path} is symmetric and its value ranges between 0 and 4 (the maximum length of a path connecting two tiles). The *cardinal distance* of two cardinal relations θ_1 and θ_2 can now be defined as the product of $d_{path}(\theta_1, \theta_2)$ and their matrix-based distance. Such a distance is then normalized to get a value between 0 and 1. Thus, as d_{path} ranges in $[0, 4]$ and d_m ranges in $[0, 9]$, the product is divided by 36. The resulting distance extends the distances presented in [12, 13] to cope with multi-tile relations.

Definition 8.5 (Cardinal Distance). *Let $\theta_1 \in CREL(d_1, d_2)$, $\theta_2 \in CREL(d_3, d_4)$. Let $\theta_1 = R_1 : .... : R_k$ and $\theta_2 = S_1 : .... : S_h$. The cardinal distance between θ_1 and θ_2, denoted by $d_c(\theta_1, (d_1, d_2), \theta_2, (d_3, d_4))$, is defined as follows:*

$$d_c(\theta_1, (d_1, d_2), \theta_2, (d_3, d_4)) =$$

$$\frac{1}{36} d_{path}(\theta_1, (d_1, d_2), \theta_2, (d_3, d_4)) \times d_m(\theta_1, (d_1, d_2), \theta_2, (d_3, d_4)). \qquad \square$$

Figure. 8.8 presents some values for $d_c(NW, (R, R), \theta_2, (d_3, d_4))$, with $d_3 \neq P$. As expected, we notice that d_c increases by increasing the number of different single-tiles in the two input cardinal directional relations, since in this case d_m increases, and by increasing the distance between the tiles corresponding to the relation supports, since in this case d_{path} increases.

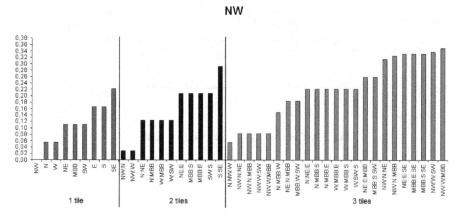

Fig. 8.8. Some distance values for the *NW* cardinal directional relationship

Example 8.3. Consider maps M_7, M_8, M_9 shown in Fig. 8.9. The target object A is a region in M_7 and becomes a line in M_8 and M_9, while the reference object B is a region in all the maps.

Intuitively, map M_7 seems more similar to map M_9 than to map M_8. However, by computing the path distance between each pair of maps, we get the following values:

- Maps M_7 and M_8: $d_{path}(N, (R, R), NW{:}N{:}NE, (L, R)) = 1$.
- Maps M_7 and M_9: $d_{path}(N, (R, R), NW{:}N, (L, R)) = 1$.
- Maps M_8 and M_9: $d_{path}(NW{:}N{:}NE, (L, R), NW{:}N, (L, R)) = \frac{3}{2}$.

Thus, the path distance alone is not sufficient to discriminate between pair M_7 and M_8 and pair M_7 and M_9. By computing the matrix-based distance between each pair of maps, we get the following values:

- Maps M_7 and M_8: $d_m(N, (R, R), NW{:}N{:}NE, (L, R)) = 2$.
- Maps M_7 and M_9: $d_m(N, (R, R), NW{:}N, (L, R)) = 1$.
- Maps M_8 and M_9: $d_m(NW{:}N{:}NE, (L, R), NW{:}N, (L, R)) = 1$.

In this case, the function discriminates between pair M_7 and M_8 and pair M_7 and M_9, but it is not able to discriminate between pair M_7 and M_9 and pair M_8 and M_9, even if map M_7 seems more similar to map M_9 than map M_8, with respect to cardinal directional relations. By combining path distances and matrix-based distances, we get the following values for the cardinal distance function:

- Maps M_7 and M_8: $d_c(N, (R, R), NW{:}N{:}NE, (L, R)) = \frac{2}{36} = 0.06$.
- Maps M_7 and M_9: $d_c(N, (R, R), NW{:}N, (L, R)) = \frac{1}{36} = 0.03$.
- Maps M_8 and M_9: $d_c(NW{:}N{:}NE, (L, R), NW{:}N, (L, R)) = \frac{3}{72} = 0.04$.

In this case, we get the expected result.

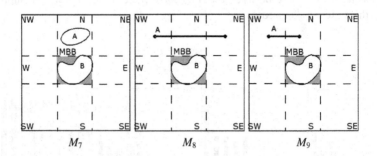

Fig. 8.9. Consistency between cardinal relations

8.6 Consistency of Spatial Relations

When two maps have a set of features in common, possibly represented with different dimensions, the problem arises of establishing whether such maps represent common objects in a consistent way. In this chapter we consider *topological and cardinal consistency*, that is, consistency with respect to topological and cardinal directional relationships existing between map objects.

Informally, two maps are consistent when, given any pair of common features, they share the same or similar spatial relationships in both maps. Different types of consistency can be defined: equality-based (eq-based) consistency and distance-based (dist-based) consistency. *Eq-based consistency* requires that the same spatial relationships exist between each pair of common features in two maps.

Definition 8.6 (Eq-based Consistency). *Let $\theta \in TREL \cup CREL$. f_1 and f_2 in map M_1 are eq-based consistent with f_1 and f_2 in map M_2, if $f_1 \theta f_2$ holds in both M_1 and M_2. M_1 and M_2 are eq-based consistent if, for any pair of features $(f_1, f_2) \in (M_1 \cap M_2)^2$, f_1 and f_2 in map M_1 are eq-based consistent with f_1 and f_2 in map M_2.*

Since topological relations are not defined for all possible pairs of dimensions (see Table 8.1), eq-based consistency cannot always be guaranteed. For example, *Overlap* is defined between pairs of regions and pairs of lines, but it is not defined between a line and a region. A similar situation arises for multi-tile relations, which are not defined when the target object is a point.

From this consideration it follows that spatial relationship equality is a too strong criterion for defining consistency. Such criteria can, however, be relaxed by considering a distance between relations. The new notion of consistency, that we call *dist-based consistency*, is always defined and requires that spatial relationships between two features in two different maps are not necessarily equal but the most similar ones. Dist-based consistency seems the most reasonable choice in real situations dealing with multiresolution, where eq-based consistency cannot always be guaranteed.

In order to formally define dist-based consistency, we rely on the distance functions we have defined in Sects. 8.4.2 and 8.5.2. More precisely, dist-based consistency can be defined by requiring that spatial relationships between pairs of features in two distinct maps must be the most similar ones, according to the introduced distance functions.

Definition 8.7 (Dist-based Consistency). *Let $\theta_1, \theta_2 \in R$, where R is either TREL or CREL. Let f_1 and f_2 be two features appearing in map M_1 with dimensions (d_1, d_2) and in map M_2 with dimensions (d_3, d_4). f_1 and f_2 in map M_1 are dist-based consistent with f_1 and f_2 in map M_2 if $f_1 \theta_1 f_2$ holds in M_1, $f_1 \theta_2 f_2$ holds in M_2, and $d_h(\theta_1, (d_1, d_2), \theta_2, (d_3, d_4))$ coincides either with*

$$min\{d_h(\theta_1, (d_1, d_2), \theta_3, (d_3, d_4)) | \theta_3 \in R(d_3, d_4)\} \text{ or}$$
$$min\{d_h(\theta_2, (d_3, d_4), \theta_3, (d_1, d_2)) | \theta_3 \in R(d_1, d_2)\}$$

where $h \in \{t, c\}$, depending on R. M_1 and M_2 are dist-based consistent if for any pair of features $(f_1, f_2) \in (M_1 \cap M_2)^2$, f_1 and f_2 in map M_1 are dist-based consistent with f_1 and f_2 in map M_2.

We denote by $r_t((d_1, d_2), \theta_1, (d_3, d_4))$ the set of all relations in $TREL(d_3, d_4)$ that are dist-based consistent with $\theta_1 \in TREL(d_1, d_2)$ and by $r_c((d_1, d_2), \theta_1, (d_3, d_4))$ the set of all relations in $CREL(d_3, d_4)$ which are dist-based consistent with $\theta_1 \in CREL(d_1, d_2)$. Dist-based consistency satisfies several useful properties.

Proposition 8.2. *Dist-based consistency satisfies the following properties:*

1. *It is a many-to-many relationship, that is, the cardinality of $r_h((d_1, d_2), \theta_1, (d_3, d_4))$ may be greater than one, $h \in \{t, c\}$.*
2. *It is symmetric, that is, $\theta_2 \in r_h((d_1, d_2), \theta_1, (d_3, d_4))$ if and only if $\theta_1 \in r_h((d_3, d_4), \theta_2, (d_1, d_2))$, $h \in \{t, c\}$.*
3. *It is reflexive, that is, $r_h((d_1, d_2), \theta, (d_1, d_2)) = \{\theta\}$ and $d_h(\theta, (d_1, d_2), \theta, (d_1, d_2)) = 0$, $h \in \{t, c\}$.*

Concerning item (1), it can be shown that, for example, $r_t((R, R), Overlap, (L, L)) = \{Overlap, Cross\}$ (see [2]).

Depending on the type of the spatial relations considered, there exist different relationships between eq-based and dist-based consistency notions, as pointed out by the following proposition.

Proposition 8.3. *The following properties hold:*

1. *For topological relationships, eq-based consistency implies dist-based consistency, that is, $\theta \in r_t((d_1, d_2), \theta, (d_3, d_4))$ except when $\theta = Touch$ and one of the following condition holds: (i) $(d_1, d_2) = (R, P)$ and $(d_3, d_4) = (P, L)$ or (ii) $(d_1, d_2) = (L, P)$ and $(d_3, d_4) = (P, R)$.*
2. *For cardinal directional relations: (a) $r_c((d_1, d_2), \theta, (d_3, d_4)) = \{\theta\}$ when $d_3 \neq P$ or θ is a single-tile relation; (b) $r_c((d_1, d_2), \theta, (d_3, d_4)) \subseteq S(\theta)$ when $d_3 = P$ and θ is a multi-tile relation.*

Proof Sketch. Item (1) directly follows from the results presented in [2]. This strange behavior is probably due to boundary information, quite relevant for the $Touch$ relationship, that are lost when transforming a region into a point. Item (2a) follows from the fact that, under the stated condition, eq-based consistency can always be defined and therefore $d_c(\theta, (d_1, d_2), \theta, (d_3, d_4)) = 0$. Moreover, it is easy to show that the distance value 0 can only be obtained when the cardinal relations coincide. Item (2b) follows from the fact that, when $d_3 = P$, $\theta \notin CREL(d_3, d_4)$ and relations in $S(\theta)$ obviously minimize the distance.

8.7 Motivating Scenarios Revisited

In the following, we complete the examples introduced in Sect. 8.3 by considering definitions and results presented in Sects. 8.4 and 8.5. To this purpose, we consider again the three distinct maps M_1, M_2, and M_3, sketched in Fig. 8.4.

8.7.1 Similarity-based Processing

Suppose the user wants to know which roads enter town T_1 and specifies the query:

$Q_1 = \{r|r$ is a road, r *Overlap* $T_1\}$.

Moreover, suppose the user decides to tolerate an error $\epsilon_t = 25\%$ in the result. If the dimension of towns and roads in the map where the query has to be executed are d_3 and d_4, the query processor can use the topology distance introduced in Sect. 8.4 to rewrite the topological predicate *Overlap* into a set of topological predicates $\{\theta_1, ..., \theta_n\} \subseteq TREL(d_3, d_4)$ such that $d_t(Overlap, (R, R), \theta_i, (d_3, d_4)) \leq 0.25, i = 1, ..., n$. Considering maps M_1 and M_2 in Fig. 8.4, according to Fig. 8.5 we have that:

- Map M_1: $d_t(Overlap, (R, R), \theta, (R, R)) \leq 0.25$, $\theta \in \{Overlap\}$.
- Map M_2: $d_t(Overlap, (R, R), \theta, (R, L)) \leq 0.25$, $\theta \in \{Cross, Touch, Cover\}$.

Thus, depending on the map where the query is executed, query Q_1 is rewritten as follows:

- Map M_1: $Q_1^1 = \{r|r$ is a road, r *Overlap* $T_1\}$ (and therefore, roads R_1, R_2, R_5 are returned).
- Map M_2: $Q_1^2 = \{r|r$ is a road, $r \theta T_1$, $\theta \in \{Cross, Touch, Cover\}\}$ (and therefore, roads R_1, R_2, R_3, R_5, R_6 are returned).

Notice that, without the use of a threshold value, no result would have been returned from map M_2.

Similarly, suppose the user wants to know which towns are NorthWest of T_1. This query can be specified as follows: $Q_2 = \{t|t$ is a town, t *NW* $T_1\}$. We notice that, differently from query Q_1, single-tile relations are always defined, whatever the dimension of the reference and target objects are. Suppose, however, the user wants to execute the query Q_2 up to an error $\epsilon_c = 6\%$. If the dimension of towns in the map where the query has to be executed is d_3, the query processor can use the cardinal distance introduced in Sect. 8.5 to rewrite the cardinal predicate *NW* into a set of cardinal predicates $\{\varphi_1, ..., \varphi_n\} \subseteq CREL(d_3, R)$ such that $d_c(NW, (R, R), \varphi_i, (d_3, R)) \leq 0.06, i = 1, ..., n$.

Considering maps M_1 and M_2 in Fig. 8.4, as towns are represented as regions in both maps, according to Fig. 8.8 we have that: $d_c(NW, (R, R), \varphi, (R, R)) \leq 0.06$ for $\varphi \in \{NW, N, W, NW:N, NW:W, W:NW:N\}$. Thus, query Q_2 is rewritten, by relaxing predicate *NW* by considering the zones around the *NW* tile, as follows:

$Q_2^1 = Q_2^2 = \{t|t$ is a town, $t \varphi T_1, \varphi \in \{NW, N, W, NW:N, NW:W, W:NW:N\}\}$.

From map M_1, towns T_2 and T_3 are returned. From map M_2, towns T_2 and T_4 are returned. Notice that, also in this second case, without the use of a threshold value the query result would have been unsatisfactory for the user, as only T_2 from map M_2 would have been returned.

8.7.2 Query Processing in GIS Mediation Architectures

In a mediation architecture, the mediator exploits the consistency concepts introduced in Sect. 8.6 to compute the most similar topological/cardinal relations to be used in order to rewrite the global user query into the local source queries.

Consider the query Q_2 introduced in Sect. 8.3:

$$Q_2 = \{r|r \text{ is a road}, r \text{ Overlap } T_1 \text{ and } r \text{ MBB:N:NE } T_1\}.$$

Suppose that maps M_1, M_2, and M_3 represent the local sources over which query Q_2 has to be executed and that, at the global level, features are represented with the maximum dimension by which they appear in the local sources. In our example, this means that, at the global level, roads, towns, and pollution areas are represented as regions.

If no rewriting is performed, from the local sources we get the following results: (i) from M_1, roads R_5; (ii) from M_2 and M_3 the empty set, since *Overlap* is not defined between lines and regions and between points and lines, even if roads R_5 and R_6 well approximate the specified condition. We notice that the cardinal directional relation *MBB:N:NE* is, however, defined in all the three maps, since no point is considered as the target object.

Under this scenario, a reasonable approach for query execution at the local level would be that of rewriting the global predicate into a local predicate which is dist-based consistent with the global one. Notice that for the multi-tile cardinal directional predicate, as the target object is not a point, no rewriting is required. For topological predicates, we consider the set of their dist-based consistent predicates, according to the results presented in [2]. Thus, query Q_1 is rewritten as follows:

- Map M_1: $Q_1^1 = \{r|r \text{ is a road}, r \text{ Overlap } T_1 \text{ and } r \text{ MBB:N:NE } T_1\}$, the result set is $\{R_5\}$.
- Map M_2: $Q_1^2 = \{r|r \text{ is a road}, r \text{ Cross } T_1 \text{ and } r \text{ MBB:N:NE } T_1\}$, the result set is $\{R_5, R_6\}$.
- Map M_3: $Q_1^3 = \{r|r \text{ is a road}, r \ \theta \ T_1, \theta \in \{Disjoint, Touch, Contain\}, \text{ and } r \text{ MBB:N:NE } T_1\}$, the result set is $\{R_5, R_6\}$.

As a second example, suppose the user is interested in the towns that are Northwest–North of town T_1. This query can be specified as follows: $Q_2 = \{t|t \text{ is a town and } t \text{ NW:N } T_1\}$. If no rewriting is performed, from the local sources we get the following results: (i) from M_1, town T_2; (ii) from M_2 and M_3, the empty set as, in map M_2, no object satisfies the *NW:N* predicate whereas, in map M_3, as discussed in Sect. 8.6, the multi-tile predicate is not defined over pairs of points. However, towns T_4 and T_5 in map M_3 well approximate the required conditions. According to Proposition 8.3, the dist-based consistent rewriting will generate the following local queries:

- Map M_1: $\{t|t \text{ is a town}, t \text{ NW:N } T_1\}$, the result is $\{T_2\}$.
- Map M_2: $\{t|t \text{ is a town}, t \text{ NW:N } T_1\}$, the result set is empty.
- Map M_3: $\{t|t \text{ is a town}, t \ \psi \ T_1, \psi \in \{NW, N\}\}$, the result set is $\{T_4, T_5\}$.

(a)

(b)

Fig. 8.10. Two snapshots of the same area taken from: (a) $ML1^{L}_{roads}$; (b) $ML1^{R}_{roads}$

8.7.3 Consistency Checking

In order to present an application example of the proposed notions to consistency checking, we evaluate the topological consistency of two maps $ML1_{roads}^{R}$ and $ML1_{roads}^{L}$, containing the streets of a municipality of Lombardy (Italy) represented as regions in $ML1_{roads}^{R}$ and as lines in $ML1_{roads}^{L}$.[6] Figure 8.10 presents two snapshots from corresponding areas of $ML1_{roads}^{R}$ and $ML1_{roads}^{L}$.

By performing a consistency check on such data sets, we find that they are neither eq-based nor dist-based consistent. For example, we discovered that roads 72 and 160 are not dist-based consistent (see Fig. 8.10). Their relation is *Touch* in Fig. 8.10(a) and *Disjoint* in Fig. 8.10(b). According to the table presented in [2], this pair violates both dist-based and eq-based consistency. On the other hand, the pair of roads 70 and 160 are eq-based consistent since their relation is *Touch* in both Fig. 8.10(a) and Figure 8.10(b).

8.8 Conclusions

In this chapter, we have identified some new approaches for processing multiresolution maps in an integrated environment by using qualitative information represented by topological and cardinal relations. In the multiresolution model considered, the same object can be represented with geometries of different dimensions (points, lines, and polygons) in different maps. The proposed approaches refer to three distinct scenarios: (i) similarity-based query processing; (ii) GIS mediation architectures; (iii) map consistency checking. In order to support the contexts presented above, two main issues have been addressed. First of all, starting from some well-known models, distance functions for topological and cardinal directional relationships have been defined. Then, such functions have been used to define two distinct consistency notions, pointing out their properties with respect to the chosen set of relationships. Finally, for each scenario considered, examples of the applications of the proposed notions have been provided. Future work includes the extension of the proposed approaches to deal with other types of spatial relations and the refinement of the proposed distance functions to cope with quantitative measures.

References

1. Belussi A, Catania B, Podestà P (2005) Towards Topological Consistency and Similarity of Multi-resolution Geographical Maps. In: Shahabi C, Boucelma O (eds) Proc. of the 13th Int. Symp. on Advances in GIS. Bremen, Germany, 220–229
2. Belussi A, Catania B, Podestà P (2006) Using Qualitative Information in Query Processing over Multi-resolution Maps Technical Report DISI-TR-06-18, Dipartimento di Informatica e Scienze dell'Informazione, University of Genova, Italy

---

[6] These data sets were provided by COGEME Spa (Rovato BS, Italy) in the context of the research project SpadaGIS funded by MIUR.

3. Boucelma O, Essid M, Lacroix Z (2002) A WFS-based Mediation System for GIS interoperability. In: Voisard A, Chen SC (eds) Proc. of the 10th Int. Symp. on Advances in GIS. McLean, USA, 23–28
4. Boucelma O, Essid M, Lacroix Z, Vinel J, Garinet J-Y, Betari A (2004) VirGIS: Mediation for Geographical Information Systems. In: Proc. of the 20th Int. Conf. on Data Engineering. Boston, USA, 855–856
5. Burns HT, Egenhofer MJ (1996) Similarity of Spatial Scenes. In: Kraak MJ, Molenaar M (eds) Proc. of the 7th Int. Symp. on Spatial Data Handling. Delft, Netherlands, 31–42
6. Clementini E, Di Felice P, Van Oosterom P (1993) A Small Set of Formal Topological Relationships Suitable for End-User Interaction. In: Abel DJ, Ooi BC (eds) Proc. of the 3rd Int. Symp. on Advances in Spatial Databases. Singapore, Lecture Notes in Computer Science, Springer, Berlin Heidelberg New York, 277–295
7. Egenhofer MJ, Al-Taha K (1992) Reasoning about Gradual Changes of Topological Relationships. In: Frank AU, Campari I, Formentini U (eds) Proc. of the Int. Conf. GIS From Space to Territory: Theories and Methods of Spatio-Temporal Reasoning. Pisa, Italy, Lecture Notes in Computer Science, Springer, Berlin Heidelberg New York, 196–219
8. Egenhofer MJ, Clementini E, Di Felice P (1994) Evaluating Inconsistency Among Multiple Representations. In: Waugh TC, Healey RG (eds) Proc. of the 6th Int. Symp. on Spatial Data Handling. Edinburgh, Scotland, 901–920
9. Egenhofer MJ, Franzosa RD (2001) Point-set Topological Spatial Relations. Int. Journal of Geographic Information Systems, 5(2):161–174
10. Egenhofer MJ, Herring J (1990) Categorizing Binary Topological Relations Between Regions, Lines and Points in Geographic Databases. Technical Report, Department of Surveying Engineering, University of Maine, USA
11. Egenhofer MJ, Mark D (1995) Modeling Conceptual Neighborhoods of Topological Line-Region Relations. Int. Journal of Geographic Information Systems, 9(5):555–565
12. Goyal RK, Egenhofer MJ (2000) Consistent Queries over Cardinal Directions across Different Levels of Detail. In: Ibrahim MT, K—ng J, Revell N (eds) Proc. of 11th Int. Conf. on Database and Expert System Applications. Greenwich, London, UK, 876–880
13. Goyal RK (2000) Similarity Assessment for Cardinal Directions Between Extended Spatial Objects. Ph.D. Thesis, Department of Spatial Information Science and Engineering, University of Maine, USA
14. OGC (OpenGeoSpatial Consortium) (1999) OpenGIS Simple Features Specification for SQL, www.opengeospatial.org/docs/99-049.pdf
15. Puppo E, Dettori G (1995) Towards a Formal Model for Multiresolution Spatial Maps. In: Egenhofer MJ, Herring JR (eds) Proc. of the 4th Int. Symp. on Advances in Spatial Databases. Portland, Maine, USA, Lecture Notes in Computer Science, Springer, Berlin Heidelberg New York, 152–169
16. Skiadopoulos S, Koubarakis M (2002) Consistency Checking for Qualitative Spatial Reasoning with Cardinal Direction. In: Proc. of the 8th Int. Conf. on Principles and Practice of Constraint Programming. Itahca, NY, USA, Lecture Notes in Artificial Intelligence, Springer, Berlin Heidelberg New York, 341–355
17. Skiadopoulos S, Koubarakis M (2004) Composing Cardinal Direction Relations. Artificial Intelligence, 152(2):537–561
18. Tryfona N, Egenhofer MJ (1996) Multi-Resolution Spatial Databases: Consistency Among Networks. In: Conrad S, Klein HJ, Schewe KD (eds) Proc. of the 6th Int. Work. on Foundations of Models and Languages for Data and Objects. Schloss Dagstuhl, Germany, 119–132

19. Tryfona N, Egenhofer MJ (1997) Consistency Among Parts and Aggregates: A Computational Model. Transactions in GIS, 1(3):189–206
20. VIRGIS: Infrastructure Logicielle pour l'Interoperabilité de Systhmes d'Information Géographique, www.telecom.gouv.fr/rntl/FichesA/Virgis.htm.

References containing URLs are validated as of September 1st, 2006.

Part III

Spatial Data Protection

9

Access Control Systems for Geospatial Data and Applications

Maria Luisa Damiani[1,3] and Elisa Bertino[2]

[1] University of Milan, Milan (Italy)
[2] Purdue University, West Lafayette (USA)
[3] EPFL, Lausanne (Switzerland)

9.1 Introduction

Data security is today an important requirement in various applications because of the stringent need to ensure confidentiality, integrity, and availability of information. Comprehensive solutions to data security are quite complicated and require the integration of different tools and techniques as well as specific organizational processes. In such a context, a fundamental role is played by the access control system (ACS) that establishes which *subjects* are authorized to perform which *operations* on which *objects*. Subjects are individuals or programs or other entities requiring access to the protected resources. When dealing with protection of information, the resources of interest are typically objects that record information, such as files in an operating system, tuples in a relational database, or a complex object in an object database. Because of its relevance in the context of solutions for information security, access control has been extensively investigated for database management systems (DBMSs) [6], digital libraries [3, 14], and multimedia applications [24]. Yet, the importance of the spatial dimension in access control has been highlighted only recently. We say that access control has a spatial dimension when the authorization to access a resource depends on position information. We broadly categorize *spatially aware* access control as *object-driven*, *subject-driven*, and *hybrid* based on whether the position information concerns objects, subjects, or both, respectively. In the former case, the spatial dimension is introduced because of the spatial nature of resources. For example, if the resources are georeferenced Earth images, then we can envisage an individual be allowed to only display images covering a certain region. The spatial dimension may also be required because of the spatial nature of subjects. This is the case of mobile individuals allowed to access a resource when located in a given area. For example, an individual may be authorized to view secret information only within a military base. Finally, position information may concern both objects and subjects like in the case of an individual authorized to display images of a region only within a military office.

There is a wide range of applications which motivate spatially aware access control. The two challenging and contrasting applications we propose as examples are the spatial data infrastructures (SDI) and location-based services (LBS). An SDI consists of the technological and organizational infrastructure which enables the sharing and coordinated maintenance of spatial data among multiple heterogeneous organizations, primarily public administrations, and government agencies. On the other side, LBS enable mobile users equipped with location-aware terminals to access information based on the position of terminals. These applications have different requirements on access control. In an SDI, typically, there is the need to account for various complex structured spatial data that may have multiple representations across different organizations. In an SDI, the access control is thus object-driven. Conversely, in LBS, there is the need to account for a dynamic and mobile user population which may request diversified services based on position. Access control is thus subject-driven or hybrid. However, despite the variety of requirements and the importance of spatial data protection in these and other applications, very few efforts have been devoted to the investigation of spatially aware access control models and systems.

In this chapter, we pursue two main goals: the first is to present an overview of this emerging research area and in particular of requirements and research directions; the second is to analyze in more detail some research issues, focusing in particular on access control in LBS. We can expect LBS to be widely deployed in the near future when advanced wireless networks, such as mobile geosensor networks, and new positioning technologies, such as the Galileo satellite system will come into operation. In this perspective, access control will become increasingly important, especially for enabling selective access to services such as Enterprise LBS, which provide information services to mobile organizations, such as health care and fleet management enterprises. An access control model targeting mobile organizations is GEO-RBAC [4]. Such a model is based on the RBAC (role-based access control) standard and is compliant with Open Geospatial Consortium (OGC) standards with respect to the representation of the spatial dimension of the model.

The main contributions of the chapter can be summarized as follows:

- We provide an overview of the ongoing research in the field of spatially aware access control.
- We show how the spatial dimension is interconnected with the security aspects in a specific access control model, that is, GEO-RBAC.
- We outline relevant architectural issues related to the implementation of an ACS based on the GEO-RBAC model. In particular, we present possible strategies for security enforcement and the architecture of a decentralized ACS for large-scale LBS applications.

The chapter is organized as follows. The next section provides some background knowledge on data security and in particular access control models. The subsequent section presents requirements for geospatial data security and then the state of the art. Afterward the GEO-RBAC model is introduced. In particular, we present the main concepts of the model defined in the basic layer of the model, the Core GEO-RBAC. Hence, architectural approaches supporting GEO-RBAC are presented. Open issues are finally reported in the concluding section along with directions for future work.

9.2 Background Knowledge on Data Security

9.2.1 Data Security

Broadly speaking, data security aims at protecting information against security breaches. Security breaches can be categorized as *unauthorized data observation, incorrect data modification, and data unavailability.*

Data unavailability occurs when information which is crucial for the functioning of the organization is not ready when needed. Information security means assuring [8]:

- confidentiality, thus protecting data against unauthorized disclosures;
- integrity, thus preventing unauthorized data modification;
- availability, thus recovering from hardware and software errors and malicious data denials.

These requirements raise practically in all application contexts. For example, Chap. 10 in this book focuses on confidentiality and integrity requirements in the context of outsourced-based architectures. In location-based applications, which is our concern, confidentiality is requested to protect various types of information which include, besides the information resources requested by the user, the location of mobile users. Integrity is needed to protect, among others, the information which is transmitted to and from the mobile devices against unauthorized modifications. Data availability is also needed to ensure the availability of the position.

Data security is supported by various tools and systems, including authentication, access control, and audit. Among these, access control is fundamental to ensure confidentiality and integrity of information resources made available to multiple users. The component in charge of access control is the ACS. When a user tries to access an information resource, a component of such system, the *access control mechanism* checks the rights of the requester against a set of authorizations, usually specified by a *security administrator.* An *authorization* states which user, or more generally, which subject can perform which operation on which objects. An authorization is defined as the triple: $< s, op, obj >$, where s, op, and obj refer to subject, operation, and object, respectively. The *subject* identifies the holder of the authorization. It can be a single or group of individuals or programs. The *object* identifies the resource. The nature and granularity of the resource depend on the application. The *operation* defines what the subject is authorized to do with the specified object. The set of authorizations defined by the organization constitutes the *access control policy.* Such a policy is defined using a *policy language* based on an *access control model.* In what follows, we present a classification of access control models.

9.2.2 Access Control Models

Access control models can be categorized into: *mandatory, discretionary,* and *role-based* access control (RBAC) models. Mandatory access control (MAC) developed in a military and national security setting; discretionary access control (DAC) has its roots in academic and commercial research laboratories; the RBAC model, since the seminal paper of [23], has gained increasingly consensus in the research community as well as in industry to finally become a standard widely adopted by organizations [13].

Mandatory Access Control

Mandatory access control models control accesses on the basis of a predefined classification of *subjects* and *objects* in the system. Objects are the passive entities storing information, such as relations in a DBMS. Subjects are active entities performing data accesses. The classification is based on a number of *access classes*, also called *labels* which are associated with every subject and object in the system. A subject is granted authorization to access a given object if and only if some relationship, depending on the access mode, is satisfied between the classifications of the subject and the object. An access class generally consists of two components: a security level and a set of categories. The security level is an element of a hierarchically ordered set. A very well-known example of such a set is the one including the levels TopSecret (TS), Secret (S), Confidential (C), and Unclassified (U), where $TS > S > C > U$. The set of categories is an unordered set (e.g. NATO, Nuclear, Army). Access classes are partially ordered as follows. An access class c_i dominates $\geq$ an access class c_j iff the security level of c_i is greater than or equal to that of c_j and the categories of c_i include those of c_j. The security level of the access class reflects the sensitivity of the information contained in the object. Categories are used to provide finer-grained security classifications of subjects and objects. Access control is based on two principles, formulated by Bell and LaPadula, which are followed by all models enforcing a mandatory security policy. The first states that a subject can read only those objects whose access class is dominated by the access class of the subject; the second states that a subject can write only those objects whose access class dominates the access class of the subject. The application of MAC policies to relational DBMSs has been extensively investigated over the past years. The introduction of such a model entails the solution of several difficult issues. Because of this complexity, the adoption of such a model in DBMSs is not as common as the next model.

Discretionary Access Control

These models are discretionary in the sense that they allow users to grant other users authorization to access the data. Specifically, DAC policies regulate the access of users on the basis of the user's identity and authorizations that specify, for each user (or group of users) and each object in the system, the access modes (e.g. read, write, or execute) the user is allowed on the object. Each request of a user to access an object is checked against the specified authorizations. If there exists an authorization stating that the user can access the object in the specific mode, the access is granted, otherwise it is denied [22]. Because of such flexibility, discretionary policies are adopted in many applications. An important aspect of DAC is related to the *authorization administration* policy. Authorization administration refers to the function of granting and revoking authorizations. It is the function by which authorizations are entered (removed) into (from) the ACS. Common administration policies include the *centralized administration* policy, by which only some privileged users may grant and revoke authorizations, and the ownership-based administration, by which grant and revoke operations on a data objects are issued by the creator of the object. The ownership-based administration is often extended with features for *administration delegation*. Administration delegation allows the owner of an object to assign other users the right to grant and revoke authorizations, thus enabling decentralized authorization administration.

Role-based Access Control

Role-based access control is centered on the notion of *role*. A role is a semantic construct which represents a job function within an organization. Specifically, the RBAC standard consists of four basic sets of elements: *users*, *roles*, *permissions*, and *sessions*.

- *User* is as a human being or an autonomous agent.
- *Role* represents the function of a user within a community. The community can be a structured organization, for example, a business enterprise or a more informal community, for example the citizens of a city. A role confers a set of permissions on the user.
- *Permission*. Permission is an approval to perform an operation on one or more objects. An object is a resource that shall be protected. An operation is an executable image of a program, which on invocation executes some function for the user over some object. The types of operations and objects depend on the application context in which RBAC is deployed. For example, in a file system, operations might include read, write, and execute; in a DBMS, operations might include insert, delete, append, and update.
- *Session*. When the user logs in, a session is established, during which the user activates some subset of roles that he or she is assigned. The permissions available to the user of the session are thus the permissions assigned to the roles that are currently active across all the user's sessions.

Over the above sets of elements, a number of relations are defined. The *user assignment* relates users to roles through a many-to-many relationship, a user can therefore be assigned multiple roles and the same role assigned to different users. The *permission-assignment* relation relates roles and permissions again through a many-to-many relationship; thus a role can be assigned multiple permissions and similarly each permission can be assigned to multiple roles. The function *SessionUser* maps each session into a user, whereas the *SessionRole* function maps a session onto a set of roles, namely the roles that are active in the session. The basic concepts are formally summarized as follows:

Definition 9.1 (Basic Concepts of RBAC). *Let* U, R, PRMS, *and* SES *denote the set of users, roles, permissions, and sessions, respectively. We define:*

- $UA \subseteq U \times R$. The user assignment relation that assigns users to roles.
- $AssignedUser : R \rightarrow 2^U$. The mapping from a role to a set of users.
- $PA \subseteq R \times PRMS$. The permission assignment relation that assigns permissions to roles.
- $PrmsAssignment : R \rightarrow 2^{PRMS}$. The mapping of a role into a set of permissions.
- $SessionUser : SES \rightarrow U$. The mapping from a session to a user.
- $SessionRole : SES \rightarrow 2^R$. The mapping from a session to a set of roles.   □

RBAC standard is defined at three levels, which are referred as Core, Hierarchical, and Constrained RBAC, respectively. *Core RBAC* defines the minimum collection of RBAC elements, element sets, and relations for a RBAC system to be defined. The Hierarchical RBAC component adds relations for supporting role hierarchies. A hierarchy is a partial order defining a seniority relation between roles, whereby senior roles acquire the permissions of their juniors. Constrained RBAC adds *separation of duty* (SoD) constraints to the RBAC [13]. SoD constraints are introduced to prevent conflicts of interest arising when a single individual can simultaneously perform sensitive tasks requiring the use of mutually exclusive duties. In the RBAC standard, enforcement of SoD policies is realized by specifying exclusive role constraints. As an additional level, administrative functions are defined to enable policy specification.

9.3 Motivations and State of the Art

In this section we discuss in detail relevant access control requirements arising in the context of geospatial applications and contexts, and then we survey related research proposals and approaches.

9.3.1 Requirements for Spatially Aware Access Control

Geospatial data have a strategic relevance in various contexts, such as emergency response, homeland security, marketing analysis tools, and environmental risks control procedures. Most applications in these areas require a fine-granularity flexible access control to geospatial data. A possible naive approach to support access control is to build ad hoc data sets (maps) for each type of access the administrator wants to grant: this has been the cartographic approach applied for many years in the past. Such an approach is not suitable when the user community is large and dynamic—which is today often the case in Web-based systems—and when the data and the access control policies dynamically change—which is the case in applications such as emergency response. Another major drawback of the conventional approach is that it does not support flexible protection granularities in the access control policies, and it does not account for various levels of representation that data may have in a geospatial database. The introduction of integrated data management systems for geospatial data, such as current integrated GIS systems, characterized by comprehensive data models, and current developments in standards for geospatial data such as GML [17] are today making possible the development of advanced ACS that go beyond such naive approaches. However, despite the importance of data protection, almost no efforts have been devoted to the investigation of access control models and systems; only very preliminary proposals exist. In order to position current research, in what follows we overview major requirements that arise in access control for geospatial data.

Richness and Multiplicity of Data Representations

Modern data management systems for geospatial data typically support multiple data representations (such as attributive, vector-based, and topological representations); additional representations are also often available, such as raster images. Not only the same entity may be represented in the data repository according to multiple representations, it can also be represented according to multiple dimensions (such as a point, 0 dimension, or a region, 2 dimensions). Also geospatial objects may be complex objects, consisting of subobjects. In some cases, one may want to hide some of the components of a given spatial object. A suitable access control model for geospatial data must thus: (i) Support the specification of authorizations against geospatial objects at a very fine granularity level. (ii) Account for the various spatial representations. This means that if a user can see an aerial image from which certain objects have been hidden, it is important that the same objects be hidden from the vector-based representation of the same area. (iii) Account for the various object dimensions and resolutions. This means that an administrator should be able to authorize a user to see a given object at 0 dimension and not at higher dimensions, thus hiding detailed information about the object shape. Similarly, in a raster representation, the administrator should be able to specify the resolutions according to which certain objects can be seen. (iv) Support various access rights. In access control models, operations that can be performed on the protected objects correspond to the access rights. This allows one to express authorizations in terms of the operations supported by the model according to which data are represented. It is thus important that in addition to access rights such as read and write, access rights that correspond to meaningful operations on geospatial objects be supported.

Dynamic and Mobile User Population

Many of the geospatial applications are characterized by a user population that constantly changes and moves. Also, in many cases, users from different administrative domains or agencies may need to access the data. A suitable access control model for geospatial data must thus perform the following: (i) Support attribute-based and profile-based user specification. Relying on user login names for grant and revoke authorization is very cumbersome and it is also a low-level approach. Recently, several attribute-based and profile-based access control mechanisms have been proposed, supported by techniques such as attribute certificates and standards such as SAML [20]. Also recent extensions to the RBAC model [7] have been developed supporting access control across different domains; under such an approach, a user authorized to use a role in a domain is automatically able to enroll in a given role in another domain. These techniques should certainly be incorporated into an access control model for geospatial data. (ii) Supporting authorizations that are dynamically enabled or disabled depending on the user location. It is important to notice that an important requirement is related to the support of mobile users and in particular to the fact that a user may be authorized to access data depending also on the user location or geographical region. For example, a taxi driver authorized to pick up passengers only in a given area of Milano cannot access passenger information when located in another area of the city. It is also important to be able to express user location in terms of the physical position or the logical position (for example "being inside of Milano train station"). Addressing such requirement also entails using secure positioning techniques.

Dynamic Application Contexts

Geospatial data are today increasingly used in various circumstances. For example, consider some data representing roads in a given region; such data can be used for planning the road maintenance schedule or for managing an emergency. It is likely that whether such data may be accessed or not by specific users depend on the current situation, and it is thus likely that different sets of access control policies may need to be activated over time, also depending on the occurrence of specific events. A suitable access control model for geospatial data must thus perform the following: (i) Support mechanisms for policy grouping and modularization. It is important that security administrators be given mechanisms according to which they can group policies into modules that are specific for handling specific situations. To better address compliance requirements, such modules should also include metadata to provide description and information about the policies inside the various modules. (ii) Support event-based activation/deactivation of policy modules. This requirement entails defining an event language and developing suitable event-monitoring techniques. Techniques such as triggers and active rules, developed in the database and the AI field, could be extended to support the automatic activation/deactivation of policies. Notice, however, that a crucial issue is represented by techniques providing high assurance event detection, to avoid an attacker preventing a relevant event from being reported or injecting a false event.

9.3.2 State of the Art

As we have seen, geospatial applications have challenging requirements of access control. Yet, such requirements have been only partially addressed by current research. The spatially aware access control models proposed in literature can be categorized as two broad classes, based on whether they are more focused on geospatial representation of data or user mobility, which are presented in what follows.

Access Control for Geospatial Data

The first access control model for geographical data has been proposed [1] to control the access to georeferenced Earth images. The purpose of such an approach is the controlled dissemination of satellite images at different levels of resolution through a DAC policy. In particular, authorizations state which images users are allowed to access and at which resolution. The system, however, is limited in that it only deals with satellite images. To overcome these limitations, an ACS has been recently developed for the protection of vector-based data available on Web [5]. The underlying model is simple but anticipates some ideas that have been more extensively developed by GEO-RBAC. The central concept in such model is that of *spatial authorization*. A spatial authorization is defined by the tuple $< u, ft, p, w >$, where u denotes the user, ft the object specified in terms of spatial feature types (such as building or house), p the operation in the form of Web service to be performed on spatial objects of the specified type (such as *InsertObject*, to introduce a new spatial element), and w is the *authorization window*. The authorization window indicates the geographical scope of the authorization, that is, the portion of space in which the authorization applies. Accordingly, one can state, for example, that user u is allowed to carry out operation p on all objects of type ft located in the authorization window w. Furthermore, the model supports the specification of administrative functions for the creation and update of spatial authorizations based on a decentralized administration policy. The notion of authorization window has also been integrated in a different access control model, which has been developed for regulating the access to a spatial database [2]. The peculiarity of this approach is that the database is based on a complex spatial data model, enabling multiple levels of spatial representation at multiple granularities. The access control model thus enables the specification of which objects at which granularity and in which portion of space can be selected and modified.

A more recent approach integrating geospatial and security standards to support controlled access to spatial information through geo-Web services has been recently developed [19]. In such an approach, a policy specification language, referred to as GeoXACML, is defined as a geospatial extension of the OASIS eXtensible access control markup language (XACML). GeoXACML supports the specification of rules which enable or deny the access to geospatial objects based on spatial criteria, such as topological relationships. As an example, consider the rule which states that an operation can be performed only on buildings which are located with the administrative boundary of Washington, DC. This approach has some similarities with the notion of authorization window, though in the window-based model [5] not only restrictions over objects can be specified but also on an administration policy.

Access Control in Mobile Applications

None of the previous models is conceived for use in a dynamic environment, which instead is the main concern of spatial and non-spatial context aware access control models.

Non-spatial context-aware access control models include generalized TRBAC (GTRBAC) [18], a RBAC-based model which incorporates a set of language constructs for the specification of various temporal constraints on roles, including constraints on role enabling, role activation, user-to-role assignments, and permission-to-role assignments. X-GTRBAC [7] is another approach which augments GTRBAC with XML for supporting the policy enforcement in a heterogeneous, distributed environment. In addition to temporal constraints, the model also supports non-temporal contextual constraints. The approach, however, is more focused on the software engineering aspects of the access control rather than on the expressivity of the policy specification language. A notable approach is the one proposed through the Generalized RBAC (GRBAC) [10]. GRBAC introduces the concept of environment roles; that is, roles that can be activated based on the value of conditions in the environment where the request has been made. Environmental conditions include time, location, and other contextual information that is relevant to access control. If compared with GEO-RBAC, the concepts of role extent and user position are close to that of context variables. However, the mechanism of contexts is very general and does not account for the specificity of spatial information, such as the multi-granularity of position and the spatial relationships that may exist between the spatial elements in space. Moreover, in GEO-RBAC a common spatial data model is adopted in order to provide a uniform and standard-based representation of locational aspects that, notably, involve not only roles but also protected objects.

The spatial dimension of access control is the basic component of the approach presented in [15]. In such work, an extension of the RBAC model is proposed based on the notion of spatial role, intended as a role that is automatically activated when the user is in a given position. The space model is however very simple and targeted to wireless network applications. It consists of a set of adjacent cells and the position of the user is the cell or the aggregate of cells containing it. The spatial granularity of the position is thus fixed while the space is rigidly structured and the position itself does not have any semantic meaning but simply a geometric value. By contrast, in GEO-RBAC the granularity of the user position may depend on the role of the user; thus no assumption is made on the space layout. Moreover, in GEO-RBAC the spatial dimension integrates geometric and semantic knowledge about the world.

A different approach which combines space and time is presented in [9]; this approach borrows from GEO-RBAC the distinction between real position and logical position and from GTRBAC the notion of temporal context. Such a model, however, does not include the notion of schema, neither supports important features of GEO-RBAC such as hierarchies of enabled roles and spatially aware SoD constraints.

9.4 An Access Control Model for Location-aware Applications: GEO-RBAC

GEO-RBAC is a comprehensive access control model specifically developed to address access control requirements of applications characterized by users that are members of mobile organizations. By mobile organization, we mean a community of individuals that, because of the role they have in the community, need to access common information resources through LBS. Mobile organizations thus consist not only of enterprises operating on field such as fleet management and resources tracking but also of health care and leisure organizations, and military and civilian coalitions created in response to a crisis, for example natural disaster and war. In these organizations, the mobile members are characterized not only by an identity but also by a functional role. Because of the different activities of the organization's members, it is reasonable to consider that LBS (services for short) are requested based on the roles of individuals and also their position.

The GEO-RBAC model addresses the above issues; it relies on RBAC standard and, like RBAC, is based on the concept of *role*. Moreover, it is structured in levels, referred as Core, Hierarchical, and Constrained GEO-RBAC, respectively. Of these levels, we describe here the Core level, which constitutes the basic component. For the additional levels, we refer the readers to [12].

Before proceeding, we introduce the running example that will be used throughout the chapter. Consider an LBS application for a university campus. The campus consists of various buildings and areas, such as departments, libraries, and recreative areas, each occupying a position in space. From an organizational point of view, people in the campus can play one or more functions, such as teachers, students, visitors, and, say, library subscribers. The members of the campus are connected to a wireless network and equipped with a location-aware PDA by using which they access various LBSs. Assume that John is a student and also a library subscriber. When located within one of the libraries of the campus, John is presented with various services, for example the *book-loan* service to request the loan of a book. Instead, when in a department, he can request the class timetable.

9.4.1 Overview of the Core Level

By adapting the definition in [13] to our model, we say that Core GEO-RBAC defines *a minimum collection of elements which are required to completely achieve a GEO-RBAC system*. The basic notion is that of spatial role, which describes a spatially bounded function for a user. Two other concepts, however, are noteworthy: (a) the *position model* which describes the position of the mobile user; (b) the notions of *role schema and role instance* which provide a representation of the role at two levels, respectively, the intensional and the extensional level. For the representation of the position of the user and the spatial properties of roles, we rely on current geospatial standards [16]. In particular, we describe the spatial objects which are located in the reference space in terms of *simple features* [16] (hereinafter, features). Features have an identity, thematic properties, and can be mapped onto a position in the reference space. A feature is denoted by its identifier. For example, *UniMi* is the identifier of the spatial object which describes the properties of a campus and its position. The position of a feature is represented through a *geometry* which can be of type point, line or polygon, or recursively a collection of geometries. Further, geometries can be related by different types of relationship. Among them, the reference set of topological relations is

{*Disjoint, Touch, In, Contains, Equal, Cross, Overlap*}. These relations are binary, mutually exclusive (if one is true, the others are false), and they are a refinement of the well-known set of topological relations [16]. Moreover, features have an application-dependent semantics that is expressed through the concept of *feature type*. A feature type captures the intensional meaning of the entity, for example, *Campus* and *Department*. The *extension* of a feature type is a set of semantically homogeneous features.

9.4.2 Basic Concepts

The Position Model

The notion of position is fundamental since it characterizes the mobile user. Unlike most proposals in mobile computing which describe the position uniquely in geometric terms, for example a point, we introduce, for the sake of flexibility, the distinction between the *real* and the *logical* position. The real position corresponds to the position of the user on Earth acquired through some positioning technology. Real positions can thus be represented as geometric elements of different types since, depending on the chosen technology and accuracy requirements, they may correspond to points or polygons. Conversely, the *logical position* is defined at a higher level of abstraction to represent positions in a way that is almost independent from the underlying positioning technology. Further, besides a geometry, the logical position has a semantics. For example, logical positions can be a house, an address number, or a road, which are represented in terms of *spatial feature types*. The logical position is computed from real positions by using a *location mapping function*. For example, a location mapping function can be defined to map a position acquired through global positioning system (GPS) onto the closer road segment. More formally, given a feature type ft, a *position mapping function for ft* is a function m_{ft} which, given a real position rp, returns a logical position corresponding to an instance of ft having rp as real position. As we will see, since the localization may respond to different application requirements, for example, with respect to accuracy, the meaning of location may vary depending on the role. For example, the position of a generic campus member may be coarsely defined in terms of campus sectors, assuming that the campus area is subdivided in sectors, whereas the position of the teacher can be represented at a higher resolution by an address.

Spatial Role

The concept of *spatial role* is the distinguishing aspect of our model. To account for the spatial context, a *spatial role* is defined not only by a name as in RBAC but also by a *role extent*. The extent of a role defines the boundaries of the region contained in the reference space. A user, who has been conferred a role r, is thus recognized to effectively play such a role only when logically located within the extent of r. For example, *CampusMember(UniMi)* is a spatial role: *CampusMember* is the role name and *UniMi* is the role extent, specifically the identifier of a spatial feature of type *Campus* denoting, among others, a polygonal space. An individual, which is a member of the campus, is recognized to play the role of *CampusMember* and thus authorized to invoke the services associated to the role, only when located in the polygonal extent of the campus.

Each role is assigned a set of *permissions*. A permission corresponds to a service. For example, the service *BookLoan* of the running example is assigned to the spatial role *LibarySubscriber(MyLibrary)*. The terms *permission* and *service* are used in this chapter in an interchangeable way. Like RBAC, when a user connects to the LBS application, a new *session* is started. Then the user selects the roles, among those which have been assigned to him/her that they wish to play. This set represents the *session roles* and the elements of the set are said to be *activated*. Unlike RBAC, however, in our model, session roles have in addition a status which is *enabled* or *disabled*. For a session role to be *enabled*, the user should be logically located within the space of the corresponding role extent. As the user moves, the status of roles change. Therefore, depending on the position of the user, only a subset of the session roles is enabled and permissions granted. As a result, the set of services which the user can access at a given point in time depends on both the session roles and the position of the user.

Role Schema and Instance

To provide a more compact representation of semantically homogeneous roles defined over different extents, we introduce the distinction between role schema and role instance. A *role instance* is a role defined over a specific extent, in compliance with the role schema. Note that the terms *roles instance* and *spatial role* are used as synonyms. A *role schema* defines common properties of roles with a similar meaning and thus simplifies the specification of roles and ultimately role engineering. Specifically, a role schema defines: (a) a common name for a set of roles; (b) the type of role extent; (c) the type of logical location; (d) the mapping function relating the real position with the logical position. For example, $CampusMember(Campus, Sector, m_{sector})$ is a schema the instances of which are roles having the following properties: the roles have the same name *CampusMember*; *Campus* is the type of the role extent; *Sector* is the type of logical position which is computed by applying function m_{sector} to the real position. Once a schema is specified, the corresponding instances can be simply created by specifying the role name and its extent, for example *CampusMember(UniMi)*. Notice that *UniMi* is a feature of type Campus and defines the boundary of the spatial role.

Because roles are assigned permissions and because of the two different levels of role representation, it seems reasonable to assign permissions to both role schemas and role instances. The permissions which are assigned to a schema are then inherited and shared by all the instances of the schema. For example, if we assign the service *getMap* to the role schema *CampusMember*, it means that such a service can be accessed by all the members of the campus. For the sake of flexibility, however, permissions can be assigned also to single-role instances.

9.4.3 Specification of Core GEO-RBAC

Now, we present the above concepts in more formal terms. Core GEO-RBAC is presented as organized in a number of logical parts, one for each major set of the RBAC model, that is roles, permissions, users, and sessions.

The general structure of the model is illustrated in Fig. 9.1.

We use the graphical representation adopted for RBAC. In defining the model, we refer to the notation introduced in the previous section and summarized in Table 9.1.

Preliminarily, we introduce the notion of partial ordering of features and feature types. Let FT be the set of feature types and $ft_i, ft_j \in FT$, with $i \neq j$, be two elements of the set. We say that ft_i is contained in ft_j, denoted by $ft_i \subseteq_{ft} ft_j$, if for each feature f_i of type ft_i, a feature f_j of type ft_j exists such that the geometry of f_i is contained in the geometry of f_j. For example, the relation of containment between *Town* and *Region* is written as $Town \subseteq_{ft} Region$. Similarly, we say that the feature f_i is contained in feature f_j, denoted by $f_i \subseteq_f f_j$, when the geometry of f_i is contained in the geometry of f_j. As we will see, such relationships will be useful in characterizing the relationships between locations and roles.

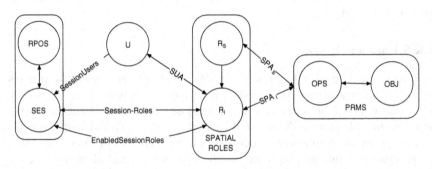

Fig. 9.1. Core GEO-RBAC

Table 9.1. Notation for the main sets used in GEO-RBAC

Notation	Meaning
FT	Feature types
F	Features
R	Role names
SES	Sessions
U	Users
$REXT\_FT$	Role extent types
$LPOS\_FT$	Logical positions types
$LPOS$	Logical positions
$RPOS$	Real positions
M	Position mapping functions
OPS	Operations
OBJ	Objects

Role Schema and Instance

A role schema defines a common name for a set of roles, the role extent type, the logical position type, and the position mapping functions relating the real position with the logical position. A role instance is defined over an extent of the type specified in the corresponding schema, while the logical position of the individual playing such a role is determined by the position mapping function specified in the schema alike. The formal definitions of role schema and instance are reported below [4]:

Definition 9.2 (Role Schema). *A Role Schema is a tuple* $< r, ext, loc, m_{loc} >$ *where:*

- $r \in R$;
- $ext \in REXT\_FT$;
- $loc \in LPOS\_FT$;
- $loc \subseteq_{ft} ext$;
- $m_{loc} \in M$ *is a location mapping function for feature type loc.*

We denote with R_S *the set of role schemas and we assume that, given a role name* $r \in R$, *r is unique in* R_S. *A role schema is also denoted as* $r(ext, loc, m_{loc})$.

Definition 9.3 (Role Instance). *Given a role schema* $r_s \in R_S$, *an instance* r_i *of* r_s *is a pair* $< r, e >$ *where r is the name of the role in schema* r_s, *thus* $r = r_s.r$ *and* $e \in F$ *is a feature of type* $r_s.ext$. *The schema of* r_i *is denoted by* $SchemaOf(r_i)$. *We denote with* R_I *the set of role instances for all role schemas. A role instance is also denoted as* $r(e)$.

Permission

A permission is associated with each service. In our model, permissions can be associated either with the role schema and inherited by all role instances of the schema or directly with the role instances. Such different granularities are formalized by introducing two functions: *S_PrmsAssignment*, relating roles schemas and permissions sets; *I_PrmsAssignment* relating spatial roles, thus role instances, to specific permissions. Function *I_PrmsAssignment** is then introduced to combine permissions directly assigned to spatial roles with permissions inherited from their role schema. Formally [4]:

Definition 9.4 (Permissions). *The set of permissions PRMS is defined as* $PRMS = 2^{(OPS \times OBJ)}$. *We also define:*

- SPA_S: $R_S \times PRMS$, *a many-to-many mapping permission-to-spatial role schema assignment relation;*
- $S\_PrmsAssignment$: $R_S \rightarrow 2^{PRMS}$, *the mapping of spatial role schema onto a set of permissions. Given a role schema* r_s, $S\_PrmsAssignment(r_s) = \{p \in PRMS \mid < r_s, p > \in SPA_S\}$;
- SPA_I: $R_I \times PRMS$, *a many-to-many mapping permission-to-spatial role instance assignment relation;*

- $I\_PrmsAssignment: R_I \rightarrow 2^{PRMS}$ the mapping of spatial role instance onto a set of permissions. Given a role instance r_i, $I\_PrmsAssignment(r_i) = \{p \in PRMS | < r_i, p >\in SPA_I\}$;
- $I\_PrmsAssignment^*: R_I \rightarrow 2^{PRMS}$ such that given a role instance r_i, $I\_PrmsAssignment^*(r_i) = I\_PrmsAssignment(r_i) \cup S\_PrmsAssignment(Sche-maOf(r_i))\}$. Hence, the permissions of a role are those assigned to its schema plus those directly assigned to the instance.

Users and Session

Spatial roles are assigned to users. The definition of the model for this part is conceptually analogous to that in RBAC. In particular, given a set of users, the following relations are defined: the many-to-many relation SUA relates users and role instances; the function $SR\_AssignedUser$ maps a role instance onto the set of users which can activate that role. Formally [4]:

Definition 9.5 (Users). *We define:*

- $SUA \subseteq U \times R_I$, a mapping user-to-spatial role instance assignment relation;
- $SR\_AssignedUser: R_I \rightarrow 2^U$, the mapping of spatial role instance onto a set of users. Formally, $SR\_AssignedUser(< r, e >\in R_I) = \{u \in U|(u, < r, e >) \in SUA\}$.

When a user logs in, a new session is activated and a number of roles are selected to be included in the session role set. Given a session s, the following two functions are defined: $SessionUser(s)$ corresponds to the user of the session; $SessionRoles(s)$ corresponds to the role that can be potentially activated in s. Formally [4]:

Definition 9.6 (Sessions). *We define:*

- $SessionUser: SES \rightarrow U$, the mapping from a session s to the user of s;
- $SessionRoles: SES \rightarrow 2^{R_I}$ with $SessionRoles(s) \subseteq \{< r, e >\in R_I|(Session-User(s), < r, e >) \in SUA\}$.

Access Control Mechanism

The session roles are the roles that the user of the session has selected. However, for a session role to be *enabled*, the user should be logically located within the space of the corresponding role extent. Therefore, depending on the user position during that session, only a subset of the session roles is enabled and permissions granted. In order to compute the logical position of a user playing a role r in a session, the location mapping function defined in the schema of r is applied to the user real position, provided by the external environment. Hence, if the logical position of the user is spatially contained in the extent of r, the role is *enabled* and thus the set of permissions assigned to the corresponding role is determined. Given a user's request, the access control mechanism determines whether the permission requested by the user belongs to the set of permission associated to the set of enabled roles *ER*. If it is the case the permission is granted otherwise it is rejected. The set of enabled roles *ER* in session *s* and real position *rp* is computed by the function *EnabledSessionRole(s, rp)*.

Definition 9.7 (Authorization Control Function). *An access request is a tuple ar =* $\langle s, rp, p, o \rangle \in SES \times RPOS \times OPS \times OBJ$. *ar can be satisfied at position rp $\in$ RPOS if:*

$$(p, o) \in \bigcup_{y \in EnabledSessionRoles(s,rp)} I\_PrmsAssignment^*(y).$$

Basic objects

$FT = \{Campus, Library, Sector, Address)\}$ with: $Sector \subseteq_{ft} Campus$
$F = \{Purdue, MyLib\}$ with: $MyLib \subseteq_f Purdue$

$PRMS = \{p_1, p_2, p_3, p_4\}$ with $\begin{cases} p_1 = GetMap \\ p_2 = ShowClassTimetable \\ p_3 = BookLoan \\ p_4 = BookSearch \end{cases}$

Schema

$R = \{Teacher, Student, LibrarySubscriber\}$

$R_S = \{St, Te, Li\}$ with $\begin{cases} St =< Student, Campus, Sector, m_{Sector} > \\ Te =< Teacher, Campus, Address, m_{Address} > \\ Li =< LibrarySubscriber, Library, Library, m_{Library} > \end{cases}$

Instances

$R_I = \{r_{St}, r_{Te}, r_{Li}\}$ with $\begin{cases} r_{St} = Student(Purdue) \\ r_{Te} = Teacher(Purdue) \\ r_{Li} = LibrarySuscriber(MyLib) \end{cases}$

Permission assignment

$SPA_S = \{(p_1, St), (p_1, Te), (p_2, St), (p_2, Te), (p_3, Li), (p_4, Li)\}$

User assignment

$U = \{Sara, John\}$

$SUA = \{s_{ua_1}, s_{ua_2}, s_{ua_3}\}$ with $\begin{cases} s_{ua_1} = \langle John, Student(Purdue) \rangle \\ s_{ua_2} = \langle Sara, Teacher(Purdue) \rangle \\ s_{ua_3} = \langle John, LibrarySubscriber(MyLib) \rangle \end{cases}$

Sessions

$SES = \{s_1\}, UserSession(s_1) = \{John\}$
$SessionRoles(s_1) = \{Student(Purdue), LibrarySubscriber(MyLib)\}$

EnabledRoles

$EnabledSessionRoles(s_1, loc_1) = \{\{Student(Purdue),$
$LibraySubscriber(MyLib)\}\}$ if John is in $MyLib$

Fig. 9.2. An example of a GEO-RBAC policy

Example

Finally, we summarize the GEO-RBAC concepts (see Fig. 9.2) by presenting the access control policy defined for the running example. The policy is built as follows. First, we define the sets of feature types, features, and permissions used in our policy. Then we specify the following role schemas: the *Student* schema specifies that an individual can play the role of student only when located within a campus. *Campus* is a spatial feature type and *Purdue* is an instance of campus. Further, the logical position of a student is represented by a sector of the campus, of feature type *Sector*. The *Teacher* schema is defined in a similar way for the individuals who are teachers in the campus. The logical position of a teacher is represented by an address. The *Library Subscriber* schema specifies that individuals can play such a role only within a library. The logical position of a subscriber is of type *library*. Notice that in this case, the logical position has the same granularity of the role extent; therefore, the actual position within the library is not relevant for the application. The role instances we consider are student at Purdue (i.e. Student(Purdue)), teacher in the same campus (i.e. Teacher(Purdue)), and subscriber of the campus library named MyLib (i.e. LibrarySubscriber(MyLib)). We then specify how permissions, namely information services, are tied to the previous schemas. For example, the service *book-loan* of the running example is assigned to the *LibrarySubscriber* schema and thus is inherited by its role instances. Now suppose that Sara is a teacher at Purdue, while John is a student in the same campus and also a subscriber of MyLib. Assume John to be the only user connected to the system. Until John is outside the Purdue campus, John is not allowed to access any information service. Conversely, when inside the library MyLib, John is recognized to be a library subscriber, besides a student, and thus can request a book loan.

9.5 A Reference Architecture for GEO-RBAC

The GEO-RBAC model defines the conceptual constructs supporting the specification of access control policies. We now complement the discussion by presenting a reference architecture for an ACS based on GEO-RBAC. The purpose is to devise the main building blocks of the access control system and at the same time the key operational aspects. We present in particular two different architectural frameworks, the first relying on a centralized architecture and the other on a distributed architecture for access control in large-scale applications.

9.5.1 A Centralized Architecture

The centralized ACS assumes that all access control functionalities, namely policy administration and enforcement, are implemented by a unique component which filters user requests accepting only those which are authorized by the current security policy. The general architectural framework that we assume is the one shown in Fig. 9.3. Such framework consists of three fundamental components [11]:

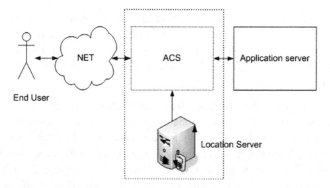

Fig. 9.3. Centralized architecture

- a set of *mobile users* equipped with mobile terminals and connected to a wireless network. Users are assumed to be identified by their terminal ID.
- the *Application Server* providing a set of location-aware information services. Such services are mapped onto *permissions* of GEO-RBAC.
- the ACS. It is a trusted component filtering the users' requests. To obtain the position of the user, the ACS accesses a *Location Server* which aggregates location data from different sources such as the network and mobile terminals equipped with GPS (connected by dotted lines in Fig. 9.3) and responds to queries such as *retrieve the position of terminal ID*. Notice that to prevent uncontrolled disclosure of location data, the ACS is the only component enabled to query the Location Server.

A request for a service is processed as follows: the user requests the service by sending a request message to the ACS. The ACS then determines whether the request can be accepted, and if it is the case the request message is properly restructured and sent to the Application Server. Finally, the requested information is then sent back to the ACS and then, through it, to the user.

In this section, we focus in particular on the key choices underlying the architecture, concerning in particular the definition of an event-driven approach to the specification of the access control mechanism.

The Access Control Mechanism

The access control mechanism applies the access control function to determine the set of enabled roles and thus the services which are accessible to the user at a given location. For the computation of the enabled roles,we need to define at what stage these roles are determined. We devise two possible strategies:

- The status of roles is computed exclusively upon user request. When the user sends a request to the ACS, the system determines which roles are enabled and then based on this information determines whether the request is accepted or rejected. The approach is thus *user-driven*.

- The status of roles is autonomously checked by an agent tracking the position of the user connected to the LBS system. As a state transition occurs for a role, that is, its status changes from enabled to disabled or vice versa, the new status is recorded and the event is notified to interested system components. The strategy is thus *event-driven*.

The simplest approach is the user-driven one because the position of the user is computed on demand and thus the status of roles is determined only when required. This approach has, however, a major drawback, in that the services which are available in a given session at a given instant are not known until the user makes an explicit access request. Therefore, it may occur that the user requires a service which is not accessible in that position or that the user is not aware of available services at a given position.

Conversely, when the event-driven strategy is adopted, the status of each role is determined asynchronously with respect to user requests. Therefore, such information needs to be recorded by the ACS. Although more complex, this approach overcomes the drawbacks of the user-driven strategy: user requests can be more efficiently processed because the current status of roles is available at the time of the request and thus need not they be computed; the users can determine the effective roles can play, before a request is made. Because it is arguably more flexible, the event-driven approach is the one we adopt.

Event-driven Architecture of the ACS

From an architectural point of view, the proposed organization for the ACS is based on the following major components:

- The *Policy DB* is a database storing the security policies specifying, among other information, the spatial roles, the services available to each role, and the roles assigned to each user.
- The *Session DB* records the *status* of sessions. The status at time t of session s is represented by the tuple $< s, t, SR, ER >$, where SR is the set of session roles, thus the roles selected by the user among those which have been assigned to her; ER is the set of roles enabled in s at time t.
- An agent called *Role Tracker*. It periodically retrieves from the Location Server the position of the users of current sessions and determines whether a state transition has occurred for the roles of each session. If this is the case, the event is communicated to the Event Manager.
- In response to a Role Tracker event, the *Event Manager* updates the status of sessions in *Session DB* and notifies the event to the corresponding terminal.

This architecture focuses on the operational meaning of access control enforcement. A number of issues however remain open. Among these, a challenging issue is related to the fact that LBS may be not only of *pull* type, like for example, directory services (e.g. Where is the closest restaurant), but also of *push* type. Under the push model, services are still requested by the user (i.e. subscribed), but the information is provided on a continuous basis. An example is the service which allows one to be automatically notified about nearby traffic jams. Because in a location-aware context the availability of the service depends on the position of the user, if the user changes position in time, it may occur that the user is no longer in a position which authorizes him/her to access the service. To our knowledge, this issue has not been addressed yet.

9.5.2 The Challenge of a Distributed Architecture for Location-based Applications

The above centralized architecture has two major drawbacks. First, the architecture is not scalable. Suppose that an organization introduces an application in which LBSs are provided by several applications servers. In such a case, a centralized policy enforcement creates a bottleneck since all requests for all application servers should be first sent to the ACS and eventually dispatched to the recipients. Similarly, the administrative operations, such as the addition or removal of a service by a service provider, should be managed by the central security administrator. The second drawback is that it does not address the authentication issue. On the other hand, as already remarked, relying on user login names for grant and revoke authorization is cumbersome and it is also a low level approach.

To overcome the limits of the centralized approach, we propose an access control framework based on a decentralized architecture. Consider the following scenario: suppose that, in our campus, services are provided by various service providers through a number of application servers $AS_1, ..., AS_n$. Moreover, roles are assigned permissions to request services from different providers. Therefore, a many-to-many relationship exists between the set of application servers and the set of roles: an individual, because of his role, can access multiple application servers, and vice versa an application server can be accessed by individuals playing different roles. To enable an efficient access to services and at the same time allow a simple interaction with the application, we propose an architectural framework based on two key design choices: (a) the access control is distributed among a set of autonomous ACS, $ACS_1, .., ACS_n$ where ACS_j with $j \in \{1, ..., n\}$ controls the access to application server AS_j. The access to each application server is then governed by a *local policy* administered by a *local security administrator*. Therefore, there are n local policies and n local security administrators. (b) The enforcement of the local policy is initiated on the client side represented by the mobile terminal and then completed on the server side.

We now describe in more detail this approach focusing in particular on the problem of user-role authentication and policy enforcement.

User–Role Authentication

We introduce the problem of user–role authentication through an example. Suppose that John is assigned the role *student*. Then when John presents such a role to the

ACS, the system is expected to grant access to the services assigned to students. Now the question which arises is how can the system trust John. In a centralized architecture, the answer is simple since the ACS maintains information on users and on all pairs user–role (user–role assignment relation). Therefore when John is first authenticated and then presents his role, the ACS verifies whether the binding < *John, student* > is specified in the user–role assignment relation. In a distributed context, however, this approach is not flexible enough: when a new user is entered into or removed from the system or a new role is assigned or deassigned to a user, such information must be replicated in every local policy. Such operation results not only are tedious and complex to manage but can also lead to inconsistent polices.

We adopt thus a different approach. The idea is that the user first obtains a digital certificate which specifies the user's role; then this certificate is presented to the ACS to certify the role assignment for the user and to subsequently grant the access to the requested service. The advantage of this approach is that the user-role assignment relation does not need to be specified at the level of local policy.

Digital certificates contain a number of attributes, such as the name, a serial number, expiration dates, and the digital signature of the certificate-issuing authority so that a recipient can verify that the certificate is real. Moreover, certificates hold spatial role information (*role certificates*). As far as we know, the idea of certificates containing spatial information has been not explored yet. The certification authority in charge of issuing role certificates is called *role provider*. User–role authentication is then performed as follows: the user downloads certificates from the role provider, one for each role, and stores them on his mobile terminal. To invoke a service, the user first selects the certificate corresponding to the role he wants to play in the interaction and then transmits it to the application server along with the service requests. In such a way, the user is authenticated along with his role.

Decentralized Policy Enforcement

Suppose now that a user, after being authenticated, requests the permission of invoking a service. Following the GEO-RBAC model, such permission is granted only if the user is located within the extent of a spatial role authorized to request that service. The question we now address is how to assess whether a relationship of spatial containment holds between the user position and the role extent. A first solution is to apply the same approach adopted in the centralized architecture, that is, the ACS, which receives the user's request, obtains the position of the user from the location server and then, based on role information, determines whether the containment constraint is satisfied. The drawback of this approach is that in large applications, it introduces a significant communication burden. We thus opt for a different solution in which policy enforcement is initiated on clients. The idea is as follows: since clients have knowledge of the roles assigned to users because of the role certificates stored on terminals, clients are potentially able to reject requests which cannot be satisfied and thus avoid sending useless requests to the servers. For this approach to be viable, however, clients must be aware of the position of users and then be able to match such position against role extents. Clients can obtain the position from the Location Server or from the positioning device eventually installed on terminal. In any case, since clients may be not trusted, it is important to ensure the integrity of position data. To that purpose, we assume position data to be provided along with a digital signature by a Secure Location Server.

Access control operations on clients are then carried out by a software component called *local access control* (LAC). Based on role certificates and user's position, the LAC determines the roles, which are enabled, and then, if there is at least an enabled role, it transmits to the application server an encrypted access request that includes the role certificate and the position. Note that the application server trusts the position sent by the client because the position has a digital signature affixed. Therefore, a malicious client cannot forge such information.

Now we consider what happens on the server side when a request is sent by LAC. The request is received by one of the ACS defined in the framework, say ACS_j. Based on the role certificate and position data sent by the client, ACS_j verifies whether the role is enabled. Notice that this operation is performed twice, first at the client and then at the server side. The reason is that the client may not be trusted, and thus, the server needs to make sure that constraints are fulfilled, with the advantage that in the case of trusted clients, unnecessary request processing at the server is avoided. Finally, the system checks whether the requested service is one of the services, which have been assigned to the specified role based on the local security policy. If this is the case, the access is permitted.

The decentralized architecture comprehensive of the LAC layer, the Role Provider, the Secure Location Server, and the set $\{ACS_1, .., ACS_n\}$ of ACS is shown in Fig. 9.4.

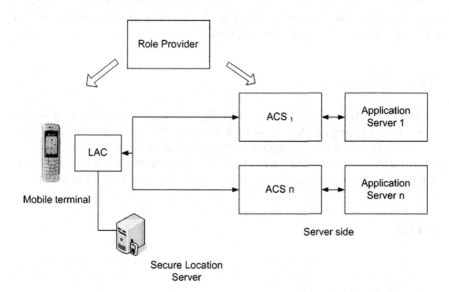

Fig. 9.4. Decentralized architecture

Concluding Remarks

It may have been noticed that in discussing the two architectural approaches, the centralized and the distributed one, we have focused on different aspects. In the centralized solution, the major concern is for the strategy to apply for the enforcement of the security policy which can be user-driven or event-driven. In the decentralized solution, the focus is on distributing the administrative and enforcement functionalities. The question which has not been addressed yet is whether in the decentralized context the enforcement strategy is user-driven or event-driven. This problem deserves a careful analysis, especially if we want to account for services that are not only of pull but also of push type. This issue will be addressed as part of our future work.

9.6 Conclusions

In this chapter, we have discussed issues related to access control in location-based applications. Such applications are characterized by a large variety of requirements affecting both the conceptual definition of access control models and the architectures of ACS. In this chapter, we have shown a rich model, GEO-RBAC, specifically tailored to geospatial applications with mobile users and then illustrated possible architectures supporting such model.

Several issues, however, are still to be addressed concerning the architectural aspects. In particular, the integration of the privacy dimension into access control, following the guidelines proposed in [11], is an important issue, given the existing privacy regulations and the increased privacy concerns by citizens and organizations. The mapping of GEO-RBAC onto an existing architectural framework, X-GTRBAC, is also crucial to obtain an ACS supporting the specification and enforcement of a rich set of context-based access control policies.

Acknowledgments

The work of M.L. Damiani has been partially supported by the European project GEOPKDD "Geographic Privacy-aware Knowledge Discovery and Delivery." The work of E. Bertino has been supported by USA NSF under the project "A Comprehensive Policy-Driven Framework for Online Privacy Protection: Integrating IT, Human, Legal and Economic Perspectives" and by the sponsors of CERIAS.

References

1. Atluri V, Mazzoleni P (2002) A Uniform Indexing Scheme for Geospatial Data and Authorizations. In: Proc. 6th Conf., on Data and Application Security, IFIP TC11/WG11.3, Cambridge, UK, 207–218

2. Belussi A, Bertino E, Catania B, Damiani M L, Nucita (2004) An Authorization Model for Geographical Maps. In: Proc. 12th Int., Symp. of ACM GIS, Washington DC, USA, 82–91
3. Bertino E, Ferrari E, Perego A (2002) MaX : An Access Control System for Digital Libraries and the Web. In: Proc. 26th Computer Software and Application Conference, Oxford, UK, 945–950
4. Bertino E, Catania C, Damiani ML, Perlasca P (2005) GEO-RBAC: A Spatially Aware RBAC. In: Proc. 10th ACM Symposium on Access Control Models and Technologies (SACMAT'05), Stockholm, Sweden, 29–37
5. Bertino E, Damiani ML, Momini D (2003) An Access Control System for a Web Map Management Service. In: Proc. 14th International Workshop on Research Issues in Data Engineering (RIDE-WS-ECEG), Boston, USA, 33–39
6. Bertino E, Sandhu R (2005) Database Security-Concepts, Approaches, and Challenges. IEEE Transactions on Dependable and Secure Computing 2(1):2-19
7. Bhatti R, Ghafoor A, Bertino E, Joshi JBD (2005) X-GTRBAC: an XML-Based Policy Specification Framework and Architecture for Enterprise-wide Access Control. ACM Transactions on Information and System Security 8(2):187–227.
8. Bishop M (2005) Introduction to Computer Security. Addison-Wesley
9. Chandran S M, Joshi JBD (2005) LoT RBAC: A Location and Time-based RBAC Model. In: Proc. 6th International Conference on Web Information Systems Engineering (WISE'05), New York, USA, 361–375.
10. Covington M, Long W, Srinivasan S, Dev AK, Ahamad M, Abowd GD (2001) Securing Context-aware Applications Using Environment Roles. In: Proc. 6th ACM Symposium on Access Control Models and Technologies (SACMAT'01), Chantilly, USA, 10–20
11. Damiani ML, Bertino E (2006) Access Control and Privacy in Location-aware Services for Mobile Organizations. In: Proc. of the 7th International Conference on Mobile Data Management, Nara, Japan
12. Damiani ML, Bertino E, Perlasca P (2005) Data Security in Location-Aware Applications: an Approach Based on RBAC. International Journal of Information and Computer Security (IJICS), in press
13. Ferraiolo D, Sandhu R, Gavrila S, Kuhn R, Chandramouli R (2001) Proposed NIST Standard for Role-Based Access Control. ACM Transactions on Information and System Security 4(2):224–274
14. Ferrari E, Adam NR, Atluri V, Bertino E, Capuozzo U (2002) An Authorization System for Digital Libraries. VLDB Journal 11(1):58–67
15. Hansen F Oleshchuk V (2003) Spatial Role-based Access Control Model for Wireless Networks. In: Proc. IEEE Vehicular Technology Conference VTC2003-Fall, Orlando, FL, USA, 2093–2097
16. ISO/TC211 (2003) 19107: Geographic information - Spatial schema
17. ISO/TC211 (2004) 19136: Geographic information - Geography Markup Language
18. Joshi JBD, Bertino E, Latif U, Ghafoor A (2005) A Generalized Temporal Role-Based Access Control Model. IEEE Transactions on Knowledge and Data Engineering, 17(1):4–23
19. Matheus A (2005) Declaration and Enforcement of Fine-grained Access Restrictions for a Service-based Geospatial Data Infrastructure. In: Proc. 10th ACM Symposium on Access Control Models and Technologies (SACMAT'05), Stockholm, Sweden, 21–28
20. OASIS SAML (2006) http://xml.coverpages.org/saml.html
21. Sandhu R, Ferraiolo D, Kuhn R (2000) The NIST Model for Role-Based Access Control: Towards a Unified Standard. In: Proc. 5th ACM Workshop on Role-Based Access Control, Berlin, Germany, 47–63

22. Sandhu R, Samarati P (1994) Access control: Principles and Practice. IEEE Communications, 32(9):40–48
23. Sandhu R, Coyne EJ, Feinstein HL, Youman CE (1996) Role-Based Access Control Models. IEEE Computer 29(2):38–47
24. Thuraisingham B (1990) Multilevel Security for Multimedia Database Systems. In: Proc. IFIP WG 11.3 Workshop on Database Security (DBSec 1990), Halifax, UK, 99–116

References containing URLs are validated as of October 1st, 2006.

Secure Outsourcing of Geographical Data Over the Web: Techniques and Architectures

Barbara Carminati and Elena Ferrari

University of Insubria, Varese (Italy)

10.1 Introduction

Service outsourcing is today a widely-used paradigm by many companies and organizations. In the same vein, in recent years a new trend has emerged, that is, *data outsourcing* [16]. Data outsourcing means moving from a traditional client–server architecture (see Fig. 10.1(a)), where the data owner directly manages the DBMS and answers user queries, to a third-party architecture (see Fig. 10.1(b)), where data owners are no longer totally responsible for data management. Rather they outsource their data (or portions of them) to one or more *publishers* that provide data management services and query processing functionalities. This paradigm has potentially many benefits. The first is the cost reduction for the owner, in that it pays only for services it uses from publishers and not for the deployment, installation, maintenance, and upgrades of costly DBMSs. Another important benefit is scalability in that the data owner could not become a bottleneck for the system since it can outsource its data to as many publishers as it needs according to the amount of data and the number of managed users.

Data outsourcing has many interesting applications in the geographical data domain. For instance, think about a geomarketing service. A data owner can outsource some of its geographical data (e.g. maps at various levels of detail) to a publisher that provides them to customers on the basis of different registration fees or different confidentiality requirements (e.g. maps of some regions cannot be distributed to everyone because they might show sensitive objectives).

However, all the benefits of data outsourcing in terms of cost reduction and better services are not enough to make data outsourcing widely adopted. One of the most serious obstacle to the widespread use of data outsourcing is related to security. If security is not considered as a primary requirement, data outsourcing can be perceived by the owner as "loss of control" over its data. The challenge is therefore how to ensure the most important security properties (i.e. confidentiality, integrity, authenticity) even if data are managed by a third-party. A naive solution is to assume the publisher is to be *trusted*, that is, to assume it always operates according to the owner's security policies. However, making this assumption is not realistic, especially for

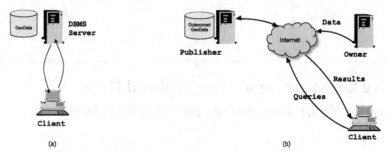

Fig. 10.1. (**a**) Two-party architecture; (**b**) third-party architecture

Web-based systems that can easily be attacked and penetrated. Additionally, verifying that a publisher is trusted is a very costly operation. Therefore, research is now focusing on the definition of techniques to satisfy main security properties even in the presence of an untrusted publisher that can maliciously modify/delete the data it manages, for instance by inserting fake records, or it can send data to unauthorized users. Several proposals can be found in the literature [13–17, 19, 22, 26], each of which focuses on a single security property and on data organized according to a specific data model (e.g. relational, XML). Such techniques, which we survey in Sect. 10.3.1, make use of encryption, to enforce confidentiality requirements and of non-standard digital signature techniques to enforce authenticity/integrity. What is still missing in these proposals is a comprehensive framework able to enforce all the above-mentioned security properties in a unified manner. For this reason, in the second part of the chapter we present our proposal for a framework providing a comprehensive solution to the problem of secure outsourcing of XML data, and we show how this framework can be successfully used in the GIS domain.

The remainder of this chapter is organized as follows. Section 10.2 discusses the requirements for secure outsourcing. Section 10.3 surveys the most important techniques that have been proposed so far for secure data outsourcing. Section 10.4 presents a comprehensive framework for the secure outsourcing of XML data, whereas Sect. 10.5 shows how this framework can be applied to geographical data. Finally, Sect. 10.6 concludes the chapter and outlines future research directions.

10.2 Security Requirements

Information security is one of the main assets of each organization, from the smallest company to international corporations. To have an idea of its key role in decision-making processes, let us consider the e-business world. Today, a company that wants to move to e-commerce has, as primary requirement, to provide a secure system. Indeed, no matter whether you are the best in your business or how cheap your prices are, today e-commerce customers do not take into account only the quality of services, rather they want to be assured also about a company's behavior with respect to security requirements. Let us now consider the outsourcing scenario. Here, we have companies/organizations/enterprises delegating their data to external entities. They basically provide another organization the core of their business, that is their data. The basic questions about security are as follows: How can a company be assured that its data are not maliciously modified by the third entity before being distributed to requesting subjects? How can a company be assured that the third entity does not provide company data to unauthorized users, including, for instance, competitors? From the user-side point of view the main security-related questions are as follows: How can a user be sure that the third-party does not maliciously modify/delete data it is authorized to access? How can the user be sure he/she has received all data it is authorized to access? Thus, it is obvious that in order to make the outsourcing paradigm well accepted and widespread, it is necessary to fulfill the main security requirements of both data owners and end users. A first step toward this is to point out the main security requirements that need to be addressed for secure outsourcing. In what follows we introduce and motivate some of them.

Confidentiality. In general, ensuring data confidentiality means that the data (or portions of it) can be disclosed only to users authorized according to access control policies stated by data owners (see Chap. 9 for issues related to access control for geographical data). However, it is obvious that when data are outsourced, confidentiality requirements are not limited to users. Indeed, in outsource-based architectures it is possible to point out two main confidentiality issues. The first, hereafter called *confidentiality with respect to users*, refers to protect data against unauthorized read operations by users. In general, confidentiality requirements with respect to users are modeled through a set of access control policies stating who can access what portions of the owner's data. In traditional client–server architectures, confidentiality with respect to users is usually enforced by means of access control mechanisms (i.e. reference monitors), which mediate each user access request by authorizing only those in accordance with the owner's access control policies. The key component is therefore the reference monitor, that is the software module in charge of enforcing access control policies. A fundamental requirement is therefore the presence of a trusted environment hosting the reference monitor. In traditional scenarios, this environment is provided by the entity managing the data, that is the owner's DBMS server. Enforcing access control in a third-party scenario would imply the delegation of the reference monitor tasks to publishers. However, since we are considering a scenario where assumptions on publisher trustworthiness cannot be made, such a solution is no longer applicable and alternative solutions should be devised (we survey some of them in Sect. 10.3.1).

Furthermore, in a scenario where data are outsourced to external publishers, confidentiality issues are also related to publishers. There thus exists a second relevant confidentiality requirement, called *confidentiality with respect to publishers*, which deals with protecting owner's data from read operations by publishers. Indeed, since publishers may not be trusted, it is necessary to devise some mechanisms to avoid their malicious use of owner's data.

Authenticity and integrity. In general, ensuring data authenticity implies that users are assured that the received data come from the intended source, whereas ensuring data integrity means ensuring that data contents are not altered during their transmission from the source to the intended recipients.[1] Enforcing authenticity/integrity requirements in outsource-based architectures implies assuring that a user receiving some data from a publisher is able to verify that the data have been generated by the owner and that the publisher has not modified their contents. In traditional architectures, a well-established approach to ensure both authenticity and integrity is based on the use of digital signatures [27]. According to this solution, before sending data to users the owner digitally signs it. Applying such a solution to third-party architectures implies that owners sign each document they send to publishers. However, this solution is not suitable for a third-party architecture in that a publisher may return to users only selected portions of a document, depending on the query they submit and/or on the access control policies in place. Thus, the user by having only selected portions is not able to validate the owner's digital signature, which has been generated for the whole document. For this reason, there is the need to investigate alternative ways to digitally sign a document, defined in such a way that they enable a user to validate the digital signature even if he/she has only received selected portions of the signed document.

Completeness. Third-party architectures introduce a further novel security requirement, that is, completeness. Since answers to user's queries are generated by possible untrusted publishers, a further important requirement is that a user receiving an answer must be able to verify the completeness of the received data, that is, the user must be able to verify that he/she receives all data answering the submitted query and satisfying owner's access control policies.

10.3 State of the Art

The most intuitive solution for secure outsourcing is to require publishers to be trusted with respect to the considered security requirements. However, as we discussed in the previous section, this solution may not always be feasible in the Web environment, since large Web-based systems cannot easily be verified to be secure and can easily be penetrated. For this reason, there is a strong need to devise strategies to ensure security properties in third-party architectures, even in the presence of an untrusted publisher. These issues have been investigated by different research groups, with several resulting proposals for the different security requirements. In this section, we survey the main results proposed so far, by grouping them according to the security requirements they address and the data model for which they are proposed. Table 10.1 summarizes, for each security requirement, the solutions discussed.

---

[1] Here we do not consider integrity with respect to the specified access control policies, since third-party architectures are mainly conceived for read accesses.

10.3.1 Confidentiality

The main efforts carried out to address confidentiality in third-party architectures have been focused on confidentiality with respect to publishers [15–17, 26] only. What is interesting to note is that all the proposals enforce confidentiality by means of cryptographic-based solutions. More precisely, the common underlying idea of all these approaches is that data owners outsource to publishers an encrypted version of the data they are entitled to manage, without providing the corresponding decryption keys. Thus, the publisher is not able to access and, as a consequence, to misuse outsourced data. Obviously, applying such a solution to a third-party scenario implies considering an interesting problem; that is how to make publishers able to evaluate queries on encrypted data, without accessing them. In recent years, this problem has been investigated in depth by several researchers, with two main resulting solutions [17, 26] that work for different data models. In what follows, we briefly introduce the method proposed in [15–17] to query encrypted relational databases, and the cryptographic scheme presented in [26] to query encrypted textual data.

Hagicumus et al. The approach proposed by Hacigumus et al. [15–17] exploits binning techniques and privacy homomorphic encryption to query encrypted relational data. More precisely, binning techniques are used to perform selection queries on encrypted relation data, whereas homomorphic encryption [25] is used to enable a third-party to perform aggregate queries over encrypted tuples. Let us start to describe how, by partitioning relation data domains, Hacigumus et al.'s approach enables third parties to evaluate selection queries. The basic idea is that for each relation R the owner divides the domain of each attribute in R into distinguished partitions, to which it assigns a different id. Then, for each tuple t in R, the owner sends the publisher the corresponding encrypted tuple t', together with the ids of the partitions to which t's attribute values belong. The publisher is able to evaluate queries by exploiting the received partition ids, without accessing the encrypted tuples. In order to do that, a user, before submitting a query to a publisher, rewrites it in terms of partition ids. Let us consider, for instance, the relation $Dept(dname, Nemployees, address)$

Table 10.1. Security requirements and possible approaches

security properties	techniques	data	approaches
authenticity/ integrity	merkle hash trees	relational data	Devanbu et al. [13]
		XML data	Bertino et al. [5]
	aggregate signatures	relational data	Mykletun et al. [19]
confidentiality	data encryption	relational data	Hacigumus et al. [15–17]
		textual data	Song et al. [26]
completeness	merkle hash trees	relational data	Devanbu et al. [13]
		XML data	Bertino et al. [5]
	aggregate signatures	relational data	Nara et al. [22]

and, for simplicity, consider only the *Nemployees* attribute representing the number of employees belonging to the department. Suppose that the domain of *Nemployees* is in the interval [1, 300], and that an equi-partition with 10 as range is applied on that domain. Moreover, suppose that a user wants to perform the following query: "SELECT * FROM Dept WHERE Nemployees =35." This query needs to be translated by the user into "SELECT * FROM Dept WHERE *Nemployees* =id_{35}," where id_{35} is the id of the partition containing the value 35.[2] Thus, the publisher is able to evaluate this query by simply looking for those tuples having the partition id associated with attribute *Nemployees* equal to id_{35}. However, we have to note that the publisher returns to the user an approximate result, in that it returns all the tuples of the *Dept* relation whose *Nemployees* attribute belongs to the range [30, 39]. Thus, further query processing has to be performed by the user to refine the received answer.

Moreover, in [17] the proposed encryption scheme is extended by enhancing it with privacy homomorphisms (PH) to enable the third-party to evaluate aggregate functions. This class of encryption functions, introduced by Rivest et al. in [25], provides the capability to calculate arithmetic operations directly on encrypted data. Indeed, the PH functions have the property that if $E(X)$ op $E(Y) = E(Z)$, then $D(Z) = X$ op Y, where $E()$ and $D()$ are the encryption and decryption function, respectively. In [17] this property is exploited to evaluate aggregate queries on encrypted tuples. More precisely, given a relation R, the proposed strategy requires to encrypt with a PH function each attribute of R, where it is expected to do some aggregations, called *aggregation attributes*. Then, the owner outsources together with the encrypted tuples, and the corresponding partition ids, also the encrypted aggregation attributes. In [17] it is shown that, by having the PH of the aggregation attributes, the third-party is able to evaluate aggregate functions over them. To better clarify how PH functions are exploited, let us consider the following query: "SELECT SUM(Nemployees) FROM Dept" which returns the total number of employees. Let us assume that the data owner has encrypted *Nemployees* with the PH function and that the publisher has been provided with PH($t.Nemployees$), for each tuple t outsourced by the owner. To calculate the SUM(), the publisher has to select all *Nemployees* attributes[3] and simply calculate the total sum. By the properties of PH functions, when a user decrypts this sum he/she will obtain the effective number of employees.

Song et al. Another relevant study carried out in the context of searching encrypted data has been done by Song et al. [26]. In particular, in [26] the authors propose an interesting cryptographic scheme that supports searching functionalities on encrypted textual data without loss of data confidentiality. The basic idea of the scheme is the following. Given a set of words, the proposed scheme first encrypts each word using a symmetric encryption algorithm with a single secret key k. Then, it generates the XOR of each encrypted word with a pseudorandom number. The resulting ciphered

---

[2] Clearly, users should receive the owner information on the techniques used to partition data and generate ids.

[3] If a WHERE condition was applied, publisher could evaluate it by means of partition ids.

words can then be outsourced to the third-party. According to this scheme, when a user needs to search for a keyword W, it generates the encrypted word $E_k(W)$ and computes $E_k(W) \oplus S$, where S is the corresponding pseudorandom number. This simple scheme allows the third-party to search for keyword W in the encrypted data, by simply looking for $E(W) \oplus S$, thus without gaining any information on the clear text. Since occurrences of the same word are combined using the exclusive OR operator with different pseudorandom numbers, by analyzing the distribution of the encrypted words, no information could be inferred regarding the clear text. According to the above-introduced basic scheme, to formulate a query, users need to know information about the pseudorandom numbers. Indeed, the scheme proposed in [26] is defined in such a way that users are able to locally compute pseudorandom numbers without any interaction with the data owner. Let us see in more detail how the scheme works. In what follows, we briefly summarize the encryption and query evaluation process proposed by Song et al., and also we refer the readers interested in the decryption process to [26]. The scheme exploits a symmetric encryption function $E()$ and two pseudorandom number generator functions, namely F and f.[4] Given as input a set of clear-text words $W_1, W_2, \ldots, W_l$, with the same length n,[5] the encryption process implies the following steps:

- Data owner generates a sequence of pseudorandom values $S_1 \ldots S_l$ of length $n-m$; parameter m can be properly adjusted to minimize the number of erroneous answers due to collision of pseudorandom number generators $F()$ and $f()$.
- For each word W_j, the outsourced ciphered word C_j is generated according to the following formula: $C_j = E_k(W_j) \oplus < S_j, F_{K_j}(S_j) >$, where $K_j = f_k(FB_j)$ and FB_j denotes the first $n - m$ bits of $E_k(W_j)$.

Let us see now how by having only a set of ciphered words, the third-party is able to search an encrypted word. When a user needs to search for a keyword W_i, he/she sends the third-party $E_k(W_i)$ and key K_i, which can be locally computed by the user. Then, for each outsourced ciphered word C_i, the third-party (1) calculates $C_i \oplus E_k(W_i)$; (2) takes the first $n - m$ bits bts of the resulting value, and computes $F_{K_i}(bts)$, where K_i is the key received by the user; (3) if the result of $F_{K_i}(bts)$ is equal to the $n - m + 1$ remaining bits, then the ciphered word C_i is returned. Indeed, if C_i contains the searched encrypted word, then $E_k(W_i) \oplus < S_i, F_{K_i}(S_i) > \oplus E_k(W_i) < S_i, F_{K_i}(S_i) >$, for the properties of the XOR operator.

10.3.2 Authenticity and Integrity

Authenticity and integrity are usually enforced through the use of digital signatures [27]—one of the most widely used techniques relying on asymmetric encryption.

---

[4] In what follows, we denote by $E_k(x)$ ($F_k(x)$, $f_k(x)$, respectively) the result of applying E (F, f, respectively) to input x with key k.

[5] This set of words can be obtained by partitioning the input clear-text into atomic quantities (on the basis of the application domain) and by padding and splitting the shortest and longest words.

Asymmetric encryption relies on the definition of a pair of keys for each user, a public and a private key. These keys are defined in such a way that a message encrypted with the public key can be decrypted only by using the private key, and vice versa. Public keys are available to anyone who needs them, whereas private keys should be kept secret. The sender computes the *digest* of the data being signed, that is, numerical summary or *fingerprint* computed using a one-way hash function. By the properties of one-way hash functions, it is infeasible to change a message digest back into the original data from which it was created. Then, the digest is encrypted with the private key. The receiver first decrypts the signature using the sender's public key, changing it back into a digest. The receiver then computes the message digest of the received data and compares it with the decrypted one. If the two coincide, then the receiver knows that the signed data has not been altered during transmission and that data has been generated by the sender, because only the sender has the corresponding private key. As pointed out in Sect. 10.2, such scheme is not suitable for third-party architectures. This is mainly due to the fact that users must be able to validate the owner's signature even if they receive a selected portion of the signed data. A naive solution to overcome this problem, but still exploiting traditional signature schemes, is to force the owner to separately sign each possible portion of data. This set of digital signatures could be outsourced together with the data, and could be properly returned to the users by the publisher. However, this solution implies an enormous overhead both in owner computation and in query answer size. For this reason, in recent years, several alternative strategies have been presented. We can group these strategies on the basis of the underlying adopted techniques. In particular, there are two main exploited techniques, that is, Merkle tree authentication [20] and signature aggregation [6]. In what follows we present them by introducing some of the related proposals.

Merkle Trees. In [20], Merkle proposed a method to sign multiple messages, by producing a unique digital signature. The method exploits a binary hash tree generated by means of the following bottom-up recursive construction: at the beginning, for each different message m, a different leaf containing the hash value of m is inserted in the tree; then, for each internal node, the value associated with it is equal to $H(h_l \| h_r)$, where $h_l \| h_r$ denotes the concatenation of the hash values corresponding to the left and right children nodes (see Fig. 10.2), and $H()$ is an hash function.

The root node of the resulting binary hash tree can be considered as the digest of all messages, and thus it can be digitally signed by using a standard signature technique and distributed. The main benefit of this method is that a user is able to validate the signature by having a subset of messages, providing him/her with a set of additional hash values. Indeed, a user, by having hash values of the missing messages, is able to locally build up the binary hash tree and thus to validate the signature.

Merkle hash trees have been used in several computer areas for certified query processing. For instance, they have been exploited by Naor and Nissim in [21] to create and maintain efficient authenticated data structures holding information about certificates validity. More precisely, [21] proposes a sorted hash tree as data

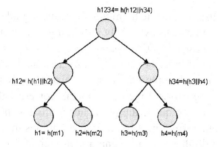

Fig. 10.2. An example of Merkle hash tree

structure, built in such a way that the tree leaves correspond to revoked certificates. Thus, verifying whether a certificate is revoked or not is equivalent to verify the existence of certain leaves in the tree. Similar schemes have also been used for micropayments [10], where Merkle hash trees are used to minimize the number of public key signatures that are required in issuing or authenticating a sequence of certificates.

Merkle hash trees have also been exploited for data outsourcing. For instance, in [13] Devanbu et al. adapt Merkle hash trees to the relational data model to prove the completeness and authenticity of query answers. In particular, in Devanbu et al.'s approach for each relation R, a different Merkle hash tree is generated, in such a way that leaf nodes represent hash values of tuples. Let us see how this enables a user to verify answer authenticity, whereas we postpone the discussion on completeness verification to Sect. 10.3.3. When a user submits a query on relation R, the publisher replies him/her with the tuples answering the submitted query and the signature generated by exploiting the Merkle hash tree generated on R, plus the hash values of the tuples of R not included in the result set. By having these additional hash values, the user is able to validate the signature of R, and thus to prove authenticity of the received tuples.

Merkle hash trees have also been investigated for third-party distribution of XML data [5, 14]. Here the challenge is how the XML hierarchical data model can be exploited in the construction of Merkle trees. A brief introduction of Merkle tree application to XML documents is given in Sect. 10.4, and also we refer interested readers to [5] for an in-depth presentation.

Signature Aggregation Schemes. Another technique recently exploited to ensure authenticity in third-party architectures is based on signature aggregation. In general, signature aggregation schemes allow one to aggregate into a unique digital signature n distinct signatures generated by n distinct data owners [6]. The validation of this unique digital signature implies the validation of each component signature. Aggregate signature schemes have also been investigated to aggregate into a unique signature n signatures generated by the same owner. This last kind of aggregation scheme can be adopted to ensure authenticity in third-party scenarios. Indeed, according to this solution, a data owner could generate a distinct signature for each distinct message[6] and then aggregate them into a unique digital signature. By having only the aggregate signature and by simply validating it, a user is able to authenticate selected messages received by third-party. Let us consider, for instance, how this solution could be applied to relational data. Let us assume that the owner generates a different signature for each tuple belonging to a relation R and then outsources to publishers all these signatures together with the corresponding tuples. When a user submits a query on relation R, the publisher evaluates the query on R. Then, it aggregates all the signatures corresponding to the tuples in the result set, and returns to the user the resulting aggregate signature, as well as the tuples answering the query. By the properties of aggregate signatures, the user, by simply verifying the aggregate signature generated by publisher, is able to validate all tuples' signatures generated by the owner, and thus to prove tuples' authenticity. Such a solution has been proposed by Mykletun et al. in [19] for relational data, where two different aggregate signature schemes, namely condensed-RSA and Boneh et al. [6], have been compared.

10.3.3 Completeness

Completeness is a novel property that is receiving growing attention due to its relevance in data outsourcing. So far completeness property has been investigated by few research groups, but three strategies have been proposed to ensure completeness of query answers in third-party architectures. In particular, two of them have been proposed for relational data [13, 22], whereas the third is for third-party distribution of XML data [5]. In what follows we introduce approaches for relational data, and we postpone completeness for XML data to Sect. 10.4. Completeness for relational data has been investigated in conjunction with authenticity and integrity requirements. The proposed solutions exploit Merkle hash tree mechanisms and aggregate signature schemes. However, it is interesting to note that despite the different exploited mechanisms they obtain a similar result, in that both the approaches presented in [13] and in [22] enable a user to prove answer completeness of range queries only. Let us introduce how this is possible by means of Merkle hash trees. The approach presented in [13] assumes that Merkle hash trees are built over tuples sorted according to a given attribute a and that the range query's predicate is against a. To enable the user to verify the answer's completeness, the third-party inserts in the answer two additional tuples, that is, a tuple preceding (subsequent) the lower (upper) bound of the result set answering the range query. By having these two values, the user can locally verify that the query answer is complete, that is, that no tuples answering the range query have been omitted. Indeed, by verifying the owner's signature, which has been generated on a Merkle hash tree built over sorted tuples, the user also verifies thatthe received lower (upper) tuple effectively precedes (follows) the ones in

---

[6] Here message has a general meaning. On the basis of the scenario, a message could be a relational tuple, a node (i.e. element or attribute) of an XML document, or more generally a data portion.

the result set. Thus, the user is ensured that the query answer is complete, that is, no tuples are missing from the result set.

By contrast, the approach presented in [22] modifies an aggregation signature scheme to include in the signature of a tuple t the hash value of tuples preceding t according to all possible sortings defined on the relation's attributes. Then, the third-party inserts into the range query's answer the boundary tuples, that is, the tuples preceding the upper and lower bound of the result set, as well as their aggregated signatures. By means of the signature chain, a user is able to prove that the third-party has not omitted any tuple.

10.4 A Comprehensive Approach to Secure Data Outsourcing

In this section we present our proposal for secure outsourcing of data, whereas in Sect. 10.5 we show how it can be applied to the outsourcing of geographical data. The framework has been developed for the protection of XML data. A key feature of our proposal is that it does not rely on trusted publishers. The framework we present assures authenticity, confidentiality, and completeness requirements of both information owners and requesting users. This is obtained through the use of non-conventional signature and encryption strategies.

The framework (see Fig. 10.3) is a classical third-party architecture. Users submit queries[7] to publishers through a *client*, which the user can download from the owner site.

Security properties enforcement requires additional information to be transmitted by the owner to both publishers and users. All such information is coded in XML and is stored in a directory server managed by owners. Keys for accessing directory entries are received by users/publishers after a mandatory registration phase. In particular, all additional information needed by publishers for confidentiality and authenticity/integrity enforcement is attached to the document sent to publishers, forming the so-called *Security Enhanced ENCryption* (SE-ENC) of the original document. All SE-ENC documents are stored by the owner in the publishers' directory entry. Similarly, all information needed by a user to verify security properties is encoded by publishers in XML and attached to the query answer, resulting in what we call the *reply document*.

In the following, we illustrate how each security property is enforced (the techniques employed are summarized in Table 10.2). More details can be found in [5, 7, 9].

10.4.1 Authenticity and Integrity

Authenticity and integrity are enforced by applying Merkle hash trees to XML documents. According to this approach, we univocally associate a hash value to the document root (i.e the digest, referred to as *Merkle hash value*) through a recursive bottom-up computation on its structure. The digest is computed by associating a *Merkle hash value* with each node n of the XML document. The Merkle hash value

---

[7] Queries are formulated in XPath [28].

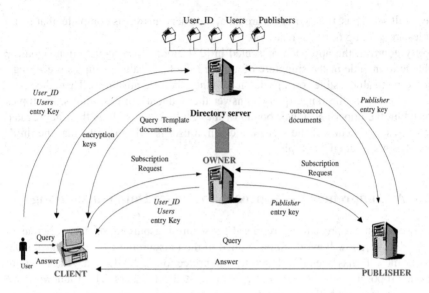

Fig. 10.3. Overall architecture

Table 10.2. Secure outsourcing: adopted techniques

security properties	techniques
authenticity/integrity	Merkle signatures, Merkle hash paths
Confidentiality:	
with respect to users	access control policies, selective encryption
with respect to publishers	encryption, encrypted queries
completeness	query templates

of a node depends on one of its children and attributes. The digest of the whole document is the Merkle hash value of the root of the document. The *Merkle Signature* is the encryption of the digest with owner's private key. The Merkle signature is inserted by the owner into the corresponding SE-ENC document by adding a `Sign` subelement to the document root. When a user submits a query, the publisher returns him/her, besides the query result, also the Merkle signatures of the documents on which the query is performed. However, this is not enough to enable a user to validate owner's signatures. What he/she needs are the Merkle hash values of the missing nodes, that is, those nodes of the original document which are not contained in the query answer. This is formalized by the notion of *Merkle hash paths*, that is, the minimum set of hash values needed by the requesting user to validate the received signature, starting from the received answer [5].

Since publishers do not operate on clear-text data (see Section 9.4.2 for more details), they cannot compute Merkle hash values. Therefore, the owner complements

SE-ENC documents with the Merkle hash value of each node. By having these hash values, the publisher is able to locally generate Merkle hash paths, without the need for accessing clear-text data.

10.4.2 Confidentiality

Confidentiality with respect to publishers is obtained through the use of symmetric encryption. The idea is quite straightforward: owners encrypt data before sending them to publishers. Publishers do not receive any decryption key and therefore owner's confidentiality requirements are enforced.

By contrast, enforcing confidentiality with respect to final users first requires the specification of a set of access control policies, stating which user can access which portions of owner's data. In our framework, these policies are specified through a credential-based access control model for XML documents proposed in [4]. To enforce confidentiality with respect to users, data managed by publishers are not encrypted with a single key, rather they are encrypted with different keys on the basis of owner's access control policies. All data portions to which the same policies apply (and therefore accessible by the same users) are encrypted with the same key. Then, each user receives from owners (in the corresponding entry of the directory server) all and only the keys corresponding to data portions he/she can access according to the specified policies.

The main weak point of this solution is that it may require the management of a great number of keys. To limit the number of keys that need to be generated, our system adopts a hierarchical key assignment scheme which requires one to permanently store, in the worst case, a number of keys linear in the number of specified access control policies. A hierarchical key assignment scheme [2] relies on the existence of a hierarchical organization (e.g. over data, security levels, roles, etc.) and it is defined in such a way that from the key associated with a level j in the hierarchy, it is possible to derive all and only the keys associated with a lower level i, where $i \leq j$, and $\leq$ is the partial order defined by the hierarchy. To limit the number of keys, we exploit the partial order that can be defined over possible access control policies configurations. Therefore, by the properties of hierarchical key management schemes, from the encryption key associated with an access control policy acp_j, we are able to derive all and only the encryption keys associated with access control policy configurations containing acp_j. These keys can be derived on the fly when needed and do not need to be permanently stored and maintained. We refer the interested reader to [3] for all the details on key generation.

Clearly, enforcing confidentiality through encryption requires enabling publishers to answer queries over encrypted data. For this purpose, we adopt an approach similar to the one proposed in [15] for relational databases (cf. Sect. 10.3.1), and we extend it to deal with XML data and data encryptions generated with multiple keys. The basic idea is to divide the domain of each document node (i.e. attribute and element) into distinguished partitions, to which a unique id is assigned. Then, the owner provides publishers the ids of the partitions corresponding to each node. The publisher is thus able to perform (approximate) queries directly on the encrypted documents, by exploiting the partitioning ids. Information on partition ids are contained in the SE-ENC document. To adapt this scheme to XML, we have, first of all, to deal with partitioning techniques for textual data, which play a key role in XML (e.g. element data content). Our approach to encrypt textual data contained in an XML document is based on two main steps. The first is *keyword extraction* and it is performed mainly to reduce the search space. Then, each keyword is considered as a distinct partition, whose id is obtained by applying a one-way hash function to the keyword itself. Properties of hash functions ensure that it is almost impossible to find two different keywords with the same id, and that it is computationally infeasible for the publisher to obtain the original keyword, given the corresponding id. The only drawback of this scheme is that the publisher could infer information on the clear- text by analyzing the distribution of hash values. Therefore, in [9] we have extended this method to be robust against data dictionary attacks.

To submit queries against encrypted data, a user must translate the query into an encrypted form which is understandable by publishers. This is done by the query translator module (see [7] for more details) that receives as input a user query and returns as output one or more ciphered queries to be submitted to publishers. Main steps done by the query translator are the completion of the user query (since the user may be allowed to see only a view of the requested document), the encryption of the tag and attribute names contained in the input query, and the translation of the query predicates in terms of partition ids. Information about partitioning functions and ids generation mechanisms is stored in the user entry of the owner's directory server.

As a final remark, note that the fact that a user submits encrypted queries to publishers ensures a certain degree of user's privacy in that publishers do not know the details of the submitted queries.

10.4.3 Completeness

Completeness verification is based on an XML-data structure called *query template*, which is generated by the owner for each outsourced document and downloaded by clients from the owner's directory the first time they query the corresponding document. The query template is generated by the owner, by applying a simple XSLT transformation [28] on the corresponding SE-ENC document, which prunes from the SE-ENC document the encrypted data contents and security-related information not necessary for completeness verification. The query template is signed by owners with a Merkle signature. Since the query template is generated from the SE-ENC document, it contains the structure of the corresponding document selectively encrypted according to the owner's access control policies, plus additional information needed for completeness verification, that is, partition ids and access control information. The idea is that by exploiting the same query processing strategy used by publishers (i.e. based on partition ids), the user is able to perform on the query template the same (encrypted) query he/she has submitted to publishers. Completeness verification can be done for all XPath queries whose conditions are based on =, <, <=, >, >= operators or the *contains*() function. Users verify the completeness of query results by comparing the node-set returned by the publisher with the node-set resulting from the query evaluation on the query template. More details on completeness verification can be found in [5].

10.5 Secure Outsourcing of Geographical Data

In this section, we show how the framework previously illustrated can be used for secure geodata outsourcing. In particular, we consider a data owner, called hereafter *GisOwner*, which is the producer of geographical data. *GisOwner* elaborates geographical data from satellites, cartographic maps, and so on and produces maps of several geographical objects, defined according to different geodata models, that is, with different features, level of details, and so on. All these maps are encoded in GML (geography markup language) [24]. We assume that *GisOwner* makes its geographical data available to customers (i.e. users) on the basis of different subscription fees. Furthermore, we suppose that it does not manage user interactions on its own, rather it outsources the maps it produces to one or more publishers, which are in charge of answering customers' queries.

In order to keep the example simple, we limit the discussion to a unique GML document, called *CambridgeCityModel*, defined according to a model for city-related geographical data and encoding information about Cambridge's rivers and roads (see Fig. 10.4). Moreover, we assume that *GisOwner* offers three kinds of subscription: *River* subscription, which allows customers to view only information on rivers, *Road* subscription, authorizing customers to access road information, and *Full* subscription, which allows access to information on both rivers and roads. In what follows, we show how, by exploiting the framework presented in Sect. 10.4.2, it is possible to avoid publisher's and user's misuses of *GisOwner*'s data. Subscriptions can be modeled by means of access control policies. For instance, the *River* subscription can be enforced by an access control policy acp_1 granting to all subscribed customers access to all and only river information (i.e. XML nodes containing river information). Whereas, *Roads* and *Full* subscriptions can be modeled by similar access control policies (referred to as acp_2 and acp_3 in what follows). Access control policies can be specified by means of GEOXACML language [18], a geospatial extension to the OASIS standard eXtensible access control markup language (XACML) [23].

According to the strategy introduced in Sect. 10.4.2, *GisOwner* selectively encrypts *CambridgeCityModel* and complements it with additional information, obtaining thus the corresponding SE-ENC document, which is then outsourced to publishers. In particular, since three different access control policies (i.e. subscriptions) are applied to the *CambridgeCityModel* document, different portions of the document are encrypted with different keys. More precisely, in the resulting encrypted

```
<?xml version="1.0" encoding="UTF-8"?>
<CityModel xmlns="http://www.opengis.net/examples"
xmlns:gml="http://www.opengis.net/gml"
xmlns:xlink="http://www.w3.org/1999/xlink"
xmlns:xsi="http://www.w3.org/2001/XMLSchema-instance"
xsi:schemaLocation="http://www.opengis.net/examples city.xsd">
    <gml:name>Cambridge</gml:name>
    <gml:boundedBy>
        <gml:Box srsName="http://www.opengis.net/gml/srs/epsg.xml#4326">
        <gml:name>Cambridge</gml:name>
    <gml:boundedBy>
        <gml:Box srsName="http://www.opengis.net/gml/srs/epsg.xml#4326">
            <gml:coord>
                <gml:X>0.0</gml:X>
                <gml:Y>0.0</gml:Y>
            </gml:coord>
        </gml:Box>
    </gml:boundedBy>
    <cityMember>
        <River>
            <gml:description>The river that runs through Cambridge.</gml:description>
            <gml:name>Cam</gml:name>
            <gml:centerLineOf>
                <gml:LineString srsName="http://www.opengis.net/gml/srs/epsg.xml#4326">
                    <gml:coord>
                        <gml:X>0</gml:X>
                        <gml:Y>50</gml:Y>
                    </gml:coord>
                    <gml:coord>
                        <gml:X>70</gml:X>
                        <gml:Y>60</gml:Y>
                    </gml:coord>
                    <gml:coord>
                        <gml:X>100</gml:X>
                        <gml:Y>50</gml:Y>
                    </gml:coord>
                </gml:LineString>
            </gml:centerLineOf>
        </River>
    </cityMember>
    <cityMember>
        <Road>
            <gml:name>M11</gml:name>
            <linearGeometry>
                <gml:LineString srsName="http://www.opengis.net/gml/srs/epsg.xml#4326">
                    <gml:coord>
                        <gml:X>0</gml:X>
                        <gml:Y>5.0</gml:Y>
                    </gml:coord>
                    <gml:coord>
                        <gml:X>20.6</gml:X>
                        <gml:Y>10.7</gml:Y>
                    </gml:coord>
                    <gml:coord>
                        <gml:X>80.5</gml:X>
                        <gml:Y>60.9</gml:Y>
                    </gml:coord>
                </gml:LineString>
            </linearGeometry>
            <classification>motorway</classification>
            <number>11</number>
        </Road>
    </cityMember>
</CityModel>
```

Fig. 10.4. GML document modeling Cambridge's rivers and roads [24]

```
< E_{k_{123}}(CityModel)>
  < E_{k_{13}}(River)>
      ...
    < E_{k_{13}}(gml:LineString)>
      < E_{k_{13}}(gml:coord)>
        < E_{k_{13}}(gml:X)>
        <Sec-Info>
          <Node-Info>
            < Query-Info > PF(0)</Query-Info >
          </ Node-Info >
        </Sec-Info >
        < /E_{k_{13}}(gml:X)>
        < E_{k_{13}}(gml:Y)>
        <Sec-Info> .... </Sec-Info >
        < /E_{k_{13}}(gml:Y)>
      < E_{k_{13}}(gml:coord)>
        < E_{k_{13}}(gml:X)>
        <Sec-Info>
          <Node-Info>
            < Query-Info > PF(70)</Query-Info >
          </ Node-Info >
        </Sec-Info >
        < /E_{k_{13}}(gml:X)>
        < E_{k_{13}}(gml:Y)>
        <Sec-Info> ....</Sec-Info >
        < /E_{k_{13}}(gml:Y)>
      < /E_{k_{13}}(gml:coord)>
      < E_{k_{13}}(gml:coord)>
        < E_{k_{13}}(gml:X)>
        <Sec-Info>
          <Node-Info>
            < Query-Info > PF(100)</Query-Info >
          </ Node-Info >
        </Sec-Info >
        < /E_{k_{13}}(gml:X)>
        < E_{k_{13}}(gml:Y)>
        <Sec-Info> ... </Sec-Info >
        < /E_{k_{13}}(gml:Y)>
      < /E_{k_{13}}(gml:coord)>
    < /E_{k_{13}}(gml:LineString)>
  < /E_{k_{13}}(River)>
< /E_{k_{123}}(CityModel)>
```

Fig. 10.5. An example of Sec-Info element

document, River elements and Road elements are encrypted with two different keys, say k_{13} and k_{23}[8] respectively.

Figure 10.5 presents a simplified version of the resulting SE-ENC document, where all additional information needed to make the publisher able to evaluate queries on the encrypted document is inserted into the Sec-Info element. We recall that this information is the partition ids defined over the data domain of elements/attributes.

Let us suppose now that a customer having the *River* subscription wants to retrieve information about all rivers in Cambridge having X coordinate equal to 100 (see Fig. 10.4). By having information on partition functions, the customer is

---

[8] We adopt a notation where the key's subscripts are stated according to the ids of the access control policies associated with that key. Thus, for instance, k_{23} denotes the key that should be delivered to users satisfying acp$_2$ and acp$_3$.

able to translate the query //River//coord/X[contains(.,100)] into an encrypted query that can be evaluated by the publisher directly on the SE-ENC document, that is, //E_{k13}(River)//E_{k13}(coord)/E_{k13}(X)//Query-Info [contains(.,'PF(100')], where PF() is the partitioning function which returns the partition id associated with value 100.

The publisher is thus able to evaluate the submitted query directly on the outsourced document (see Fig. 10.5), and returns to the user only those nodes answering it (that is, only the first E_{k13}(River) element). Only the users provided with the proper key (i.e. k_{13}) are able to decrypt and access the data. Since encryption keys are distributed according to access control policies satisfied by users, this ensures confidentiality with respect to the users.

It is important to note that the proposed framework also allows the evaluation of more complex queries. For instance, if *CambridgeCityModel* stored the *X,Y* coordinates into attributes instead of elements, we could evaluate XPath predicates, whose conditions are based on =, <, <=, >, >= operators. Thus, for instance, it could be possible to perform queries like "Retrieve all rivers contained in a given rectangle." This can easily be evaluated by an XPath predicate on the attribute storing *X* coordinate (*Y* coordinate, respectively), which verifies that the coordinate is included between the *X* coordinates of the rectangle.

10.6 Conclusions

The chapter dealt with a new and promising paradigm for data management, that is, data outsourcing. The chapter, besides illustrating the basic concepts of this paradigm and its possible applications on the GIS domain, focused on security issues arising when data are outsourced to a third-party. Enforcing security requirements of both final users and data owners is a primary need to make data outsourcing widely accepted. The chapter discussed main security requirements and analyzed the related literature in view of these requirements. Then, it presented a comprehensive framework for secure outsourcing of XML data and illustrated its application to geographical data.

Data outsourcing is a new and emerging area, as such interesting research issues still need to be addressed. For instance, a possible extension is considering privacy as a further security requirement. This implies investigating several issues. Indeed, besides the protection of user queries that is achieved in our framework by query encryption, there is the need to protect user personal data as well as to consider concerns on possible data mining operations performed by publishers. Additionally, an interesting issue is how techniques for intellectual property protection (see Chap. 11) can be integrated into the proposed framework. Finally, the investigation of encryption strategies and related query processing for more complex queries, such as the ones supported by Web feature service interface standard (WFS) [29], is a further interesting research direction.

References

1. Agrawal R, Kiernan J, Srikant R, Xu Y (2002) Hippocratic Databases, In: Proc. 28th International Conference on Very Large Databases (VLDB'02), Hong Kong, China, 143–157
2. Akl SG, Taylor PD (1983) Cryptographic Solution to a Problem of Access Control in a Hierarchy. ACM Transaction in Computer Systems, 1(3):239–248
3. Bertino E, Carminati B, Ferrari E (2002) A Temporal Key Management Scheme for Broadcasting XML Documents. In: Proc. 9th ACM Conference on Computer and Communications Security (CCS'02), Washington, VA, 31–40
4. Bertino E, Ferrari E (2002) Secure and Selective Dissemination of XML Documents. ACM Transactions on Information and System Security, 5(3):290–331
5. Bertino E, Carminati B, Ferrari E, Thuraisingham B, Gupta A (2004) Selective and Authentic Third-Party Distribution of XML Documents. IEEE Transactions on Knowledge and Data Engineering, 16(10):1263–1278
6. Boneh D, Gentry C, Lynn B, Shacham H (2003) Aggregate and Verifiably Encrypted Signatures from Bilinear Maps. Advances in Cryptology - EUROCRPYT 2003: International Conference on the Theory and Applications of Cryptographic Techniques, Warsaw, Poland, Lecture Notes in Computer Science, Springer, Berlin Heidelberg New York, 416–432
7. Carminati B, Ferrari E, Bertino E (2005) Securing XML Data in Third-Party Distribution Systems. In: Proc. ACM 14th Conference on Information and Knowledge Management (CIKM'05), Bremen, Germany, 99–106
8. Carminati B, Ferrari E, Bertino E (2005) Assuring Security Properties in Third-party Architectures. In: Proc. 21st International Conference on Data Engineering (ICDE'05), Tokyo, Japan, 574–548
9. Carminati B, Ferrari E (2006) Confidentiality Enforcement for XML Outsourced Data. In: Proc. 2nd EDBT Workshop on Database Technologies for Handling XML Information on the Web (DATAX'06), Munich, Germany
10. Charanjit S, Yung M (1996) Paytree: Amortized Signature for Flexible Micropayments. In: Proc. 2nd Usenix Workshop on Electronic Commerce, Oakland, CA, 213–221
11. Cockcroft S, Clutterbuck P (2001) Attitudes Towards Information Privacy. In: Proc. 12th Australasian Conference on Information Systems, Coffs Harbour, NSW, Australia
12. Chor B, Goldreich O, Kushilevitz E, Sudan M (1999) Private Information Retrieval. Journal of the ACM, 45(6):965–982
13. Devanbu P, Gertz M, Martel C, Stubblebine SG (2000) Authentic Third-party Data Publication. In: Proc. 14th Annual IFIP WG 11.3 Working Conference on Database Security, Schoorl, the Netherlands, 101–112
14. Devanbu P, Gertz M, Kwong A, Martel C, Nuckolls G, Stubblebine SG (2001) Flexible Authentication of XML Documents. In: Proc. 8th ACM Conference on Computer and Communications Security (CCS'01), Philadelphia, PA, 136–145
15. Hacigumus H, Iyer B, Li C, Mehrotra S (2002) Executing SQL over Encrypted Data in the Database Service Provider Model. In: Proc. ACM SIGMOD'2002, Madison, Wisconsin, 216–227
16. Hacigumus H, Iyer B, Mehrotra S (2002) Providing Database as a Service. In: Proc. 18th International Conference on Data Engineering (ICDE'02), San Jose, CA, 29
17. Hacigumus H, Iyer B, Li C, Mehrotra S (2004) Efficient Execution of Aggregation Queries over Encrypted Relational Databases. In: Proc. 9th International Conference on Database Systems for Advanced Applications (DASFAA'04), Jeju Island, Korea, 125–136

18. Matheus A (2005) Declaration and Enforcement of Fine-grained Access Restrictions for a Service-based Geospatial Data Infrastructure. In: Proc. 10th ACM Symposium on Access Control Models and Technologies (SACMAT'05), Stockholm, Sweden, 21–28
19. Mykletun E, Narasimha M, Tsudik G (2006) Authentication and Integrity in Outsourced Databases. ACM Transactions on Storage, 2(2):107–138
20. Merkle RC (1989) A Certified Digital Signature. In: Proc. of Advances in Cryptology, Santa Barbara, CA, 218–238
21. Naor M, Nissim K (2000) Certificate Revocation and Certificate Update. IEEE Journal on Selected Areas in Communications, 18(4):561–570
22. Narasimha M, Tsudik G (2005) DSAC: Integrity of Outsourced Databases with Signature Aggregation and Chaining. In: Proc. ACM Conference on Information and Knowledge Management (CIKM'05), Bremen, Germany, 235–236
23. OASIS Consortium, eXtensible Access Control Markup Language (XACML) (version 1.1), http://www.oasis-open.org/committees/xacml/
24. OGC Open Geospatial Consortium, Geography Markup Language (version 3.1.1), http://portal.opengeospatial.org/files/?artifact_id=4700
25. Rivest R, Adleman L, M Dertouzos (1978) On Data Banks and Privacy Homomorphisms. In: Foundations of Secure Computation, Academic Press Inc., 171–179
26. Song DX, Wagner D, Perrig A (2000) Practical Techniques for Searches on Encrypted Data. In: Proc. IEEE Symposium on Security and Privacy, Oakland, CA, 44–55
27. Stallings W (2000) Network Security Essentials: Applications and Standards. Prentice Hall
28. World Wide Web Consortium, http://www.w3.org
29. Web Feature Service Interface Standard, http://www.opengeospatial.org/docs/02-058.pdf

Information Hiding for Spatial and Geographical Data

Maria Calagna and Luigi V. Mancini

University of Rome "La Sapienza", Rome (Italy)

11.1 Introduction

A *geographical information system* (GIS) represents geographical and spatial data through digital maps that are built according to a specific representation model. A digital map represents a set of features that are stored as couples < *attribute*, *value* > for each position in the map. These features may be organized into different layers in the map. Each layer logically represents a set of elementary features that can be used in a specific working activity, or they can be used in addition to other features belonging to other layers. Figure 11.1 illustrates the abstract representation of a GIS as a collection of different layers. In this example, the digital map is composed of four layers: a spatial coordinate system layer, a regions layer, a rivers layer, and a towns layer. All of them may be visualized together or separately. Also, features that belong to different layers may be selected by the user and some operations can be performed over them in order to obtain a new digital map. This is a distinguishing property of GIS. Thus, digital maps can be visualized according to GIS users' preferences, that is, GIS users may select some features to proceed with their working activity, or they can compose new layers starting from the ones that they have. For example, if a user is interested in visualizing the rivers, which go through the eastern towns, first, the user should select the features from the towns and rivers layers properly, and then the user should be able to create a new map that visualizes a new layer containing the rivers features that correspond to the user query.

In recent years the development of communications in terms of technology and applications has influenced several domains in the public sector, including finance, marketing, arts, and administrative services. The main relevant events include the spreading of satellite and mobile communications. Consequently, Web-based applications are emerging to make information available to public and private clients. The technology evolution improves the cooperation between the public agencies and the efficiency of public and private services. However, while the technological advances help people to access public information and public services more easily, on the other hand, new problems about privacy and security arise. For example, a typical risk concerns the fact that the data could be captured by unauthorized users. Thus,

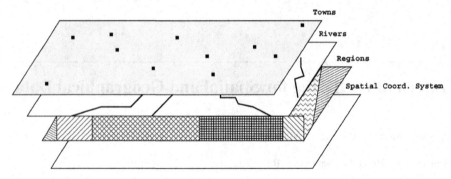

Fig. 11.1. A digital map is built upon a set of layers. Each layer logically represents a kind of information that can be used as it is or in conjunction with information belonging to other layers

we need to make additional efforts in order to achieve the security and the protection requirements for the new emergent applications.

Digital watermarking is a common solution that is used in order to realize the above requirements. It is classified as an information hiding technique, since it embeds a secret message, the watermark, in some valuable data in order to protect the intellectual property. In the GIS field, it is employed in order to address the problem of intellectual property protection, including *copyright protection* and *fingerprinting* of digital maps. The former concerns the protection of the owner digital rights on maps, while the latter can actually identify who is authorized to receive a digital map in a distributed GIS. Then, if the receivers try to circumvent the system policies, by re-distributing illegal digital maps to unauthorized entities, it is possible to identify who is the *traitor*. Section 11.3 describes the GIS application scenarios that can benefit from watermarking.

A related information hiding technique is steganography that is used in order to conceal the secret message's very existence. We observe that steganography may be used with the aim to communicate secretly among authorized users. The authorized users may retrieve the hidden content, while unauthorized users are unable to detect that any hidden content is exchanged between the authorized entities. The addition of steganography to digital maps is more powerful than its application to other kind of representation, including, simple images, for example. In fact, according to the GIS description that was given above, many GIS operations are available to the users who are able to combine geographical and spatial data in several ways. Section 11.5 illustrates a possible application that integrates steganography with digital mapping. The chapter is structured as follows. Section 11.2 introduces the fundamentals of security and protection mechanisms, particularly concerning information hiding techniques. Further details on watermarking and steganography are given in Sects. 11.2.1 and 11.2.2, while Sect. 11.2.3 introduces the main properties of information hiding systems. The structure of the chapter continues with Sects. 11.3–11.4 and 11.5–11.6 introducing the watermarking and steganography techniques for spatial and geographical data, respectively. In particular, Sect. 11.4 illustrates a proposal for a watermarking scheme, developed within the SPADA@WEB project [27], that is based on the use of the SVD (*singular value decomposition*) transform by blocks, while Sect. 11.6 illustrates a proposal for a steganography GIS model that is based on the use of public keys with the aim to transmit sensitive content inside digital maps, making the hidden content available only to authorized users. Finally, some related works are presented in Sect. 11.7, and we conclude with Sect. 11.8.

11.2 Information Hiding: A Background on *Watermarking* and *Steganography*

Security can be added to existent technology (hardware and software) in order to minimize the risks that are related to the novel scenarios, improving robustness of the system. Security concerns three main requirements: confidentiality, integrity, and authentication. These properties were well addressed in Chap. 9 and 10. Cryptography and digital signatures can be used to achieve these requirements, thus making the communication secure. When used in conjunction, they can be considered in terms of a public-key infrastructure (PKI): each communicating entity owns a pair of keys (public-key, secret-key); the entities are able to securely communicate with each other with the help of a *trusted* certification authority (CA). Cryptography helps people to achieve data confidentiality. Ciphered data are accessible to only legitimate receivers that use a secret-key to decrypt the ciphered data and retrieve the original data. The problem is that once data are deciphered, the original data may be captured by adversaries that could use the captured information maliciously. Attacks are classified as *passive* or *active*, whether they leave the captured information unchanged or modify it. Generally, active attacks are more dangerous than passive attacks since they involve the integrity and the authenticity requirements, in addition to the confidentiality requirement. Moreover, legitimate users would not be able to recognize that data are compromised. Legitimate users continue to work normally, while they are actually feeding a malicious adversary. Another drawback related to the adoption of cryptography is the fact that if cipher-text is easily recognizable, then, a passive adversary (*eavesdropper*) that intercepts the communication traffic is acknowledged about the use of cryptography.

So, we need additional protection mechanisms that can be used in conjunction with the security mechanisms. Protection mechanisms should help content providers and distributors to preserve the digital rights in the whole value chain, by regulating access and usage of the resources. A digital rights management (DRM) system is a framework that is used to control and manage digital rights, including the intellectual property ownership. Basic security and protection components are cryptography and watermarking. DRM rules are enforced by policies that may be expressed by rights expression languages. Two main approaches are the MPEG REL [23] and the ODRL [24]. MPEG REL [23] expresses licences in terms of the following features: *principals*, identifying who is able to perform some actions; *rights*, specifying actions; *resources*, identifying the objects that may be used by principals in order to exercise some rights; and *conditions*; that should be validated before any action is performed. Similarly, ODRL [24] is based on the following features: *assets*, identifying any physical or digital content; *rights*, identifying permissions on the assets; and *parties*, including end-users and rights holders. Both the approaches may be integrated as shown in [8].

In the following, we describe in detail two information hiding techniques, watermarking and steganography, that may be applied to the GIS field.

11.2.1 Watermarking

Given a *cover* that represents some valuable data (a song, a video, a map image) it is possible to embed a digital *watermark* inside them [16, 17, 19]. After embedding, the watermark and the cover are inseparable and they are used to generate the *watermarked content*. Often, the watermark is invisible, so it is imperceptible in the watermarked content. According to applications, watermarking systems can be classified into two categories: *fragile* and *robust*. In a fragile watermarking scheme, the watermark is designed to be fragile so as to detect and localize possible modifications made to the watermarked content. Typical applications of fragile watermarking systems include *content authentication* (the watermark is used to detect modifications applied to the original content), *copy control* (the watermark indicates whether the content can be copied or not) and *transaction tracking* (the watermark records transaction history of the content, typically identifying first authorized user). Robust watermarking schemes are used for copyright protection: the watermark is designed to be robust against attacks so as to *prove ownership* of the image (the watermark is used to prove ownership in a court of law). In addition, according to the detection phase, watermarking systems may be classified into two categories: *non-blind* or *blind*, based on the use of the original cover to extract the watermark by the detector. The term *oblivious* is also used as a synonym for blind. Blind watermarking systems are preferable since they can easily be used in a distributed environment where the original cover is not required and multiple watermarked copies may be delivered to the users.

11.2.2 Steganography

Steganography (literally, *covered writing*) is an ancient art to hide traditional writing. Since Greek and Roman times, it has been used to convey a secret without raising any suspicion [16]. The main feature of classical steganography consists of the encoding system secrecy. Once it is discovered, it makes all future communications useless. Modern steganography is based on the secrecy of a shared secret-key that the entities use to communicate with each other. As for watermarking systems, steganography systems embed some secret information inside a cover object, producing a stego-object. Based on the underlying cryptographic techniques, steganography systems are classified into secret-key and public-key, based on the use of a shared secret-key or a pair of secret-key and public-key for secure communication. Secret-key stegosystems use a shared secret-key in embedding and extraction phases. A typical issue in secret-key steganography concerns the difficulty of exchanging the secret-key between the communicating entities. This problem is overcome by public-key steganography: the secret is embedded in the public-key of the receiver. Then, the receivers can extract the secret from the stego-object by applying their secret-key. In Sect. 11.6, we apply a public key stegosystem to describe a possible application to geospatial maps.

While the main requirements for a watermarking system are the imperceptibility and the robustness against possible attacks, the goal of steganography systems is the undetectability of the secret message. In the following section, some properties concerning watermarking and steganography systems are illustrated.

11.2.3 Properties of Information Hiding Systems

We refer to an information hiding system as a system that embeds a secret into a cover-object, producing a stego-object. In the following, we use this terminology in order to illustrate some properties of these systems.

Based on the redundancy of the coding, some information may be embedded in a cover-object. The total number of bits that are required for encoding the secret message is called "data payload." For images, the data payload is equivalent to the number of bits required to encode the secret message; for audio files, the data payload is equivalent to the number of bits of the secret message per second and, for video files, it is equivalent to the number of bits of the secret message that can be transmitted per second or on a frame-by-frame basis.

The *Imperceptibility* Requirement

The main requirement for the information hiding systems is *imperceptibility*, that is, the stego-object and the cover-object maintain the same level of quality. This means that embedding the secret is not intrusive, thus it does not degrade the original quality. A common metric for the quality difference is the peak-to-signal-noise ratio (PSNR) that is based on the minimum squared error (MSE). For images, the *MSE* is given by

$$MSE = \sum_{i=1}^{M} \sum_{j=1}^{N} [I_1(i, j) - I_2(i, j)]/MN \qquad (11.1)$$

where I_1 and I_2 represent two images of $M \times N$ pixels. Then, the *PSNR* is given by

$$PSNR = 20 \log(max\_intensity/RMSE) \qquad (11.2)$$

with *max_intensity* indicating the maximum intensity value in the images and, RMSE indicating the squared root of the MSE.

When the PSNR value is higher, the quality difference is worse. This metric provides an estimate of the quality difference between two different objects (we are interested in the quality difference between the cover-object and the stego-object). Then, the relative values are more meaningful than the absolute values. The use of the PSNR metric helps to analyze different information hiding systems and the distortion due to the embedded information.

The *Undetectability* Requirement

Steganography is used by entities whose aim is to communicate secretly through a public channel, without making the communication suspicious. An important requirement of such a communication is *undetectability* of the secret message. This property may be under attack according to different criteria. Commonly, attacks against stegosystems are referred to by the term "steganalysis." A stegosystem is insecure if an adversary can differentiate between cover-objects and stego-objects. Steganalysis includes the following attacks: *stego-only* attack (only the observed stego-object is available to the adversary), *known cover-object*, and *known secret* attacks (if the cover-object is available or the possible hidden secret is known, then the adversary may search for distinguishing patterns in the observed stego-object), *known stego* attack (the steganography technique is available to the adversary), and *chosen secret* attack (the steganography technique and the possible hidden secret are available to the adversary, who may suspect a steganography technique was applied). In [14], the undetectability requirement is defined in terms of the impossibility for a polynomial time adversary to differentiate between cover-objects and stego-objects. Other steganalysis attacks assume that the distribution of cover-objects is known [4], that is, an adversary is able to discern if the observed object corresponds to a cover-object or a stego-object. In fact, an important class of steganalysis includes the statistical correlation-based attacks. These attacks can be used to discover the hidden information since they are based on the idea that the hidden information may be more random than the information in the cover it replaces. The amount of randomness may be computed by the chi-square test. Given $e(i)$, the number of times an event occurs (the event of a bit 1 or 0, for instance), and $E(i)$, the expected number of times the pure random event should occur, the randomness is computed by

$$\chi^2 = \sum \frac{e(i) - E(i)}{E(i)}$$

The lower the χ^2 value, the higher the amount of randomness, implying hidden information in a file with high probability.

11.3 Digital Watermarking for Spatial and Geographical Data

So far, little effort has been made to make spatial and geographical data publicly available in a digital way. Their acquisition and maintenance are very expensive, especially if data are integrated from different cooperating sources. Thus, GIS organizations are not interested in distributing them without the support of an efficient and effective DRM system that will be able to protect data throughout the value chain: from the creator to the distributor to the final user. Recently, some technical solutions to protect digital rights of digital maps, including digital watermarking, have been proposed. Also, the creation of the GeoDRM group [22] has the aim of taking into consideration the expertise in the GIS field and to provide standardized procedures to protect distributed data throughout the value chain.

In the following, we illustrate two possible scenarios where watermarking is applied to digital mapping: *map copyright protection* and *map fingerprinting*. In both situations, we adopt the concept of third-party architecture that has been explored in Chap. 10. However, we distinguish the publisher as two different entities that play an important role in the GIS field: the map producer and the mapping services provider.

Map Copyright Protection. As a solution for copyright protection, watermarking can be employed for assessing the digital rights on maps. People that benefit from copyright protection are the map owners, the map producers, and the mapping service providers. The following example, shown in Fig. 11.2, illustrates a possible scenario in the GIS field. Suppose that the local public agency has to collect geospatial data concerning a particular region for different kind of activities: urban growth monitoring, industrial pollution monitoring, analysis of the rain course in a specific year, homeland security, and so on. It may ask specialized organizations to acquire and maintain the large amount of geospatial data, that is, they outsource this complex task to the map producer, who has more expertise, in order to provide high-quality services at reduced costs (see Chap. 10 for further details). In this scenario, the DRM should take into account the owner digital rights (in the example, the public agency) and the producer digital rights (in the example, the outsourcing organization). Moreover, some map servers (e.g. Yahoo Maps, MapQuest, etc.) can be authorized to provide other mapping services for multipurpose uses. Then, the DRM should take into account the provider digital rights as well. Typically, exported data concern simple routing queries (e.g. how to reach location B from location A) or displaying information through satellite images. In some cases, exporting such data may be subject to additional fees.

Map Fingerprinting. As a digital object, a digital map may be copied, modified, and retransmitted. Attention is focused on preventing illegal use of digital maps by both the map producer and the mapping service providers. In literature, this issue is known as "traitor tracing": if someone circumvents the original security policy of the system, it should be possible to show that an information leakage happened and to identify who leaked information (traitor).

Watermarking may be employed to solve this issue. In this case, we refer to the technical solution as "fingerprinting." A fingerprinting system for a GIS embeds

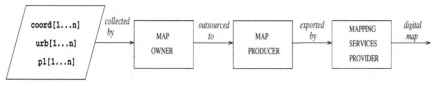

Fig. 11.2. This scheme illustrates a real-world scenario in the GIS field. Watermarking can be added to digital maps in order to protect the digital rights throughout the value chain: the map owner is responsible for collecting geographical and spatial data (*e.g. public agencies employees*), the map producer is responsible for acquiring and maintaining the digital map, finally, the mapping services provider offers further functionalities to the final users

a code into each delivered map. This code identifies the receivers univocally such that if one of them circumvents the system, it is possible to identify who leaked information by monitoring the delivered maps and tracing the fingerprinting code that was embedded by the map owner. The map producer may be interested in embedding another fingerprinting code too. Figure 11.3 illustrates a possible scenario with the map owner and the map producer employing a *GIS fingerprinting system* (GFS) to monitor digital maps delivery. Geospatial information has a very high degree of precision. So, digital watermarking should be applied with high accuracy in order to minimize the risk of changing or destroying the original values in the map. A typical approach consists of adjusting the level of precision used for storing the geographical coordinates. The digital watermark may be added to the original content according to two different approaches, as follows:

1. It may be embedded to the digital content, *directly*.
2. It may be embedded in the transform domain of the original content, according to some criterion.

11.3.1 Related Work on Watermarking for Spatial and Geographical Data

The most common representation models for digital maps are the *raster* and the *vector* model. In the raster model, a digital map is represented as a matrix, where each element represents a point on the surface. In the raster model, only one attribute is represented by a digital map. Whereas, in the vector model, a digital map is stored in a database table with the primary key composed of a unique identifier and the coordinates of the geometric primitives forming the map. One or more attributes are associated to this key.

Thus, visualization of a raster map corresponds to a standard image, where each pixel is characterized by an intensity value, related to the feature value at a specific position in the map. Vector maps are visualized by means of geometric features (*point, line, polygon*), where each geometric feature is associated to one or more attributes.

However, the raster representation of a digital map has some peculiarities that differ from a standard image. First of all, digital maps have very high resolution and they contain some large homogeneous areas that makes the application of digital watermarking a little difficult. In fact, embedding a watermark produces some

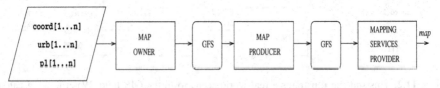

Fig. 11.3. This scheme extends the basic scheme of Fig. 11.2. It illustrates a possible scenario with the map owner and the map producer employing a *GIS fingerprinting system* (GFS) in order to monitor digital maps delivery

distortions that could be perceived in terms of quality degradation of the resulting watermarked digital map against the original one. Similarly, the vector maps show this feature; in fact, maps are represented in terms of geometric features and embedding a watermark may change the geometry of the map.

In [18], this issue is analyzed and a blind watermarking algorithm is provided: the watermark is imperceptible, so the resulting map is of high quality. The basic idea is to add local small noise signals at the boundaries between the homogeneous regions in order to obtain the watermarked map. Also, this watermarking algorithm is robust against cropping and shifting attacks since the watermark is added locally.

In the following, we illustrate how this watermarking algorithm works. The region boundaries are extracted by means of a segmentation algorithm. The watermark is chosen from the set $\{-1, 0, 1\}$ to avoid possible boundary points shifting in the resulting watermarked map. Indicating with L_w, the watermark length, the following formula states that both the endpoints of the original and the modified boundaries will coincide.

$$\sum_{n=0}^{L_w-1} w_n = 0 \tag{11.3}$$

In general, it is possible to constrain the endpoints of both the original and the modified boundary to coincide at regular intervals, as follows:

$$\sum_{n=m}^{n+l-1} w_n = 0 \tag{11.4}$$

with $m = 0, l, 2l, \ldots, L_w$. The watermark detection is performed by means of correlation between the watermark and the watermarked map. The watermark is assumed to be present if the correlation value is over a specified threshold.

According to the experimental results, a watermark of 75 elements is added to a color-mapped chart image of resolution 10000×3000, which represents hydrographic resources. Even if the watermark is embedded *unobtrusively*, the watermarking algorithm is shown to be robust against main attacks, including the JPEG compression and the addition of Gaussian noise. Regarding the JPEG compression attack, the watermark can be detected when the quality factor is over 20; this result is acceptable since digital maps should preserve their quality, otherwise they are useless.

The watermark embedding comes at the price of little distortions in the digital map. Watermarking developers try to reduce these distortions to make the watermark imperceptible. Another requirement of watermarking is the *usability* of watermarked content. This requirement is very important in those applications where the watermarked content serves as input for further operations. In GIS applications, watermarked maps may be visualized or processed by common GIS operations (selection, buffering, overlay, vector model conversion, classification) to extract the desired geographical and spatial information.

In [13], the authors take into consideration the impact of watermarking on digital maps for further processing, especially for classification of satellite images. GIS allow users to integrate different sources of geographical and spatial data. The remotely sensed satellite images are an important source of this kind of data. They consist of multispectral images that capture different kinds of information through different bands, or layers, where each pixel represents a specific position on the Earth and the pixel brightness value represents the remotely sensed value of some information, such as water, vegetation, urban areas, and so on.

Usually, features from different bands may be integrated to perform homeland reasoning and decision making. In general, classification of scientific images, including satellite and medical images, is very important in order to perform high-level tasks that significantly help field people to elaborate additional views that are created upon available data.

In [13], the impact of different methods of invisible watermarking against typical classification algorithms (the *migrating means* among the iterative optimization algorithms and, the *nearest neighbor* among the agglomerative hierarchical clustering algorithms) is analyzed. The watermark may be embedded in the spatial domain of the original image, directly or through a transform domain. According to experimental results, watermark embedding in the spatial domain performs better than transform domain embedding, since the misclassification percentage is less in the first case. One possible interpretation concerns the way watermarking distortion is introduced; the first approach tends to distribute changes uniformly, while the last one tends to distribute changes at those positions that are near the edges in the image.

11.4 A Proposal for a SVD-based Digital Watermarking Scheme

Within the SPADA@WEB project [27], we developed a *blind* image watermarking system [5]; only the embedded watermark and the singular values of the cover image are required in the detection phase, the original cover is not required. This approach is suitable for Internet applications; in this scenario, the original cover cannot be available to receivers. It works on a block-by-block basis and makes use of SVD compression to embed the watermark. The watermark is embedded in all the non-zero singular values according to the local features of the cover image so as to balance embedding capacity with distortion. The experimental results show that our scheme is robust, especially with respect to high-pass filtering. To the best of our knowledge the use of the *block-based* SVD for watermarking applications has not yet been investigated.

The following section introduces the basic fundamentals of the SVD transform, then the SVD-based watermarking algorithm is presented.

11.4.1 The SVD Transform

A common representation of images is in terms of matrices. In the spatial domain, images are usually represented as matrices of $M \times N$ pixels, while in the transform domain, images are represented as matrices of coefficients. According to SVD, a matrix may be factorized in three matrices to achieve the goal of compression [15]. Given a generic matrix A with M rows and N columns, $A^{M \times N}$ ($M \leq N$), then the

$SVD(A) = U\Sigma V^T$, where U and V are orthogonal matrices ($UU^T = VV^T = I$, with I the identity matrix) and Σ is a diagonal matrix, $\Sigma diag(\sigma_1, \sigma_2, ..., \sigma_M)$, whose diagonal entries $\sigma_i(i = 1, ..., M)$ are in decreasing order and they are called *singular values* [12]. The singular values and their distribution vary with different images. The image energy spreads over all the singular values in the presence of random texture, while only the first singular value dominates all the others in the presence of smooth regions [25]. The rank of A is defined as the maximum number of linearly independent columns or, alternatively, as the maximum number of linearly independent rows. An interesting property of the SVD of a matrix is that the rank of a matrix is equal to the number of the non-zero singular values [12]. With r denoting the rank of the matrix A, we have $A = U_r \Sigma_r V_r^T$, where U_r and V_r are obtained, respectively, by picking the first r singular vectors of U and V, and Σ_r is a diagonal matrix of order r ($r \leq M$), whose diagonal entries are $\sigma_1, \sigma_2, ..., \sigma_r$. Applying this result to the matrix representation of images, we have the remaining singular values $\sigma_i = 0$, for $i > r$; thus, these singular values do not contribute to the energy of the image. Then, the *SVD* compression may be considered for lossless compression schemes. A higher compression ratio is achieved by adaptive rank selection [11]: in some applications, including Web publication or multimedia streaming, it is acceptable to compress the original data with some loss according to the requirement that a percentage of the original information is maintained in the compressed data. Thus, fewer ranks than the rank r may be enough.

The *SVD* may be applied to the whole image or, alternatively, to small blocks of it. In image processing, the analysis of many problems can be simplified substantially by working with block matrices [15]. Processing small blocks is suitable to capture local variations that are addressable in a given block and may disappear at a coarser level of representation, such as the whole image. Other advantages of the block-oriented approach include reduced computation and parallel processing [15]. The use of SVD by blocks is suitable in our watermarking scheme, since we embed the watermark in each block according to their rank and the magnitude of singular values. Typical values for the block size are 4, 8, 16, 32, 64.

In the following, we develop a new watermarking system based on block-based SVD compression. This is a novel approach with respect to previous solutions that apply the SVD to the entire image, without considering how singular values vary with different blocks of the same image. Also, embedding the watermark in the most significant singular values of each block makes the watermarking system more robust to possible attacks.

11.4.2 The Watermarking Embedding

We decided to embed the watermark in the most significant singular values of each block to prevent possible removal attacks, including lossy compression. The model of our watermarking system is discussed in the following. Let $A^{M \times N}$ be the matrix representation of the original image to be protected by watermarking. First, it is divided into blocks H_i ($i = 1, ..., B$), whose size $m \times n$ is small enough to capture the local features. The watermark to be embedded into the ith block is represented by the non-zero entries of a diagonal matrix $W_{k_{r_i}}$. The watermark is embedded in the first k_{r_i} singular values of H_i, and then the new singular value matrix is used to obtain the watermarked block H_i'. Embedding a watermark consists of the following steps:

1. Choose a proper block size and divide the original image that is represented by the matrix $A^{M \times N}$ into B blocks.
2. For each block H_i ($i = 1, \ldots, B$),
   - Apply the SVD to the matrix H_i and calculate its rank, r_i. According to the rule $\sigma_j = 0$, for $j > r_i$ we represent it as $H_i U_{r_i} \Sigma_{r_i} V_{r_i}^T$.
   - Select the most significant singular values of each block $(\sigma_1, \ldots, \sigma_{k_{r_i}})$ for watermark embedding. We may represent the ith block as

$$H_i \left(H_{k_{r_i}} \middle| H_{r_i - k_{r_i}} \right)$$

   where $H_{k_{r_i}}$ is limited to the first k_{r_i} singular values and $H_{r_i - k_{r_i}}$ is related to the remaining singular values in the ith block.
3. The watermark is embedded as follows:

$$\Sigma_{k_{r_i}}' = \Sigma_{k_{r_i}} + \alpha \Sigma_{k_{r_i}} W_{k_{r_i}}$$

where $\Sigma_{k_{r_i}}$ represents the original singular values matrix of order k_{r_i} and, $\Sigma_{k_{r_i}}'$ is the updated singular values matrix after the watermark embedding. The remaining non-zero singular values σ_j ($k_{r_i} < j \leq r_i$), which are the diagonal entries of $\Sigma_{r_i - k_{r_i}}$, remain unchanged. Hence, the reconstructed watermarked block H_i' is represented as

$$H_i' = \left(H_{k_{r_i}}' \middle| H_{r_i - k_{r_i}} \right) \tag{11.5}$$

where the sub-blocks $H_{k_{r_i}}'$ and $H_{r_i - k_{r_i}}$ are given, respectively, by

$$H_{k_{r_i}}' = U_{k_{r_i}} \Sigma_{k_{r_i}} V_{k_{r_i}}^T + \alpha \cdot U_{k_{r_i}} \Sigma_{k_{r_i}} W_{k_{r_i}} V_{k_{r_i}}^T \tag{11.6}$$
$$= H_{k_{r_i}} + \alpha \cdot U_{k_{r_i}} \Sigma_{k_{r_i}} W_{k_{r_i}} V_{k_{r_i}}^T$$

$$H_{r_i - k_{r_i}} = U_{r_i - k_{r_i}} \Sigma_{r_i - k_{r_i}} V_{r_i - k_{r_i}}^T \tag{11.7}$$

In other words, the ith block is reconstructed by updating its first k_{r_i} singular values according to step 3 of the watermark embedding algorithm and updating the sub-blocks of H_i according to (11.5)–(11.7). Thus, the watermarked image is obtained by replacing all the original blocks H_i with the watermarked blocks H_i', for $i = 1, \ldots, B$.

11.4.3 The Watermarking Extraction

In the extraction phase we take into consideration only the singular values σ_j, $j = 1, \ldots, k_{r_i}$, of each sub-block $H_{k_{r_i}}'$, described by (11.6). Given a possible counterfeit image, whose sub-blocks are indicated with $H_{k_{r_i}}'$ and the singular value matrix $\Sigma_{k_{r_i}}$ of the correspondent sub-blocks in the cover image, the watermark elements w_j' in each sub-block are extracted according to the following rule: $w_j' = \frac{\sigma_j' - \sigma_j}{\alpha \sigma_j}$, where w_j' are the diagonal elements of the extracted watermark $W_{k_{r_i}}$, while σ_j and σ_j' are the singular values of $H_{k_{r_i}}$ and $H_{k_{r_i}}'$ respectively, for $j = 1, \ldots, k_{r_i}$. The extracted watermark elements w_j' are compared to the embedded ones w_j by means of correlation.

Our threat model consists of two main categories of attacks: *removal* (an adversary attempts to discover the embedded watermark and may remove it from the watermarked image) and *distortive* (an adversary applies some quality-preserving transformations, including signal processing techniques, to the watermarking image making the watermark undetectable). We observe that our scheme is robust against both categories of attacks. As for the removal attacks, an adversary is unable to remove the watermark, since we assume the adversary has no access to the singular values of the cover image. Recall that the removal attack is more dangerous, especially in the *GIS* field, as the digital maps have a great value and then they should be protected from malicious attacks. As for the distortive attacks, Sect. 11.4.4 shows that the attacks included in the Checkmark benchmarking tool [20] do not alter the embedded watermark, which can be detected correctly.

Our scheme has very low false-positive rates. The reader may find further details in [5].

11.4.4 Experimental Results

In this section, we show the robustness of the block-based SVD watermarking scheme against possible attacks included in the Checkmark benchmarking tool [20]. We applied our scheme to several images. For space reasons, we summarize in Table 11.1 only the experimental results on the 512×512 *Aerial* grayscale picture, which is illustrated in Fig. 11.4 together with the watermarked image. In the following, we compare our SVD scheme to the Cox scheme that is implemented according to the original paper on the spread spectrum algorithm [7]. The watermarks are drawn from a Gaussian distribution $N(0, 1)$ in both schemes. We set the *block size* $m_i = 8, i = 1, \ldots, B$ and the watermark strength $\alpha = 0.3$ in the SVD scheme, to obtain an acceptable quality difference between the cover image and the watermarked image as for the Cox algorithm (the PSNR is 30 dB for the SVD scheme and 32 dB for the Cox scheme). In addition, we observe that by applying our scheme, we may embed a longer watermark in the image with respect to the Cox scheme, without decreasing the quality difference too much. By our settings, we embed a watermark of about 8000 elements in the original image. Embedding 8000 elements in the Cox scheme produces images of lower quality, with PSNR = 25.30 dB. The robustness of the watermarking scheme is evaluated by means of the correlation measure between the *extracted* watermark from a possible counterfeit image and the *embedded* watermark.

For each attack, we obtain different levels of distortion, by changing the corresponding parameter, that is indicated in the second column of Table 11.1. The third and the fourth columns show the correlation value between the extracted watermarks from the attacked image and the embedded one for both the watermarking

Fig. 11.4. The cover aerial picture is on the left, while the picture on the right represents the watermarked aerial picture obtained by applying the SVD scheme with block size 8 and watermark strength 0.3

schemes. The last column shows the quality difference between the attacked image and the original cover when our scheme is applied. These experiments give evidence that our watermarking scheme is robust against that kind of attacks, including the high-distortive ones. Actually, it depends on the specific application if the level of distortion is acceptable or not. We illustrate the worst case in Table 11.1 in order to show the robustness properties of our scheme that is applicable in other domains that could be different from the GIS field.

Regarding filtering techniques, we can see that the larger the window size is, the more distortive the attack is. Regarding downsampling/upsampling attack, the choice of the first pair of downsample/upsample factor leads to lower distortion, while the choice of the second pair leads to higher distortion. Regarding JPEG compression and the addition of white Gaussian noise (AWGN) we considered the parameters that lead to more distortion in the attacked image: the quality factor is 10 in the JPEG compression, while the noise strength is 20 in the AWGN attack. Our scheme is robust against contrast enhancement techniques that improve the visual quality of the image and are considered high-frequency preserving [10]. They include histogram equalization and intensity adjustment as well. Finally, we show the robustness results against clipping attack. To summarize, our scheme is more robust than the Cox scheme against median filtering, downsampling/upsampling, histogram equalization, intensity adjustment, and clipping attacks, while it is comparable to that scheme against some other attacks, including JPEG compression and AWGN.

Regarding geometrical attacks, including rotation, translation, and scaling, our claim is that our algorithm is robust against this kind of attacks, since the following properties hold:

- *Rotation*. Given an image A and its rotated A^r, both have the same singular values.
- *Translation*. Given an image A and its translated A^t, both have the same singular values.
- *Scaling*. Given an image A and its scaled A^s, if A has the singular values σ_i, then A^s has the singular values $\sigma_i * \sqrt{L_r L_c}$ with L_r the scaling factor of rows and L_c

Table 11.1. Robustness of the SVD watermarking scheme against attacks

attack name	parameter	correlation (Cox)	correlation (*SVD*)	PSNR (SVD attacked image)
Gaussian filter	window size = 5	0.95	0.98	29.00
median filter	window size			
	3	0.65	0.80	25.50
	5	0.80	0.80	27.00
midpoint filter	window size			
	3	0.90	0.94	26.13
	5	0.58	0.60	24.15
trimmed mean filter	window size			
	3	0.95	0.95	29.18
	5	0.82	0.89	27.77
downsampling/ upsampling	downsample factor/ upsample factor			
	0.75/1.33	0.87	0.98	24.00
	0.50/2.00	0.74	0.77	23.05
denoising with perceptual remodulation	window size			
	3	0.88	0.90	24.28
	5	0.77	0.77	24.67
hard thresholding	window size			
	3	0.91	0.94	24.95
	5	0.90	0.93	25.82
soft thresholding	window size			
	3	0.90	0.90	27.00
	5	0.80	0.83	27.40
JPEG compression	quality factor = 10	0.77	0.77	29.20
sharpening	window size = 3	0.77	0.76	28.50
AWGN	noise strength = 20	0.72	0.88	22.00
histogram eq.	-	0.33	0.65	20.00
intensity adjustment	intensity values/ gamma correction			
	$[0, 0.8] \rightarrow [0, 1] / 1.5$	0.65	0.70	21.15
clipping	clipped portion			
	75%	0.31	0.55	12.78

the scaling factor of columns. If rows (columns) are mutually scaled, A^s has the singular values $\sigma_i * \sqrt{L_r}$ ($\sigma_i * \sqrt{L_c}$).

11.5 Steganography for Spatial and Geographical Data

Steganography can be used to transmit secret information through digital maps in which it is possible to slightly modify some elements, while preserving the quality of the original map. Given such a capability, potential users of digital mapping with the help of steganography can exchange information by means of maps, without arising suspicion by unauthorized users. Also, they can agree on the representation model and they are able to encode and decode information based on some criterion. The strength of steganography is that the secret message is undetectable by those who do not know the secret key; it was reported on USA Today 2001 that terrorists embed information concerning possible targets in apparently innocuous images that are exchanged in chat rooms.

In past years, the problem of transmitting secret messages bypassing conventional schemes has emerged. Research efforts have focused on the following two different approaches:

1. Create a secure steganography scheme that can be employed to communicate secretly (e.g. for military purposes).
2. Apply steganalysis techniques to detect illegal use of steganography.

We remind the reader that the aim of steganography is to conceal the transmission of extra information through innocuous vehicles. The robustness requirement is not necessary since a potential adversary does not know about the existence of any extra information that is carried secretly. Therefore, it is not necessary to prevent removal or distortion attacks that are directed at making the secret information useless. Satellite images are becoming interesting for different classes of people. An interesting approach concerns the addition of steganography to traditional satellite or aerial images to make available different kinds of information to different classes of users. What we have in mind is the possibility of adding some hidden content that could be visualized properly only by authorized people. For example, a digital map, produced from satellite or aerial observation, could be transferred by public channel. Some people, such as public agencies employees, could be authorized to receive highly detailed information that should be forbidden to unauthorized users. For instance, consider a digital map with disease data or income amounts of families living in a specific area associated to it. Distribution of the digital map through a public channel should prevent unauthorized users from having access to this kind of information. In the military field, the addition of steganography to digital maps may be employed to communicate secretly meeting points, base stations, the planned route for traveling, and generally, additional information that should be maintained as secret.

In the following section, we introduce a new conceptual GIS model that integrates public-key steganography with digital mapping. Some related work concerns the modeling and security of public-key steganography systems [1, 2]. However, to the best of our knowledge, the use of this technique in conjunction with digital mapping is a novel application.

11.6 A Proposal for a Steganography GIS Model

In technical terms, public-key steganography is employed to realize a GIS that allows authorized users to have access to secret information. The basic idea is that potential users u_i are associated with a couple of secret-key and public-key, $< SK_i, PK_i >$, that are employed to exchange information with the other entities. The secret information is embedded in the digital map with the public-key of the authorized users. Then, the digital map may be delivered to all the users. However, the hidden content can be extracted only by those users that own the corresponding secret-key that is used to extract this information.

The main advantage of this approach is to convey different levels of strategic information in an innocuous way without raising suspicion in other potential users who cannot be authorized to access some kind of information. In the scenario of a public channel this solution is more appropriate than the use of cryptographic mechanisms because encrypted content can easily be recognized. On the other hand, steganography content is indistinguishable from other content that does not contain hidden information inside. In terms of security, one important requirement is the confidentiality of sensitive information.

Figure 11.5 shows the diagram of a GIS employing public-key steganography to make sensitive content available only to authorized people. In the following, we describe the functionality of this approach. As illustrated in Fig. 11.5(a), the sensitive information (hidden content) is embedded in the digital map with the public-key PK_i of the potential users class u_i. This produces a stego-map that can be visualized differently according to the users that receive it. In fact, as illustrated in Fig. 11.5(b), users visualize the stego-map by applying the extraction application. The extraction application is composed of two extraction functions: $Extract_{DM}()$ and $Extract_{HC}()$. By applying the $Extract_{DM}()$ function, users retrieve the public digital map (cover). By applying the $Extract_{HC}()$ function, users that own the corresponding secret-key SK_i are able to retrieve the hidden content, while users that do not own the right secret-key are unable to notice any hidden content in the stego-map.

The embedding and extraction functions are illustrated in Table 11.2. A detailed description of these functions is given in the following sections. Our aim is to provide a GIS conceptual model that employs public-key steganography. We do not address technical details about the implementation of this model, including, for example, the number of GIS users or the possibility to employ multiple public-keys per user. Such details strongly depend on the specific scenario; however, this issue goes outside the scope of the chapter.

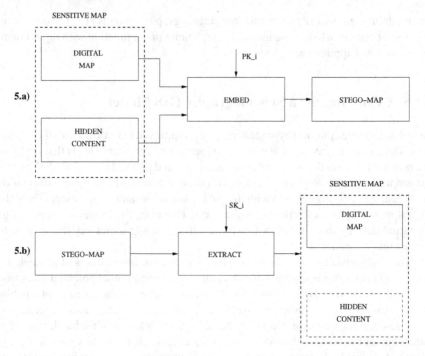

Fig. 11.5. GIS employs steganography to transmit sensitive information (*hidden content*) inside a digital map and makes this sensitive information available only to authorized people: (**a**) illustrates how hidden content is embedded in a digital map in order to create a stego-map; (**b**) illustrates how the users can extract the digital map, and eventually the hidden content, from a stego-map

Table 11.2. The *Embedding* application is used to create a stego-map composed of a digital map and possible sensitive content. The *Extraction* application retrieves all possible information from a stego-map, thus making the digital map, and eventually the hidden content, accessible

Procedure *Embedding*		Procedure *Extraction*	
Input:	DM, HC, PK_i;	Input:	SM, SK_i;
	$Embed(DM, HC) = SM$;		$Extract(SM)_{SK_i} = HC$;
Output:	SM	Output:	DM, HC

11.6.1 The Embedding Application

The *Embedding* function takes a digital map DM, a content HC that represents the secret to be hidden in the digital map ,and the public-key PK_i of the class of users i authorized to see the hidden content inside the digital map. The *Embedding* function produces a stego-map SM as the final output. This map will be available to different users, but it will be visualized differently by the recipients, as said before. The users that own the right secret-key SK_i will be able to retrieve both the digital map DM and the sensitive content HC.

Let us describe both the *Embed* and the *Extract* functions in detail. These functions are the core functions applied by the *Embedding* and the *Extraction* applications, respectively. Let us start with the *Embed* function that is illustrated in Table. 11.3. It takes a digital map DM, the content HC that should be hidden in the digital map, the secret-key of the Embedder SK_E, and the public-key PK_i of the class of users i that will be authorized to visualize the hidden content. By applying the $Embed_{DM}()$ function, the digital map is stored in the GIS with the help of the secret-key of the Embedder SK_E, so that each user may verify and visualize it by applying the corresponding $Extract_{DM}()$ function with the public-key of the Embedder PK_E that everyone knows. The resulting publicly available digital map is dm.

By applying the $Embed_{HC}()$ function, the sensitive content is hidden in the digital map with the help of the public-key PK_i of the class of users i that will be authorized to visualize the hidden content. The secret content that will be available only to this class of users is hc.

Then, the resulting stego-map is composed of dm and hc, and it may be transmitted to different users in a secure way. The security of this application is based on the security of the public-key steganography method that is used to produce the stego-map. Some security results were given in Sect. 11.2.3.

11.6.2 The Extraction Application

In this section, we introduce the *Extraction* application that is based on the $Extract()$ function, given in Table. 11.4.

The $Extract()$ function takes a stego-map SM, the secret-key of a generic user SK_m, and the public-key PK_i that was used by the *Embedding* application to embed the sensitive content in a digital map. The digital map may be visualized by every user by extracting it from the stego-map. In fact, each user knows the public-key of the *Embedder*, so they can have access to the digital map dm. Also, if they own the right secret-key, that is, the secret-key SK_m associated to the public-key PK_i in the key space K_{GEN}, then, they will be authorized to have access to

Table 11.3. *Embed* is the core function of the *Embedding* application. It is used to create a stego map that is composed of the digital map dm and some possible sensitive content hc that is embedded inside dm with the public-key PK_i of the class of authorized users

Procedure *Embed*	
Input:	DM, HC, SK_E, PK_i;
	$Embed_{DM}(DM, SK_E) = dm$;
	$Embed_{HC}(DM, HC)_{PK_i} = hc$;
	$SM = dm + hc$;
Output:	SM

Table 11.4. *Extract* is the core function of the *Extraction* application. The digital map may be extracted from the stego-map by applying the public-key of the *Embedder* that every user knows. Also, if users own the right secret-key, then they will be authorized to have access to the sensitive content that was hidden in the stego-map

$$
\begin{aligned}
&\text{Procedure } Extract \\
&\text{Input:} \quad SM, PK_E, SK_m, PK_i; \\
&\qquad\qquad Extract_{DM}(SM, PK_E) = dm; \\
&\qquad\qquad \text{if } (SK_m, PK_i) \in K_{GEN} \text{ then} \\
&\qquad\qquad Extract_{HC}(SM)_{SK_m} = HC; \\
&\qquad\qquad hc = HC; \\
&\qquad\qquad \text{else} \\
&\qquad\qquad Extract_{HC}(SM)_{SK_m} = \emptyset; \\
&\qquad\qquad hc = \emptyset; \\
&\qquad\qquad \text{end} \\
&\text{Output:} \quad dm, hc
\end{aligned}
$$

the embedded sensitive content. The *Extract*() function releases dm and eventually the hidden content hc. The application of the $Extract_{HC}$ is transparent to the users, so the knowledge of the class of users that are authorized to have access to the sensitive hidden content is not revealed to those users that do not belong to that class. These users will simply notice that no hidden content is available to them.

11.7 Related Work on Information Hiding

In this section, due to the space limits, we chose to describe only a related work on information hiding, though it was not designed specifically for the watermarking or steganography purpose. Among different data hiding proposals [10] [26], a recent work [3] shows that the SVD is a powerful tool to be used for information hiding. According to the scheme that is introduced in [3], the orthogonal matrices U and V are used as vessels for the secret message. While the singular value matrix Σ is uniquely determined by the original matrix, the orthogonal matrices are not univocally determined [12]. In fact, an important requirement of the algorithm is that the original matrix has a uniquely determined SVD decomposition that is assured if its singular values are pair-wise distinct and non-zero. The secret message bits p_k are embedded in some elements u_{ij} of the U matrix according to the formula:

$$ u_{ij}' = p_k|u_{ij}| \tag{11.8} $$

The updated orthogonal matrix U', whose elements are u_{ij}', replaces U in the SVD decomposition of the original matrix and the resulting stego is:

$$ A' = U'\Sigma V^T \tag{11.9} $$

This scheme works on a block basis, thus the original cover is divided into a series of $n \times n$ blocks, the SVD is applied to each block, and then the secret message bits are embedded in some elements of the resulting orthogonal matrix U. It could be a good practice to embed the secret message only in a few u_{ij} elements in order to address the invisibility issue. In particular, the elements of the first rows and columns are not used for embedding, as changing them may introduce noticeable distortions in the stego image. Figure 11.6 illustrates this approach for a block of size 8; the gray elements remain unchanged, the central portion represents the elements of U where the secret message is embedded according to (11.8), while the other elements (those marked with '−') are updated according to the rule that the new matrix U' should be orthogonal. In fact, let U_i' denote the ith column of the matrix U', then according to the orthogonality rule it is required that $U_i' \Delta U_j' \delta_{ij}$. In the example shown in Fig.11.6, the orthogonality rule is satisfied if:

$$U_1' \Delta U_3' = 0$$
$$U_2' \Delta U_3' = 0$$
$$U_1' \Delta U_4' = 0$$
$$U_2' \Delta U_4' = 0$$
$$U_3' \Delta U_4' = 0$$

and so on, until the lower entries in $U_5', \ldots, U_8'$ are determined. Then, the matrix U' is orthogonal and the stego image is obtained according to (11.9).

Advantages of this approach are its simplicity and the embedding capacity. However, it could be found that it is a very difficult task to have a unique SVD of a given matrix. This requirement is not valid when singular values are not pairwise distinct, or they differ slightly. Then, the basic algorithm can be extended by forcing the

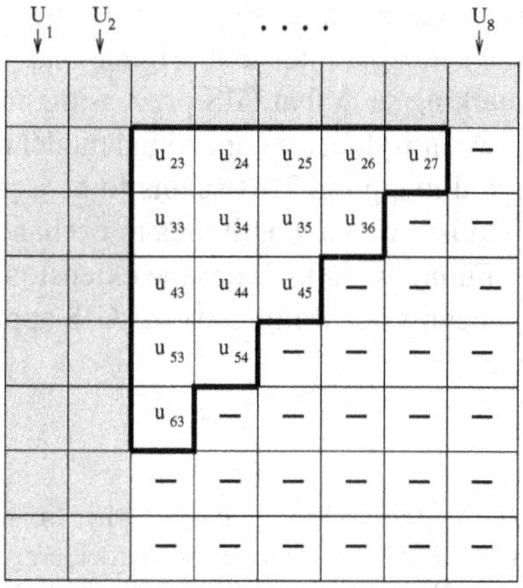

Fig. 11.6. SVD unitary embedding, block size = 8

singular values to be pairwise distinct. One possible strategy is to space them according to the following formula

$$\sigma_{k+1} \longrightarrow \sigma_k - h$$
$$\sigma_{k+2} \longrightarrow \sigma_k - 2h$$
$$\dots$$
$$\sigma_n \longrightarrow \sigma_k - (n-k)h$$

where $h = (\sigma_k - \sigma_n)/(n-k)$.

Another algorithm enhancement could be the use of an error correction code to detect most of the watermarks correctly.

Even if this approach is proposed to address the issue of data hiding, we observe that further studies are needed. Basically, the goal of steganography is imperceptible data hiding, that is, the secret message is undetectable to innocuous observers. We observe that proof of undetectability is missing. Similarly, since goal of watermarking is robustness against common manipulations of the watermarked content, proof of robustness is required.

11.8 Conclusions

Novel applications integrate digital mapping with information hiding. Information hiding techniques are employed to embed secret information inside data without noticeable distortion. Two different approaches are possible: *watermarking* to address the intellectual property issue of digital maps and *steganography* to convey secret information to authorized users by taking advantage of digital mapping. In Sect. 11.4.4, we have illustrated some experimental results concerning the robustness of our watermarking scheme based on SVD by blocks. Future work concerning this scheme includes the analysis of the watermarking impact on further processing of digital maps. In fact, a relevant topic is to guarantee that original map features may be identifiable clearly after watermarking such that GIS processing may lead to correct results. As for steganography, we introduced a conceptual model that integrates public-key steganography with digital mapping. This seems to be a promising application that may be used in different GIS working activities to exchange secret information among cooperating entities. Future work includes the extension of the model and the implementation of technical details according to some GIS applications.

References

1. von Ahn L, Hopper NJ (2004) Public-Key Steganography. In: Advances in Cryptology (Eurocrypt 2004), LNCS vol. 3027, Springer Berlin Heidelberg, 323–341
2. Backes M, Cachin C (2005) Public-Key Steganography with Active Attacks. In: Proc. Theory of Cryptography Conference (TCC'05), Cambridge, MA, USA, LNCS vol. 3378, Springer-Verlag, Berlin, Germany, 210–226

3. Bergman C, Davidson J (2005) Unitary Embedding for Data Hidig with the SVD. In: Proc. SPIE: Security, Steganography and Watermarking of Multimedia Contents VII, San Jose, CA, USA, 619–630
4. Cachin M (2004) An Information-Theoretic Model for Steganography. Information and Computation, 192(1):41–56
5. Calagna M, Guo H, Mancini LV, Jajodia S (2006) A Robust Watermarking Scheme based on SVD Compression. In: Proc. ACM SIGAPP Symposium on Applied Computing (SAC'06), Dijon, France, 1341–1347
6. Checcacci N, Barni M, Bartolini F, Basagni S (2000) Robust Video Watermarking for Wireless Multimedia Communications. In: Proc. IEEE Wireless Communications and Networking Conference (WCNC'00), Chicago, IL, USA, 1530–1535
7. Cox IJ, Kilian J, Leightont T, Shamoon T (1997) Secure Spread Spectrum Watermarking for Images, Audio and Video. IEEE Transactions on Image Processing, 6(2):1673–1687
8. Delgado J, Prados J, Rodrriguez E (2005) A New Approach for Interoperability between ODRL and MPEG-21 REL. In: Proc. 2nd International Workshop on the Open Digital Rights Language (ODRL'05), Lisbon, Portugal, 9–20
9. Döllner J (2005) Geospatial Digital Rights Management in Geovisualization. The Cartographic Journal, 42(1):27–34
10. Ganic E, Eskicioglu A M (2004) Robust DWT-SVD Domain Image Watermarking: Embedding Data in all Frequencies. In: Proc. ACM Multimedia and Security workshop (MM-SEC'04), Magdeburg, Germany, 166–174
11. Goldstein JS, Reed IS (1997) Reduced-Rank Adaptive Filtering. IEEE Transactions on Signal Processing, 45(2):492–496
12. Golub GH, Van Loan CF (1996) Matrix Computations, Johns Hopkins University Press, Baltimore
13. Heileman G L, Yang Y (2003) The Effects of Invisible Watermarking on Satellite Image Classification. In: Proc. 3rd ACM workshop on Digital Rights Management (DRM'03), Washington, DC, USA, 120–132
14. Hopper NJ, Langford J, von Ahn L (2002) Provably Secure Steganography. In: Advances in Cryptology (CRYPTO 2002), 77–92
15. Jain AK (1989) Fundamentals of Digital Image Processing. Prentice Hall
16. Johnson NF, Duric Z, Jajodia S (2001) Information Hiding Steganography and Watermarking - Attacks and Countermeasures. Kluwer Academic Publishers, Boston/Dordrecht/London
17. Kwok SH, Yang CC, Tam KY (2000) Watermark Design Pattern for Intellectual Property Protection in Electronic Commerce Applications. In: Proc. 33rd Hawaii International Conference on System Sciences (HICSS'00), 6037
18. Masry M A (2005) A Watermarking Algorithm for Map and Chart Images. In: Proc. SPIE: Security, Steganography and Watermarking of Multimedia Contents VII, San Jose, CA, USA, 495–503
19. Memon N, Wong PW (1998) Protecting Digital Media Content. Communications of ACM, 41(7):35–43
20. Peng ZR, Tsou MH (2003) Internet GIS - Distributed Geographical Information Services for the Internet and Wireless Networks. John Wiley and Sons, Hoboken, NJ
21. Sion R (2004) Proving Ownership over Categorical Data. In: Proc. 20th International Conference on Data Engineering (ICDE'04), Boston, MA, 584–596
22. The GeoDRM working group, http://www.opengeospatial.org/groups/?iid=129
23. The MPEG-21 REL (ISO/IEC 21000-5:2004), http://rightsexpress.contentguard.com
24. The ODRL Initiative, http://odrl.net

25. Tian B, Shaikh MA, Azimi-Sadjadi MR, Vonder Haar TH, Reinke DL (1999), A Study of Cloud Classification with Neural Networks using Spectral and Textural Features. IEEE Transactions on Neural Networks, 10(1):138–151
26. Tseng YC, Pan HK (2002), Data Hiding in 2-Color Images. IEEE Transactions on Computers, 51(7):873–880
27. Web-based Management and Representation of Spatial and Geographic Data, http://homes.dico.unimi.it/dbandsec/spadaweb
28. Zhang XP, Li K (2005) Comments on "An SVD-Based Watermarking Scheme for Protecting Rightful Ownership". IEEE Transactions on Multimedia: Correspondence, 7(3):593–594

References containing URLs are validated as of October 1ˢᵗ, 2006.

Part IV

Innovative Applications for Mobile Devices

12

Geographical Data Visualization on Mobile Devices for User's Navigation and Decision Support Activities

Stefano Burigat and Luca Chittaro

University of Udine, Udine (Italy)

12.1 Introduction

Users who operate in the field (e.g. maintenance personnel, geologists, archaeologists, tourists, first responders) often bring with them paper sheets (e.g. city maps, forms, technical plans, object descriptions) containing data needed for their activities. Even when these data are available in digital form, the mobile condition of these users makes them prefer more portable and manageable solutions (such as paper) to potentially more powerful and flexible ones (such as laptop computers). For example, it is easier to handle and look at a paper map rather than a digital laptop map while on the move. However, the increasing availability of small and powerful mobile devices (PDAs and Smartphones) is opening up new opportunities. At first devoted to manage user's personal information, these devices can now be employed to assist users in carrying out different tasks in the field.

Unfortunately, the design of effective applications for mobile scenarios cannot rely on the traditional techniques devised for desktop applications due to a number of issues:

- Data presentation and exploration on mobile devices are heavily affected by the small size and resolution of displays. For example, displaying a map in its entirety to assist users in navigating a geographical area typically provides only an overview without sufficient detail, while zoomed-in views provide more detail but lose the global context. Thus, users are forced to perform panning and zooming operations, a process that is cognitively complex, disorienting, and tedious.
- The limited processing power of mobile devices restricts the amount of data that can be managed locally and prevents the use of computationally expensive algorithms. For example, 3D representations of geographical data (e.g. virtual reproductions of terrain) are still uncommon on mobile devices, while they are widely used in some desktop scenarios.
- Input peripherals of mobile devices heavily constrain the set of possible interface designs. For example, keyboards (either physical or virtual) are limited in size and/or number of keys, making it difficult to manually insert data.

- User's mobility affects the design of applications; factors such as user's speed, different and changing activities, distractions, device autonomy, and environmental conditions (e.g. weather, lighting, traffic) must be taken into consideration. Moreover, users are typically involved in other tasks in the field and cannot focus their attention primarily on the mobile device.
- Mobile phones are used by a larger population than traditional computers. Therefore, usability, which has already proven to be an important factor for the acceptance of desktop applications by final users, becomes even more important in mobile applications.

However, the power of mobile graphics hardware is increasing and this makes it possible to provide users on the move with more sophisticated and flexible visualizations [11]. In the following sections, we will survey research that has investigated the visualization of geographical data on mobile devices. Indeed, a wide range of user's activities in the field depends on obtaining and exploiting data on the geographical area where users are located. We will first deal with research on map visualization, since maps are the most commonly used means to provide users with geographical information on an area, thus representing a key building block for applications that aim at supporting users in the field. Then, we will focus on more specific topics, concerning navigation of a geographical area and the support of user's decisions in the field.

12.2 Map Visualization on Mobile Devices

In recent years, research on geographical data visualization on mobile devices has mainly focused on how to best represent and interact with maps on small displays. Indeed, maps are the most efficient and effective means to communicate spatial information [23]. They simplify the localization of geographical objects, reveal spatial relations and patterns, and provide useful orientation information in the field.

As highlighted in Sect. 12.1, using maps on small displays differs from using paper maps in a mobile setting or viewing maps on a desktop computer. This motivated researchers to propose various solutions for the *adaptation* of map contents for mobile scenarios. In particular, since a map is always strongly related to the context of its use, various contextual elements (e.g. user, location, task, device) have been considered to perform the adaptation so that users could get what is most suitable for their needs.

For example, Fig. 12.1 shows a detailed map of an area of the city of Munich (on the left) and the same map adapted for a specific mobile scenario (on the right). The adapted map provides the users with information on the route to follow to reach their hotel from their current position near a pub, passing nearby a shopping area they are interested in (identified with the shopping cart icon). Only relevant information along the route, such as street names and points of interest (POIs), is highlighted and the map has been rotated to be oriented in the main direction of the route.

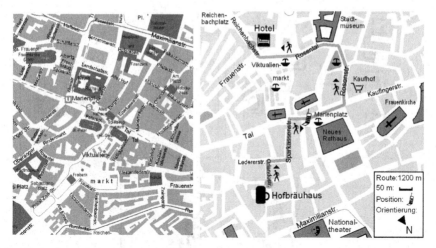

Fig. 12.1. Comparison between an original map (*on the left*) and the same map adapted for a specific mobile scenario [30]. Image courtesy Tumasch Reichenbacher, Technical University Munich, DE. Base data copyright: Städtisches Vermessungsamt München

Reichenbacher [31] claims that the most important task in adapting maps is highlighting relevant map features. He proposes various graphical means to put visual emphasis or focus on a feature that are as follows:

- Highlighting the feature using colors
- Emphasizing the outline of the feature
- Enhancing the contrast between the feature and the background
- Increasing the opacity of the feature while decreasing the opacity of other map contents
- Focusing the feature while blurring other map contents
- Enhancing the level of detail of the feature with respect to other features
- Animating the feature by means of blinking, shaking, rotating, increasing/ decreasing the size

Nivala and Sarjakoski [27] employ some of these principles to adapt the symbols to use in a map according to the current usage situation and user preferences, so that more fluent map reading and interpretation processes could be obtained. Figure 12.2 illustrates an example of map symbol adaptation according to season and user's age. Map (a) at the top left is a summer map for the age group 46 and above, map (b) at the top right is a winter map for the age group 18–45, and map (c) at the bottom is a winter map for young people under 17 years old. The most obvious difference among the maps are the different symbols. The two maps at the top differ in information content: during the winter and summer different kinds of POIs are relevant for the user (for instance, swimming places in summer time and skiing tracks in winter time). There is another difference between the first two maps: pictograms with a white background are provided in map (a) to improve the contrast of the symbol

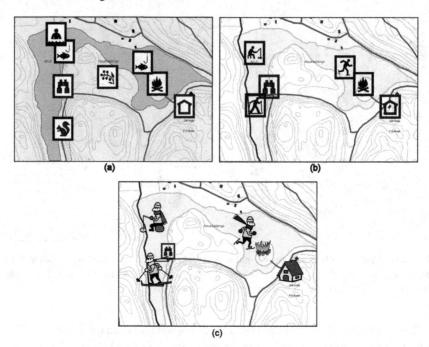

Fig. 12.2. Maps of the same area for different seasons and age groups [27]. Image courtesy Nivala, A.M. and Sarjakoski, L., Finnish Geodetic Institute, FI

for elderly people, while map (b) uses transparent symbols to prevent as much of the information as possible from being hidden, which is critical with small screens. Maps (b) and (c) also use different symbols for the same POIs for adults and for teenagers. Indeed, traditional map symbols may not be familiar to young people and more illustrative symbols were designed to more accurately reflect how they perceive the world.

Some authors propose general approaches for the adaptation of maps to mobile devices. For example, Chalmers et al. [10] studied how to enable map-based applications to adapt to variations in display specifications, network quality, and user's current task. Their approach is based on explicitly considering variants of the features represented on maps (e.g. variants of a road may represent the road at different levels of detail) and on modeling user's preferences for each feature. Users specify their preference for features of a particular type by associating a weight to the type, for instance, to describe a preference for displaying roads rather than rivers. When performing the adaptation, the system determines the content to display according to both user's preferences and overall goals that must be met (e.g. downloading all needed content within a predefined time). This approach was found to be useful in deriving adapted maps with reduced content for transmission with different bandwidths and with different map download time requirements, degrading the data presented while providing users with as much relevant detail as possible.

Zipf [38] provides a comprehensive overview of the design steps involved in adaptive map generation, considering a wide range of variables such as user preferences and interests, tasks, cultural aspects, communicative goals, and current context and location. For example, the orientation of a map can be adapted so that the map is aligned in the direction the user is walking, thus simplifying navigation of an environment, while the meaning of colors can be taken into specific account when generating maps for different cultures.

One of the design steps identified by Zipf and investigated by different authors is shape simplification through *generalization*. Generalization is a graphic- and content-based simplification of the data presented on a map that aims at abstracting irrelevant details to reduce the cognitive effort of the user, and at simplifying the process of creating a lower scale map from a detailed one. As reported in Chaps. 4 and 5, generalization techniques can also be used to support progressive transmission of vector data through wired or wireless networks, albeit studies in this direction for mobile scenarios are still at an early stage.

Agrawala and Stolte [1] developed some techniques for the generalization of cartographic data that improve the usability of maps for road navigation on mobile devices. Standard computer-generated maps are difficult to use because their large, constant scale factor hides short roads and because they are usually cluttered with extraneous details such as city names, parks, and roads that are far away from the route. The techniques proposed by Agrawala and Stolte are based on cognitive psychology research showing that an effective route map must clearly communicate all the turning points on the route and that precisely depicting the exact length, angle, and shape of each road is less important. By distorting road lengths and angles and simplifying road shape, it is possible to clearly and concisely present all the turning points along the route in less screen space. The generalized maps that are obtained exaggerate the length of short roads to ensure their visibility while maintaining a simple, clean design that emphasizes the most essential information for following the route. These generalized maps can fit to the display size of a PDA by rotating the entire route so that the largest extent of the map is aligned with the vertical axis of the page, thus providing extra space in the direction the route needs it most.

Generalizing map features is a useful approach to simplify the display of maps on the small screen of mobile devices, but maps can still be too large to fit into the available screen space. Thus, several techniques have been proposed in the literature to visualize large maps on mobile devices.

A basic approach is to display only a portion of the map and to let users control the portion shown by conceptually moving either a '*viewport*' on top of the map, or the map under the viewport. Scrollbars are typically used to support this interaction, providing separate vertical and horizontal viewport control. Another mechanism is *panning* allows users to drag the map in any direction without any constraint on the movement. It is also common to provide users with a *zooming* function that allows one to increase or decrease the size of the visible portion of the map [16]. Alternative interaction techniques have been developed to simplify these operations on mobile devices. Jones et al. [19], for example, present a technique that combines zooming and scrolling into a single operation, depending on how much users drag the pointing device on the screen with respect to the starting position. Figure 12.3 illustrates the technique. When users start an action by tapping on the map with a pen, two concentric circles are drawn and their center is the location of the action. As the user drags the pointing device, a direction line is drawn between the starting position and its current location, indicating the direction of travel. If the pointer remains within the inner circle, the user is free to scroll within the map in any direction. As the pointer moves further away from the starting position, the scroll rate increases. When the pointer moves beyond the inner circle (threshold A in Fig. 12.3), both zooming and scrolling operations take place. As the user moves closer to the outer circle (threshold B in Fig. 12.3), the map progressively zooms out and the scroll speed is modified to maintain a consistent visual flow. When the pointer reaches the outer circle no further zooming occurs, while scrolling remains active. The rectangle indicates to the user the area of the map that will be displayed once the navigation operation is completed. Its size changes proportionally to the current zoom value. An experimental evaluation showed that the proposed technique reduces the physical navigational workload of users with respect to a standard technique based on the use of scrollbars, panning, and zoom buttons.

Other techniques to display maps on small screens are based on combining panning and zooming with compression or distortion and can be classified into *Overview&Detail* and *Focus&Context* techniques.

Overview&Detail techniques provide one or multiple overviews (usually at a reduced scale) of the whole map, simultaneously with a detailed view of a specific

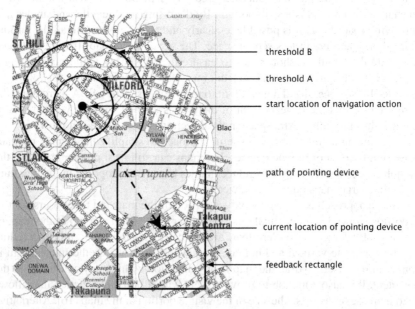

Fig. 12.3. Illustration of the system proposed by Jones et al. [19] to combine panning and zooming into a single operation, including control feedback cues (*which are emphasized for clarity*). Image courtesy Steve Jones, University of Waikato, NZ

portion of map. The Large Focus-Display [21] is an example of such a technique where the overview is a downscaled version of a map that highlights the currently displayed region as a rectangular viewfinder (Fig. 12.4). Users can drag and resize the viewfinder to perform panning and zooming operations. By examining the size and position of the viewfinder, users are also able to derive useful information for the browsing process, such as the scale ratio between the displayed portion and the whole map. Despite these advantages, the overview covers parts of the detailed view and its content is hard to understand because of the scale that needs to be used for it on a small screen.

As a possible alternative, Rosenbaum and Schuman [32] allow users to interact with a grid overlaid on the currently displayed map area to perform panning and zooming operations. The grid is proportional to the whole map and each grid cell can be tapped to display the corresponding portion of map. Cells can also be merged or splitted to provide users with different zoom levels.

Unlike Overview&Detail techniques, Focus&Context techniques are able to display a map at different levels of detail simultaneously without separating the different views. To achieve this result, only a specific area of a map, called "focus area," is represented in full detail, embedded in surrounding context areas distorted to fit the available screen space. These techniques are generally based on the assumption that the interest of the user for a specific map region decreases with the distance from this region. A typical example of Focus&Context technique is given by the Rectangular FishEye-View [29], where a rectangular focus is surrounded by one or more context belts, appropriately scaled to save screen space. The scaling factor for each context belt is usually chosen in such a way that less detail is displayed with increasing distance from the focus.

Variable-scale maps [17] apply the same principle used in the Rectangular FishEye-View but show in full detail a circular area surrounding a specific point

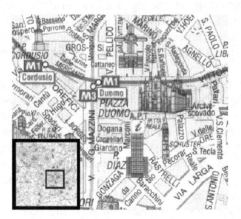

Fig. 12.4. Large Focus-Display: a detailed view of a map area is complemented by an overview of the whole map (*displayed in the corner*) that highlights the currently displayed portion as an interactive rectangular viewfinder

Fig. 12.5. Variable-scale maps: a circular area (*in the center*) is shown in full detail, while the remaining area is generalized and distorted to fit the available space [17]. Image courtesy Lars Harrie, Lund University, SWE

(not necessarily the center of the map) while using a small scale and applying generalization and distortion operations to fit the remaining map area in the available space (Fig. 12.5).

Unlike variable-scale maps, focus maps [39] are not based on distortion but on subdividing a map into different regions of interest and displaying each region with a different amount of detail according to its degree of interest. This is achieved by using generalization and color. Map features lying inside regions with a high degree of interest are less generalized than those inside regions with a low degree of interest. Moreover, bright and shiny colors are used for the former regions, while softer and duller shades of the same colors are used for the latter. As an application example, regions of interest may comprise the region a user is currently in and, if the current task involves movement, the regions the user is about to encounter. In this way, user's attention is directly drawn toward those regions that are currently most relevant, but the other regions can still be used, for example, to help the users to locate and to orient themselves.

Despite their capability to improve map visualization on small screens, Focus&Context techniques are unsuitable for users who need to use large undistorted maps to perform spatial tasks involving distance measurements, such as first responders who need to identify locations of potential hazards in a building or view the real-time location of other team members. To better support these users, one can provide them with information to locate relevant objects even when they are off-screen. This is the approach followed by Baudisch and Rosenholtz [4] who propose Halo a technique to visualize off-screen objects locations by surrounding them with circles that are just large enough to reach into the border region of the display window. From the portion of the circle visualized on-screen, users can derive the off-screen location of the object located in the circle center. A user study has shown that Halo enables users to complete map-based route planning tasks faster than a technique based on displaying arrows coupled with labels for distance indication. In a subsequent work [9], we compared Halo with two other techniques based on exploiting size and body length of arrows, respectively, to inform about the distance of objects. In our study, arrows allowed users to order off-screen objects faster and more accurately according to their distance, while Halo allowed users to better identify the correct location of off-screen objects.

12.3 Supporting User's Navigation in the Field

Navigation can be generally defined as the process whereby people determine where they are, where everything else is, and how to get to particular objects or places [20]. Helping users navigate the geographical area they are in is a typical goal of systems supporting activities in the field. For example, it is a key service of *mobile guides* [5], applications that exploit information such as user position, place, current time, and task, to provide users with information and services related to a specific geographical area. A number of the proposed techniques are based on the visualization of 2D maps representing the considered geographical area but alternative solutions have also been investigated, especially to provide users with directions to reach specific objects or places.

12.3.1 2D Map-based Techniques

2D maps provide information about the geographical area the users are in. By exploiting positioning technologies such as GPS (global positioning system), they can highlight the user's current position by means of a graphical symbol. Furthermore, the positions of objects and other people can be presented. Additionally, maps can show routes and *landmarks* (i.e. distinctive features of an environment, such as churches and squares, that can be used as reference points during navigation) for reaching specific objects or places in a geographical area.

Most of the research results presented in Sect. 12.2 are also significant in the design of maps for navigation. For example, by investigating the effect of map generalization on user's performance in route-following tasks in a geographical area, Dillemuth [13] found that a generalized map was more effective than an aerial photograph (Fig. 12.6). Indeed, users took less time to complete tasks and performed less zooming operations in the former rather than in the latter condition. However, Dillemuth also points out that missing or erroneous information in a map cause confusion and errors in navigating an area, thus suggesting that an accurate aerial map with a lot of detail would be preferable to a generalized but outdated map.

Baus et al. [6] studied how to perform map adaptation for pedestrian navigation according to user's walking speed and accuracy of positional information. Figure 12.7(a) presents an example map for a slowly moving user and unprecise positional information, whereas Fig. 12.7(d) shows a map for exact positional information at higher speed. The precision of positional information is encoded in the size of the dot that represents user's current position on the map. A decreasing positional

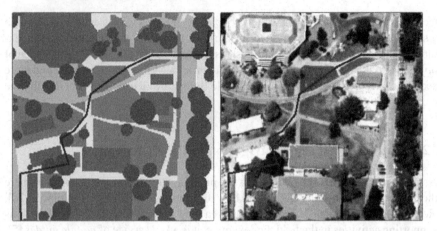

Fig. 12.6. Examples of generalized map (*on the left*) and aerial photograph (*on the right*) used in the route-following study by Dillemuth [13]. Image courtesy Julie Dillemuth, University of California, Santa Barbara, USA

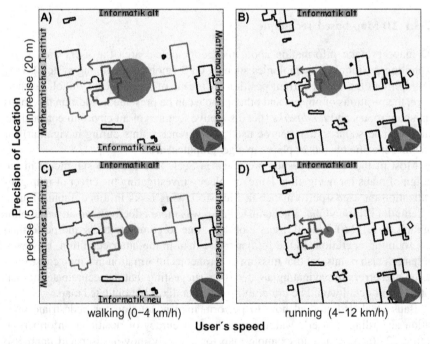

Fig. 12.7. Map adaptation according to user's moving speed and precision of location information [6]. Image courtesy Jörg Baus, Saarland University, DE and Antonio Krüger, University of Muenster, DE

information results in a bigger dot. In addition, if the user moves fast, a greater portion of the map is presented to help the users orient themselves and the amount of information about buildings is reduced.

In general, using a map for orientation implies a mental effort to switch between the egocentric perspective of the viewer and the geocentric perspective of the map. The effort is smaller if the map is *forward-up* (the top of the map shows the environment in front of the viewer) rather than *north-up* (the top of the map always shows the northern part of the environment). Indeed, as shown by various studies (e.g. [18, 34]), the number of navigational errors is lower with a forward-up map compared to a north-up map. Moreover, forward-up maps allow the users to better understand their orientation and to reach targets faster [33]. A further investigation on the orientation of mobile maps has been carried out by Winter and Tomko [37] who argue that it is more intuitive for the users to find their actual position at the bottom of the map rather than at its center, where their positions are typically shown. Indeed, by looking down at the device in their hand, the users perceive the bottom of the map as the closest part to their body, and the map as showing all features in the space ahead of them. This suggests a map design that moves the user's position to the bottom of the mobile map to reduce the cognitive workload of map reading.

12.3.2 Perspective Views and 3D Map-based Techniques

In recent years, some attempts have been made to explore the use of perspective maps and 3D graphics to communicate geographical information on mobile devices for navigation purposes. Perspective views (Fig. 12.8) are based on showing maps with an inclination that should make it easier for users to match what they see in the display with their view in the real world.

Fig. 12.8. Perspective views: maps are inclined to make it easier for users to match what they see in the display with their view in the real world. Source: iGO Website

Exploiting 3D graphics can add further possibilities of visual encoding, can significantly increase the quantity of data displayed on the same screen, and can take advantage of users' natural spatial abilities. However, 3D approaches often suffer from problems such as graphic occlusion and difficulties in comparing heights and sizes of graphical objects. Moreover, designing interfaces for visualizing and manipulating 3D data on mobile devices is much more complex than in desktop applications because of the limitations highlighted in Sect. 12.1.

One of the first investigations on using 3D graphics on mobile devices to support user's navigation has been carried out by Rakkolainen et al. [28] who proposed a system combining a 2D map of a urban environment with a 3D representation of what users see in the physical world (Fig. 12.9). By evaluating the system in the field with a mockup implementation on a laptop computer rather than a PDA, they found that 3D models help users to recognize landmarks and find routes in cities more easily than using a 2D map only.

However, user evaluation of mobile 3D maps is still in its infancy and the first results are not fully consistent. For example, Laakso et al. [25] reached different conclusions with respect to Rakkolainen et al. and found that 3D maps were slower to use both in initial orientation and in route finding compared to 2D maps. Their evaluation concerned a system, called "TellMarisGuide," that supports tourists when they are visiting harbors by visualizing 3D maps of the environment along with more classical 2D maps (Fig. 12.10). 3D maps are used to support navigation in a city and route finding to POIs, such as city attractions or restaurants.

Fig. 12.9. Combining a 2D map of a urban environment with a 3D representation of what users see in the physical world [28]. Image courtesy Ismo Rakkolainen, Tampere University of Technology, FI

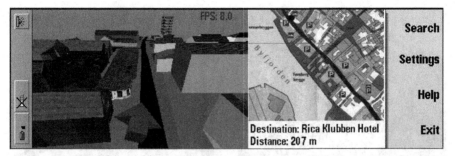

Fig. 12.10. The TellMarisGuide system combining 2D and 3D maps [25]. Image courtesy Katri Laakso, Nokia, FI

Besides providing navigation support, 3D maps can also simplify the access to information related to a geographical area. For example, our LAMP3D system [7] is meant to support the location-aware presentation of 3D content on mobile devices. LAMP3D provides users with a 3D representation of a geographical area, synchronized with the physical world through the use of GPS data, and allows them to request information on the objects they see in the world by directly tapping on their virtual reproduction on the screen (see Fig. 12.11). Using this approach, content is filtered according to user's position and the information about the closest POIs is easier to get.

12.3.3 Alternative Approaches: Text, Audio, and Route Sketches

Besides 2D and 3D maps, other approaches can be used to provide users with information to navigate a geographical area. These approaches are mostly used to support route-following tasks, that is, to provide users with instructions to correctly move along a route to reach some place or object. A very simple approach consists in providing users with textual instructions that are usually easily understood by users and that need few technical resources. However, long descriptions are often needed to give directions (because the context must be explicitly described) and this may quickly increase the cognitive load on the user, reducing the usefulness of the approach.

Another approach consists in providing audio directions. An important advantage of audio is that it does not require users to look at the screen of the mobile device to obtain navigation information, thus simplifying interaction. Unfortunately, audio instructions suffer from the same limitations of textual instructions when complex descriptions must be provided. In an experimental study, Goodman et al. [15] found that text, speech, and text+speech are equally effective in presenting landmark information to people for navigation purposes. Audio can also be used to enable blind users to build a mental model of a geographical area. This can be obtained, for example, by representing important map features, such as POIs, with distinct and unique sounds, called "hearcons" [22]. With hearcons, a representation of the real world with various POIs is given by a virtual auditory environment around

Fig. 12.11. In the LAMP3D system, the user selects objects with a stylus to obtain information on them [7]

the user. The distance between the user's position and each POI is mapped directly onto the loudness of the hearcons. Moreover, by using different sound families for different types of information, the sources can be distinguished through the sense of hearing.

Finally, an additional approach to support navigation consists in the use of *route sketches*, that is, graphical abstractions of a route that provide users with essential information about it. Arrows are a frequently used abstraction: they are familiar to users of car navigation systems and can be ideal for users with limited orientation and map-reading abilities. As reported in [24], the main advantage of route sketches is also their limitation since the high level of abstraction may also take away information that would help a user to find her way. Moreover, this approach is highly dependent on the accuracy of information on user's orientation. If this information is inaccurate, users may be provided with wrong directions, thus compromising their navigation effort.

12.3.4 Combining Audio and Visual Directions

In [12] we carried out a user study to compare different ways of improving users' navigation abilities by combining visual and audio directions on location-aware mobile guides.

We implemented three interfaces that provide the same audio directions but differ in the way they provide visual directions; the first interface adopts a traditional map-based solution, the second combines map indications with pictures of the environment, and the third combines arrow indications and pictures.

As shown in Fig. 12.12(a), the first interface (Map Interface) visualizes the path the user has to follow as a (blue) line and the path the user has already completed as a (gray) bold line. With this interface, the user has to determine the direction by interpreting the map with respect to the physical environment. However, due to the rich information it usually provides, the map can be used even when GPS provides inaccurate data. The map is forward-up and includes street names, POIs (represented by red flags) along the path and its starting and ending points (represented by green flags). The users can pan and zoom the map to look at specific areas and a special button allows them to center the visualization of their current position.

The second interface (Map + Pictures) combines the Map Interface shown in Fig. 12.12(a) with the visualization of pictures as shown in Fig. 12.12(b). The map is visualized when users are walking, while the pictures appear when users are in proximity of relevant choice points such as crossroads. They provide specific views

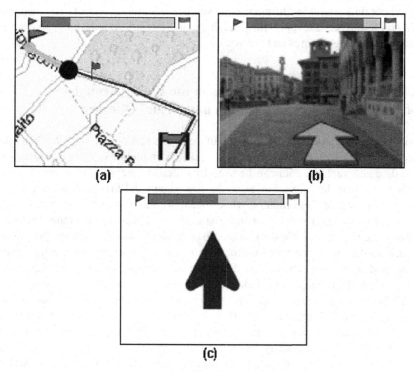

Fig. 12.12. The interfaces compared in the study by Burigat and Chittaro [12] to provide visual directions to users: (a) map; (b) pictures; (c) arrows

of the traveled geographical area and contain yellow 2.5D arrows that are meant to simplify the user's understanding of the direction to follow.

In the third interface (Arrows + Pictures), the map has been replaced by large black arrows (see Fig. 12.12(c)) that indicate the direction to follow. The behavior is similar to the Map + Pictures interface. To determine which direction the user is facing when moving we exploit the succession of position points provided by the GPS and present the appropriate arrows according to the path the user should follow.

The results of the evaluation (described in detail in [12]) show that combining a map with pictures that indicate the direction to follow or removing the map completely and replacing it with a combination of directional arrows and pictures significantly improve user's navigation times with respect to the traditional map condition. This is likely due to the fact that it is more difficult to understand the correct direction to follow with a map (even if the map is forward-up) than with more explicit picture-based indications. Using pictures provides quite good navigational support, because it simplifies the visual recognition of landmarks, and it has the additional advantage of depending only weakly on the actual direction of the user, thus being useful even when position and orientation information are inaccurate. This result is also confirmed by users' comments about the feeling of disorientation sometimes due to the use of the map. Moreover, while both approaches exploiting pictures allowed users to obtain a similar performance, the solution combining pictures and map was highly preferred because it provided a higher amount of information compared to the solution exploiting pictures and arrows.

12.4 Supporting User's Decisions in the Field Through Geographical Data Visualization and Visual Queries

Navigation is only one, albeit important, of the user's tasks in the field that can take advantage of the visualization of geographical data on mobile devices. In particular, up-to-date geographical data can be used by different categories of users to properly support their decisions. For example, firefighters and first-responders can use accurate geographical information while managing the impact of disasters to take decisions to evacuate residents, change management tactics, inform other crews by updating the set of available data on the disaster. Ecologists can employ geographical data to determine the best location to perform observations of animal or plant species and collect data about individuals. Utilities maintenance personnel may accurately locate equipment in the field and update information about its status.

All these activities require users to gain access to geographical data visualizations in the field as well as to manipulate them by modifying features, collect new data, take geo-referenced measurements. Specific mobile GIS (geographical information system) applications are usually devoted to this purpose.

With respect to geographical data visualization, mobile GIS applications are able to display both raster and vector data and manage different geographical features associated with a geographical area (e.g. roads, buildings, boundaries, trees) as separate layers so that users can display only the data they are interested in. Figure 12.13

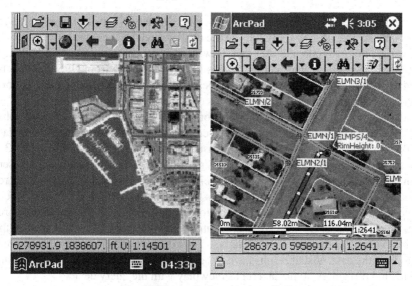

Fig. 12.13. ArcPad screenshots [14]

shows the mobile version of a well-known GIS (ESRI's ArcPad [14]). Users can navigate maps with standard tools such as pan and zoom, and can display map features and their associated attributes, including photographs, documents, video, or sound recordings. Interactive functions allow users to measure distance, radius, and area on-screen and create, delete, and move point, line, and polygon features.

Mobile GIS applications are usually tailored to specific needs by creating custom forms for data entry and by integrating tools to solve specific field-based problems. In [36], for example, a mobile GIS is used to update maps of archaeological areas with the indication of interesting locations and to rapidly collect data about them and the artifacts found therein.

Geographical data can be characterized by an extraordinarily rich number of different attributes, and users operating in the field often need to explore such an information space to support their decisions. While mobile GIS is able to collect, manipulate, and display geographical data in the field, specific visual interfaces are still needed to provide users with complementary exploration and analysis tools. In particular, while mobile GIS allows users to display on a map only the data they are interested in by activating or by deactivating specific layers, solutions that provide users with the capability to visually explore this data, for example, by querying its attributes, are still uncommon.

An approach to support mobile users in visually accessing and querying GIS databases is presented by Lodha et al. [26]. They offer users a variety of queries (e.g. how far, where, closest) for many different types of geometric primitives (e.g. points, lines, polygons) and objects (e.g. buildings, metro stops). Users perform queries by directly interacting with the displayed geographical data. For example, a user can select a building on an aerial map, query for buildings in that area, and receive information such as the building's name in a schematic view. The user can paint points, lines, and arbitrary polygons on the display, and use these primitives as input to queries. For example, the user can draw two polygonal regions to find buildings contained within their intersection. User's location obtained from GPS can also be used as input to queries, for example to locate the nearest metro stop and telephone at the end of a path, highlighting buildings close to the user's path.

In [8], we have presented an application, called "*mobile analysis of geographic data*" (MAGDA), aimed at supporting users in the analysis of georeferenced data on PDAs. The approach we followed is based on exploiting *dynamic queries* [3, 35] that are typically used in desktop scenarios to explore large data sets, providing users with a fast and easy-to-use method to specify queries and visually analyze their results. The basic idea of dynamic queries is to combine input widgets (called "query devices" [2]), such as sliders or check buttons, with graphical representations of results, such as maps. By directly manipulating query devices, users can specify the desired values for the attributes of elements in a data set and can easily get different subsets of the data. Visual results have to be rapidly updated to enable users to learn interesting properties of the data set as they play with the query devices.

MAGDA allows users to select different categories of geographical objects (as with mobile GIS layers) and displays these objects as icons superimposed on the map of the considered geographical area (see Fig. 12.14) in their georeferenced position.

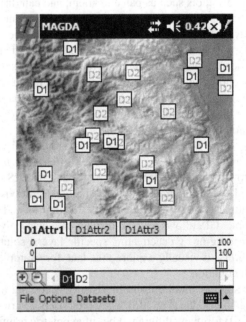

Fig. 12.14. The map displays all elements of the selected categories. A tabbed panel contains all query devices related to the currently explored category, which is highlighted in the toolbar at the bottom of the screen [8]

To define queries, users interact with query devices contained in a tabbed interface, where each tab allows users to specify values for a single attribute of the considered elements. Figure 12.14 shows an example where elements of categories called "D1" and "D2" are shown on the map, and it is possible to specify values for the attributes "D1Attr1," "D1Attr2," and "D1Attr3" of category "D1" by accessing the appropriate tabs. Elements belonging to the currently explored category are highlighted (by shading all other elements) to improve their visibility and reduce visual clutter.

As is typical of dynamic query systems, MAGDA can display all and only the elements that satisfy the current query (we call this *on-off visualization*). Thus, while users are manipulating query devices to specify attribute values, icons representing elements that satisfy (or not) the query are switched on (or off), providing a perceptual visual cue to make it easy for users to understand the effects of changes in the query (see Fig. 12.15).

However, the on–off visualization does not allow one to quickly determine how elements are distributed in an area according to how much they satisfy a query. This prevents users, for example, from easily identifying suboptimal results (which may be particularly important when a query produces no results) or finding out interesting patterns, trends, or anomalies in the data that may prompt further investigations in the field (such as elements fully satisfying a query surrounded by elements that do not satisfy it at all).

A possible solution to this problem consists in exploiting graphical properties of icons to highlight the state of the objects. Reichenbacher [31], for example, suggests varying the opacity of icons to show qualitative or quantitative differences between objects. This method catches user's attention and directs it to the important and more

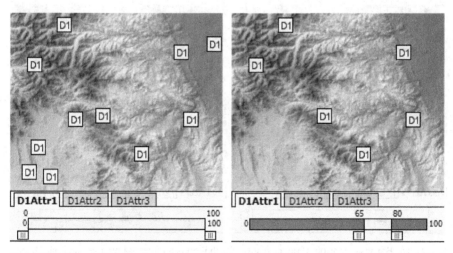

Fig. 12.15. On–off visualization of query results. On the left, no condition has been specified for attributes and all elements are visualized. On the right, a specific range of values has been specified for one attribute and only elements satisfying this condition are visualized [8]

relevant information without completely neglecting other information that could become important.

The solution we adopted in MAGDA, called "bar visualization," augments all icons on the map with a vertical bar that is used to represent how much each element satisfies the user's query (Fig. 12.16). By default, we fill the bar associated with each element with a green area whose size is proportional to the number of satisfied conditions, while the remaining area gets filled in red. Using this visualization, it is easy for users to track elements that fully satisfy a query (i.e. those with a completely green bar) as well as to visually compare how much different elements satisfy the specified set of conditions (the less an element satisfies the query, the bigger the red area in the bar).

Figure 12.17 shows a scenario where MAGDA is used as a tool to analyze georeferenced probes of soil characterized by three continuous attributes: pH, salinity, and cation-exchange capacity (CEC). By properly manipulating query devices associated with these attributes and by looking at the results on the map of the considered geographical area, users are able to visually analyze the data to identify patterns, outliers, and clusters. For example, by exploiting the correlation between soil chemical properties and vegetation it is possible to identify which probed areas are the most suitable for different plant species. In the figure, a 60–200 interval has been specified for the CEC property (these are typical CEC values for an organic soil type), a 5.5–7.5 interval for the pH property (optimal pH values for plant growth), and a 0–4 interval for the salinity property (values above 4 may restrict the growth of many plants). As shown by the bar visualization, areas on the left are more suitable for most plants compared with areas on the right. However, while in the field, users may be

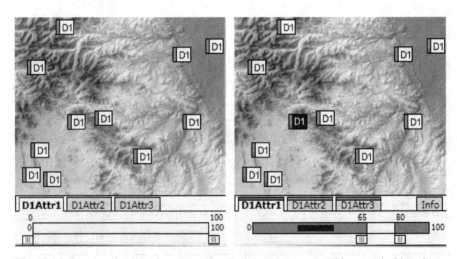

Fig. 12.16. Bar visualization of query results. Each icon is augmented by a vertical bar showing how much the corresponding element satisfies the user's query. Users can visually perceive the effects of their queries by observing changes in the color-filled areas of bars while manipulating query devices [8]

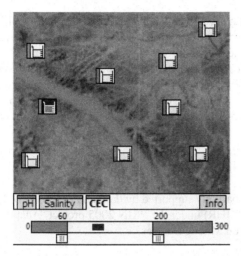

Fig. 12.17. Analyzing correlations between soil chemical properties and vegetation [8]

interested in exploring areas that are only partially suitable for plants (e.g. to acquire data on plant species that are able to survive in those areas). Furthermore, if interesting phenomena are identified (e.g. a fully unsatisfied result surrounded by fully satisfied ones, or a sudden variation in how much a query is satisfied between bordering areas), they may prompt researchers to perform more accurate investigations.

12.5 Conclusions

Mobile technologies have tremendous potential in supporting users in the field. In particular, various tasks, from searching for specific objects in an area to field data acquisition, can benefit from the possibility of exploiting devices such as PDAs and Smartphones to get and store georeferenced data. Designing solutions for the visualization, exploration and use of geographical data on mobile devices is thus of fundamental importance.

This chapter has given an overview of research on geographical data visualization on mobile devices, with emphasis on supporting user's navigation and user's decisions in the field. In recent years, the first of these two topics has received a lot of attention and various techniques have been proposed. However, there is still a need to compare these techniques and assess their effectiveness in different situations. For example, how much perspective and 3D map-based techniques are suitable to support navigation tasks is still unclear and further studies are necessary to identify the best presentation techniques for different classes of users.

The constant evolution of mobile technologies is also introducing the opportunity to use mobile devices as interactive tools to analyze geographical data and obtain the most appropriate information to support user's decisions where and when needed. However, current solutions, such as mobile GIS applications, while suitable for collecting and displaying geographical data in the field, are still inadequate as mobile analysis tools and additional investigations are thus needed to provide users with more powerful and flexible tools.

In the near future, a growing demand for mobile solutions to support user's activities in the field is likely. As new needs and new issues emerge, research on geographical data visualization on mobile devices will remain crucial to produce applications that can be easily and effectively used while on the move.

Acknowledgments

This work has been partially supported by the Italian Ministry of Education, University and Research (MIUR) under the PRIN 2005 project "Adaptive, Context-aware, Multimedia Guides on Mobile Devices."

References

1. Agrawala M, Stolte C (2001) Rendering Effective Route Maps: Improving Usability Through Generalization. In: Proc. Conference on Computer Graphics and Interactive Techniques (SIGGRAPH 2001). ACM Press, New York pp. 241–249
2. Ahlberg C, Truvé S (1995) Exploring Terra Incognita in the Design Space of Query Devices. In: Proc. Working Conference on Engineering for Human Computer Interaction (EHCI 95). Chapman & Hall, Boca Raton pp. 305–321
3. Ahlberg C, Williamson C, Shneiderman B (1992) Dynamic Queries for Information Exploration: an Implementation and Evaluation. In: Proc. Conference on Human Factors in Computing Systems (CHI 92). ACM Press, New York pp. 619–626
4. Baudisch P, Rosenholtz R (2003) Halo: a Technique for Visualizing Off-screen Locations. In: Proc. Conference on Human Factors in Computing Systems (CHI 2003). ACM Press, New York pp. 481–488
5. Baus J, Cheverst K, Kray C (2005) A Survey of Map-based Mobile Guides. In: Zipf A, Meng L, Reichenbacher T (eds) Map-based mobile services – Theories, Methods, and Implementations. Springer, Berlin Heidelberg New York pp. 197–216
6. Baus J, Kruger A, Wahlster W (2002) A Resource-adaptive Mobile Navigation System. In: Proc. Conference on Intelligent User Interfaces (IUI 2002). ACM Press, New York pp. 15–22
7. Burigat S, Chittaro L (2005) Location-aware Visualization of VRML Models in GPS-based Mobile Guides. In: Proc. Conference on 3D Web Technology (WEB3D 2005). ACM Press, New York pp. 57–64
8. Burigat S, Chittaro L (2005) Visualizing the Results of Interactive Queries for Geographic Data on Mobile Devices. In: Proc. Symposium on Advances in Geographic Information Systems (ACM GIS 2005). ACM Press, New York pp. 277–284
9. Burigat S, Chittaro L (2006) Visualizing Locations of Off-screen Objects on Mobile Devices: a Comparative Evaluation of Three Approaches. In: Proc. Conference on Human-Computer Interaction with Mobile Devices and Services (MobileHCI 2006). ACM Press, New York pp. 239–246

10. Chalmers D, Sloman M, Dulay N (2001) Map Adaptation for Users of Mobile Systems. In: Proc. World Wide Web Conference (WWW 2001). ACM Press, New York pp. 735–744

11. Chittaro L (2006) Visualizing Information on Mobile Devices. IEEE Computer 39(3): 40–45

12. Chittaro L, Burigat S (2005) Augmenting Audio Messages with Visual Directions in Mobile Guides: an Evaluation of Three Approaches. In: Proc. Conference on Human-Computer Interaction with Mobile Devices and Services (MobileHCI 2005). ACM Press, New York pp. 107–114

13. Dillemuth J (2005) Map Design Evaluation for Mobile Displays. Cartography and Geographic Information Science 32(4):285–301

14. ESRI. http://www.esri.com/software/arcgis/about/mobile.html

15. Goodman J, Brewster S, Gray P (2005) How Can We Best Use Landmarks to Support Older People in Navigation? Behaviour & Information Technology 24(1):3–20

16. Gutwin C, Fedak C (2004) Interacting with Big Interfaces on Small Screens: a Comparison of Fisheye, Zoom, and Panning Techniques. In: Proc. Conference on Graphics Interface (GI 2004). A K Peters, Wellesley pp. 145–152

17. Harrie L, Sarjakoski L T, Lehto L (2002) A Mapping Function for Variable-scale Maps in Small-display Cartography. Journal of Geospatial Engineering 2(3):111–123

18. Hermann F, Bieber G, Duesterhoeft A (2003) Egocentric Maps on Mobile Devices. In: Proc. International Workshop on Mobile Computing (IMC 2003). IRB Verlag, Stuttgart pp. 32–37

19. Jones S, Jones M, Marsden G, Patel D, Cockburn A (2005) An Evaluation of Integrated Zooming and Scrolling on Small Screens. International Journal of Human-Computer Studies 63(3):271–303

20. Jul S, Furnas G W (1997) Navigation in Electronic Worlds. SIGCHI Bulletin 29(4): 44–49

21. Karstens B, Rosenbaum R, Schumann H (2004) Presenting Large and Complex Information Sets on Mobile Handhelds. In: Candace Deans P (ed) E-Commerce and M-Commerce Technologies. IRM Press, Hershey pp. 32–56

22. Klante P, Krosche J, Boll S (2004) AccesSights - a Multimodal Location-aware Mobile Tourist Information System. In: Proc. Conference on Computers Helping People with Special Needs (ICCHP 2004). Springer, Berlin Heidelberg New York pp. 287–294

23. Kraak M J (2002) Current Trends in Visualisation of Geospatial Data with Special Reference to Cartography. Indian Cartographer 22:319–324

24. Kray C, Laakso K, Elting C, Coors V (2003) Presenting Route Instructions on Mobile Devices. In: Proc. Conference on Intelligent User Interfaces (IUI 2003). ACM Press, New York pp. 117–124

25. Laakso K, Gjesdal O, Sulebak J R (2003) Tourist Information and Navigation Support by Using 3D Maps Displayed on Mobile Devices. In: Proc. Mobile HCI Workshop on HCI in Mobile Guides pp. 34–39

26. Lodha S K, Faaland N M, Wong G, Charaniya A P, Ramalingam S, Keller A P (2003) Consistent Visualization and Querying of GIS Databases by a Location-Aware Mobile Agent. In: Proc. Computer Graphics International (CGI 2003). IEEE Press, Los Alamitos pp. 248–253

27. Nivala A M, Sarjakoski L T (2005) Adapting Map Symbols for Mobile Users. In: Proc. 22nd International Cartographic Conference (ICC 2005) July 9–16 La Coruna, Spain, CD-ROM

28. Rakkolainen I, Vainio T (2001) A 3D City Info for Mobile Users. Computers & Graphics 25(4):619–625

29. Rauschenbach U, Jeschke S, Schumann H (2001) General Rectangular FishEye Views for 2D Graphics. Computers and Graphics 25(4):609–617

30. Reichenbacher T (2001) Adaptive Concepts for a Mobile Cartography. Journal of Geographical Sciences 11:43–53

31. Reichenbacher T (2004) Mobile Cartography – Adaptive Visualisation of Geographic Information on Mobile Devices. PhD thesis, Technischen Universitat Munchen

32. Rosenbaum R, Schumann H (2005) Grid-based Interaction for Effective Image Browsing on Mobile Devices. In: Proc. of the SPIE, Volume 5684. SPIE - The International Society for Optical Engineering, Bellingham pp. 170–180

33. Sas C, O'Grady M, O'Hare G (2003) Electronic Navigation – Some Design Issues. In: Proc. Conference on Human-Computer Interaction with Mobile Devices and Services (MobileHCI 2003). Springer, Berlin Heidelberg New York pp. 471–475

34. Schilling A, Zipf A (2003) Generation of VRML City Models for Focus Based Tour Animations - Integration, Modeling and Presentation of Heterogeneous Geo-Data Sources. In: Proc. Conference on 3D Web Technology (WEB3D 2003). ACM Press, New York pp. 86–97

35. Shneiderman B (1994) Dynamic Queries for Visual Information Seeking. IEEE Software 11(3):70–77

36. Tripcevich N (2004) Flexibility by Design: How Mobile GIS Meets the Needs of Archaeological Survey. Cartography and Geographic Information Science 31(3):137–151

37. Winter S, Tomko M (2004) Shifting the Focus in Mobile Maps. In: Mortia T. (Ed.) Proc. International Joint Workshop on Ubiquitous, Pervasive and Internet Mapping UPIMap 2004, Tokyo, pp. 153–165

38. Zipf A (2002) User-adaptive Maps for Location-based Services (LBS) for Tourism. In: Proc. Conference for Information and Communication Technologies in Travel & Tourism (ENTER 2002). Springer, Berlin Heidelberg New York pp. 329–338

39. Zipf A, Richter K F (2002) Using Focus Maps to Ease Map Reading: Developing Smart Applications for Mobile Devices. Kunstliche Intelligenz 4:35–37

References containing URLs are validated as of October 1st, 2006.

13

Tracking of Moving Objects with Accuracy Guarantees

Alminas Čivilis[1], Christian S. Jensen[2], and Stardas Pakalnis[2]

[1] Vilnius University, Vilnius (Lithuania)
[2] Aalborg University, Aalborg Ost (Denmark)

13.1 Introduction

In step with the increasing availability of an infrastructure for mobile, online location-based services (LBSs) for general consumers, such services are attracting increasing attention in industry and academia [9, 18]. An LBS is a service that provides location-based information to mobile users. A key idea is to provide a service that is dependent on positional information associated with the user, most importantly the user's current location. The services may also be dependent on other factors, such as the personal preferences and interests of the user [3].

Examples of LBSs abound. A service might inform its users about traffic jams and weather situations that are expected to be of relevance to each user. A friend monitor may inform each user about the current whereabouts of nearby friends. Other services may track the positions of emergency vehicles, police cars, security personnel, hazardous materials, or public transport. Recreational services and games, as exemplified by geocaching [8], the Raygun [7] game, may be envisioned. In the latter type of game, individuals catch virtual ghosts (with geographical coordinates) that are displayed on the screens of their mobile phones.

Services such as these rely to varying degrees on the tracking of the geographical positions of moving objects. For example, traffic jams may be identified by monitoring the movements of service users; the users that should receive specific traffic jam or weather information are identified by tracking the users' positions. Some services require only fairly inaccurate tracking, for example, the weather service, while other services require much more accurate tracking, for example, location-based games. In contrast to Chap. 12, this chapter does not consider issues to do with user interface design, but rather concerns support for fundamental functionality that may be exploited by mobile services.

We assume that the users are equipped with wireless devices (e.g. mobile phones) that are online via some form of wireless communication network. We also assume that the GPS [20] positions of the users are available.

To accomplish tracking with a certain accuracy, an approach is used where each wireless device, termed "a moving object," monitors its real position (its GPS position) and compares this with a local copy of the position that the server-side database assumes. When needed in order to maintain the required accuracy in the database, the object issues an update to the server. The database may predict the future positions of a moving object in different ways.

The challenge is then how to predict the future positions of a moving object so that the number of updates is reduced. This in turn results in reduced communication and server-side update processing. The chapter initially covers three basic techniques for predicting the future positions of a moving object. The first two are point- and vector-based tracking, where an object is assumed to be stationary and to move according to a velocity vector, respectively. In the third approach, segment-based tracking, the future movement of an object is represented by a road segment drawn from a representation of the underlying road network and a fixed speed. A road segment is a polyline, that is, a sequence of connected line segments. So, this representation assumes that a moving object moves along a known road segment at constant speed.

As explained above, a moving object is aware of the server-side representation of its movement. The server uses this representation for predicting the current position of the moving object. The client-side moving object uses the representation for ensuring that the server's predicted position is within the predefined accuracy.

The chapter also covers techniques that aim to improve the basic segment-based approach. The chapter considers modifications of the segments that make up the representation of the road network. The chapter covers the use of anticipated routes for the moving objects, which are represented as (long) polylines, instead of individual segments drawn from the road network representation. The chapter explores the use of acceleration profiles instead of modeling the speed of an object as being constant in between updates.

In summary, the chapter covers three types of techniques that aim to reduce the communication and the update costs associated with the tracking of moving objects with accuracy guarantees, and it reports on empirical evaluations of these techniques and the best existing tracking techniques based on real data.

Chapter 9 concerns access control for LBSs – the reader is referred to that chapter for further information on this highly relevant aspect of tracking.

The coverage of tracking techniques is primarily based on proposals by Čivilis et al. [4, 5] and Jensen et al. [11]. These works share the general approach with Wolfson et al. [22, 24]. The chapter offers results of new empirical performances studies, based on two real GPS data sets, of the techniques presented. A more detailed coverage of related studies is given in Sect. 13.8.

The presentation is organized as follows. Section 13.2 describes the general approach to tracking and describes the data sets used in experiments. Section 13.3 describes point-, vector-, and segment-based tracking. Section 13.4 covers improvements in the segment-based approach using road network modifications. Sections 13.5 and 13.6 present techniques for update reduction using routes and acceleration profiles, respectively. Section 13.7 is a summary, and Sect. 13.8 covers commercial developments and points to further readings.

13.2 Background

We first describe the general approach to tracking that we use. Then we describe the real-world GPS and road network data that we use for evaluating the different tracking techniques.

13.2.1 Tracking Approach

We assume that moving objects are constrained by a road network and that they are capable of obtaining their positions from an associated GPS receiver. Moving objects, also termed "clients," send their location information to a central database, also termed "the server," via a wireless communication network. We assume that disconnects between client and server are dealt with by other mechanism in the network than the tracking techniques we consider. When a disconnect occurs, these mechanisms notify the server that may then take appropriate action.

After each update from a moving object, the database informs the moving object of the representation, or prediction function, it will use for the object's position. The moving object is then always aware of where the server thinks it is located. The moving object issues an update when the predicted position deviates by some threshold from the real position obtained from the GPS receiver. We term this the "shared prediction-based approach" to tracking.

Figure 13.1 presents a UML activity diagram for this tracking approach (activity diagrams model activities that change object states). The object initially obtains its location information from its GPS receiver. It then establishes a connection with the server and issues an update, sending its GPS information and unique identifier to the server.

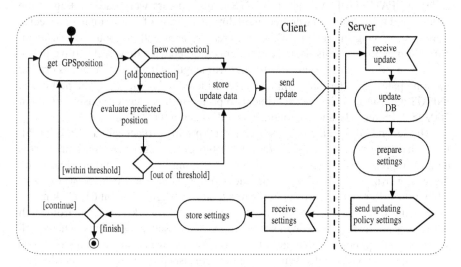

Fig. 13.1. Tracking scenario diagram

Having received this update, the server determines which tracking technique and threshold to be used for the object (these are predefined), and it stores the information received from the object in the database. If segment-based tracking is to be used, the server also uses map matching to determine on which road segment the object is moving. The server then sends its representation, or prediction, of the object's current and future position to the object.

Having received this information from the server, the object again obtains its actual, current location information from the GPS receiver. It then calculates its predicted position using the representation received from the server, and it compares this to the GPS position. If the difference between these two exceeds the given threshold, the client issues an update to the server. If not, a new comparison is made. This procedure continues until it is terminated by the object. Although the server may also initiate and terminate the tracking, we assume for simplicity that the object is in control. This aspect has no impact on the chapter's exposition.

13.2.2 Data Description

As mentioned, we assume that GPS is used for positioning of the moving objects. In making this assumption, we note that Galileo-based positioning [19] and hybrid GPS/Galileo-based positioning are likely to work even better when they become available. In experiments, the results of which will be reported upon throughout the chapter, we used two data sets of GPS logs. Both were obtained by installing GPS receivers together with small computers in a number of vehicles. The positions of the vehicles were recorded every second while the vehicles were driving. Positions were not recorded for a vehicle when its engine was turned off.

The first GPS data set stems from a Danish intelligent speed adaptation project called "INFATI" [10]. A total of 20 GPS equipped cars were participating in the project, and their positions were recorded during a period of approximately 8 weeks. Cars were driving in Aalborg, Denmark area, an area with a population of about 140,000 inhabitants. This data set represents the behavior of vehicles traveling in semiurban surroundings. Here, the average trip length is 9.5 km (continuous driving ignoring pauses shorter than 5 minutes is considered to be one trip). The part of the INFATI data set used in the experiments reported in this chapter consists of about 500,000 GPS records, and the total trip length is about 9,000 km.

The second GPS data set stems from a road-pricing project called "AKTA" [16]. Here, the participating cars were driving in the Copenhagen, Denmark area. This data set represents the behavior of vehicles traveling in a larger urban area. Here, the average trip length is 17.9 km, and the part of the data set used consists of about 4,000,000 GPS records, corresponding to a total trip length of about 67,000 km.

For the experiments, we also used digital road networks obtained from both projects. The initial road networks were composed of sets of segments, where each segment corresponds to some part of a road between two consecutive intersections or and intersection and a dead end. A segment consists of a sequence of coordinates, that is, it is a polyline. Further, the road networks are partitioned into named roads or streets, meaning that each segment belongs to precisely one road or street. Each segment identifies its road or street by means of a street code. Chap. 2 offers additional detail on more comprehensive modeling of road networks.

13.3 Fundamental Tracking Techniques

We proceed to describe three tracking techniques that follow the scenario described in Sect. 13.2.1 but differ in how they predict the future positions of a moving object. These were covered by Čivilis et al. [4]; minor variations of the first and third of these were also studied by Wolfson and Yin [24] (see Sect. 13.8 for additional discussion).

13.3.1 Point-based Tracking

Using this technique, the server represents an object's future position as the most recently reported position. An update is issued by an object when its distance to the previously reported position deviates from its current GPS position by the specified threshold. Thus, the movement of an object is represented as a "jumping point." This technique is the most primitive among the techniques presented, but it may well be suitable for movement that is erratic, or undirected, with respect to the threshold used. An example is the tracking with a threshold of 200 m of children who are playing soccer.

The algorithm for point-based tracking, **PP** (Predict with Point), is simple.

Algorithm 13.3.1 PP(mo)

(1)   **return** $mo.p$

As the prediction is constant, the predicted position is the same as the input position. Here $mo.p$ is the position of the moving object.

13.3.2 Vector-based Tracking

In vector-based tracking, the future positions of a moving object are given by a linear function of time, that is, by a start position and a velocity vector. Point-based tracking then corresponds to the special case where the velocity vector is the zero vector.

A GPS receiver computes both the speed and the heading for the object it is associated with — the velocity vector used in this representation is computed from these two. Using this technique, the movement of an object is represented as a "jumping vector." Vector-based tracking may be useful for the tracking of "directed" movement.

Algorithm **PV** (Predict with Vector) predicts the location of the given object mo at a given time t_{cur}.

Algorithm 13.3.2 PV(mo, t_{cur})

(1) $p_{pred} \leftarrow mo.p + mo.\bar{v}(t_{cur} - mo.t)$
(2)   **return** p_{pred}

The result of the algorithm is the location of mo at time t_{cur}. The predicted location is calculated by adding the time-dependent traveled distance ($t_{cur} - mo.t$) to the starting point in the direction of vector $\bar{v}$.

13.3.3 Segment-based Tracking

Here, the main idea is to utilize knowledge of the road network in which the objects move. A digital representation of the road network is thus required to be available. The server uses the GPS location information it receives from an object to locate the object within the road network. This is done by means of map matching, which is a technique that positions an object on a road network segment, specified as a distance from the start of that segment, based on location information from a GPS receiver.

In segment-based tracking, the future positions of an object are given by a movement at constant speed along the identified segment that is represented as a polyline. The speed used is the speed most recently reported by the client. When or if a predicted position reaches the end of its segment, the predicted position remains there from then on. In effect, the segment-based tracking switches to point-based tracking.

Special steps are needed to ensure robustness when segment-based tracking is used. In particular, if for some reason, a matching road segment cannot be found when a moving object issues an update, the segment-based approach switches temporarily to the vector-based approach that is always applicable. On the next update, the server will again try to find a matching road segment in the database. This arrangement ensures that segment-based tracking works even when map matching fails. Map matching may fail to identify a segment for several reasons. For example, the map available may be inaccurate, or it may not cover the area in which the client is located.

Using segment-based tracking, the movement of an object is represented as a set of road segments with positions on them, and as jumping vectors in case map matching fails. This technique takes into account the shape of the road on which an object is moving – an object thus moves according to the shape of the road.

Algorithm **PS** (Predict with Segment) is defined as follows.

Algorithm 13.3.3 PS(mop, t_{cur})

(1) $m_{pred} \leftarrow mop.m + mop.plspdt_{cur} - mop.t)$
(2)   **if** $m_{pred} >= \mathcal{M}mop.pl, p_n)$ **then return** $mop.pl.p_n$
(3)   **elsif** $m_{pred} <= 0$ **then return** $mop.pl.p_0$
(4)   **else**
(5)   **return** $\mathcal{M}^{-1}(mop.pl, m_{pred})$
(6)   **end if**

Here, an object's position is given by a polyline *mop.pl* and a measure *mop.m*. The predicted location is given as a new measure that is equal to the old measure (line 1) plus the distance traveled since the last update, $t_{cur} - mop.t$. The traveled distance is negative if the object moves against the direction of the polyline. If the new measure is outside the polyline (lines 2 and 3), the result is one of the end points of the polyline. In this way, the position prediction stops at a boundary point of the polyline (the first point $pl.p_0$ or the last point $pl.p_n$). Function $\mathcal{M}$ calculates the measure value on a given polyline of a given coordinate point. Otherwise, the coordinate point of m_{pred} is calculated with the inverse function $\mathcal{M}^{-1}$ that calculates the coordinate point of a given measure value on a given polyline.

13.3.4 Comparison of Tracking Techniques

Results of experimental studies with the three tracking techniques are presented in Fig. 13.2. Here both the INFATI and AKTA data sets, described in Sect. 13.2.2, are used. For the experiment, the INFATI data set contributed with approximately 500,000 GPS positions collected from five cars. The AKTA data set contributed with 4,000,000 GPS positions obtained from five cars. Labels for results obtained using the INFATI data start with 'i' while labels for results obtained using the AKTA data start with an 'a.'

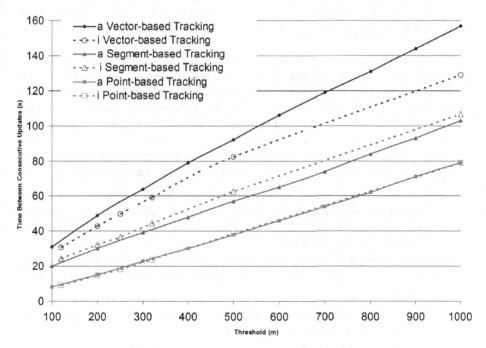

Fig. 13.2. Comparison of tracking techniques

The results were obtained by simulating the scenario described in Sect. 13.2.1 with thresholds ranging from 100 to 1,000 m. Specifically, the movement of each car was simulated using the log of GPS positions for the car. So a client program and a server program interact, and a simple experiment management system is in charge of the bookkeeping needed to obtain the performance results. Instead of obtaining GPS positions from a GPS device in real time, the client program utilizes the GPS logs, which of course makes the simulation much faster than the reality being simulated. The bookkeeping involves the counting of updates sent from the client program to the server program and keeping track of time.

All performance studies reported in this chapter follow this pattern. The studies differ in the specific GPS data and road networks used, and in the tracking policies used.

In Fig. 13.2, accuracy threshold values in meters are on the x-axis. The client obtains a GPS position from the GPS device every second and performs a comparison between the GPS position and the predicted position. The y-axis then gives the average number of seconds between consecutive updates sent from the client to the server to maintain the required accuracy.

It is seen that the time between updates increases as the accuracy threshold increases, that is, as the required accuracy decreases. Point-based tracking shows the worst performance. The largest improvement of the segment- and vector-based techniques over the point-based technique is for smaller thresholds, while for larger thresholds the improvement is smaller. For thresholds below 200 m, segment- and vector-based tracking policies are more than two times better than point-based tracking.

Segment-based tracking is outperformed because the road segments in the underlying road network are relatively short, having an average length of 174 m. For example, this means that a relatively straight road is represented by several segments. In this case, vector-based tracking may need less updates. So, although vector-based tracking is simpler and performs slightly better, we find it likely that it is possible to improve segment-based tracking to become the best.

In addition, segment-based tracking, by relating the location of a moving object to the underlying road network, offers other advantages that are as follows:

- Buildings, parking places, traffic jams, points of interest, traffic signs, and other road-related information that is mapped to the road network can easily be associated with the location of a moving object.
- Road network-based distances can be used in place of Euclidean distances.
- Acceleration profiles, driver behavior on crossroads, and other road-related data that increase the knowledge about the future positions of moving objects can be exploited.

Consequently, a promising direction for obtaining improved tracking is to continue in the direction of segment-based tracking.

13.4 Modifying the Road Network Representation

Recall that with segment-based tracking, the predicted position of an object moves at constant speed along a segment drawn from the underlying representation of the road network until it reaches the end of the segment, at which time the predicted position remains at the end of the segment. The experimental study reported in Sect. 13.3 indicates that the numbers of updates in segment-based tracking are closely correlated with the numbers of segment changes. This motivates attempts to modify the underlying road network representation so that less segment changes occur.

We proceed to cover several modifications of the road network representation. The main idea is to connect road segments in such a way that moving objects would have to change segments as few times as possible as they travel in the road network. We first describe three modification approaches and then report on experimental studies with these approaches.

13.4.1 Modification Approaches

The general idea of the road network modification is to iterate through all segments in the road network to be modified according to some specified ordering. During each iteration, the modification algorithm orders all available segments and then tries to extend the topmost, or current, segment with other segments. To do this, the algorithm identifies all the existing segments that start or end at the start or end of the current segment and extends the current segment with the most attractive of such segment(s) according to some other specified ordering. A current segment that has been extended is considered in the next iteration, but the segment(s) that were used for the extension are disregarded. A current segment that has not been extended becomes part of the final result and is not considered any further. Several different ordering strategies can be considered. Details on the modification approaches beyond what is reported next are given elsewhere [5].

Street code-based Approach

As described at the end of Sect. 13.2.2, a named street (e.g. "Main Street") is represented by many segments, and the segments have street codes that identify the named street that they represent part of.

The ordering in the street code-based (SSC) approach exploits the availability of street codes for the segments. The idea is to give priority to connecting polylines with the same street code. In this way, longer road segments are constructed that tend to correspond to parts of the same street. In cases where there are several candidates with the same street code, priority is given to the shortest polyline. This strategy reduces the probability that unconnected polylines will be short.

Tail Disconnection Approach

The SSC does not distinguish between main roads and side streets. The observation that motivates the tail disconnection approach (TSC) is that moving objects can be assumed to be moving on main roads most of the time. This approach thus first connects polylines while disregarding side streets, termed "tails," and it only subsequently takes the tails into account.

To be more precise, a polyline *pl* in a set of polylines *rn* is a *tail* if at least one delimiting point of *pl* is not connected to any delimiting point of any other polyline in *rn*. Tails are also termed "first-level tails." The *i*th level tails in *rn* are those polylines that are tails in the set obtained by subtracting all tails at lower levels than *i* from *rn*.

A few observations are in order. If a road network has a purely hierarchical structure, each polyline is a tail at some level. Polylines that belong to a circular structure in a road network, that is, a structure where each constituent polyline is connected at both ends, are not tails. A highest tail level is assigned to all non-tail polylines. The ranking is based on the tail level.

Direction-based Approach

The last approach takes into account the directions of the candidate polylines at the connection points. The idea is that moving objects are expected to prefer to be moving as directly as possible toward their destinations, which means that they will tend to move as straight as possible and by making as few turns as possible.

This approach thus gives preference to polylines that continue in the same direction as much as possible when extending a polyline. Put differently, preference is given to polylines with a direction at a connection point that has as small an *angle* as possible with respect to the direction of the polyline to be extended, again at the connection point.

Figure 13.3 explains this further. The property *angle* denotes the angle between the polyline being extended and a candidate for use in the extension. The figure contains two such candidates and thus has two angles. Specifically, the line segment at the connection point of the polyline to be extended is itself extended toward the candidate polyline. This extension corresponds to a straight extension of the polyline's line segment at the connection point. The property *angle* is then the angle between the extended line segment and the line segment of the candidate polyline at the connection point. A small angle is thus preferable.

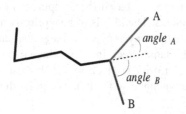

Fig. 13.3. The *angle* property between polylines

13.4.2 Comparison of Approaches

The goal of the road modification approaches described in the previous section is to connect the polylines of road segments into longer polylines, so that moving objects travel on fewer polylines. In doing this, we assume that objects in a road network move mostly along the main roads. We proceed to evaluate the results of the road network modifications in terms of how well the constructed polylines correspond to the main roads.

All policies succeeded in connecting short polylines into longer ones. The polylines created by the SSC were the worst in connecting the main roads. In residential and other areas, it is common for a main road and its side streets to have the same street code. (Or put differently, a road may have many side roads.)

Figure 13.4 offers a comparison of the update performance for segment-based tracking using the unmodified road network and the road networks resulting from the

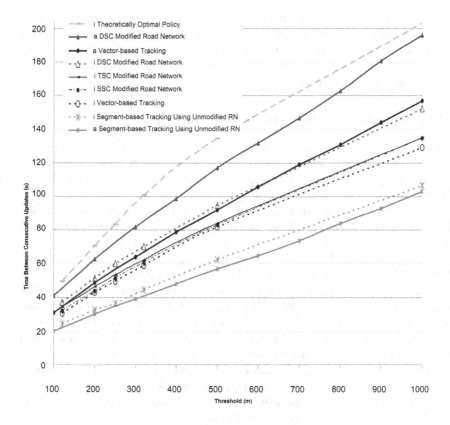

Fig. 13.4. Comparison of road network modification approaches

application of SSC, TSC, and DSC. Vector-based tracking is also included. To avoid clutter, only the most interesting techniques are illustrated for the AKTA data.

In the comparison, 568,307 GPS records were used from the INFATI data set and about 4,000,000 GPS records from the AKTA data set. The curves show experimental results using thresholds ranging from 100 to 1,000 m.

All three road network modifications increase the performance of segment-based tracking, which then outperforms vector-based tracking. Segment-based tracking has the best performance when using the road network resulting from the direction-based modification.

The performance of a theoretical form of tracking that is optimal under the assumption that the speed of an object is modeled as being constant between updates is also included in Fig. 13.4. This technique is explained in Sect. 13.5.

13.5 Update Reduction Using Routes

The focus of this section is on the use of the routes of moving objects for update reduction. We first describe the theoretical, constant-speed optimal form of tracking mentioned in the previous section. Then we consider the use of an object's routes, which are 'long' segments, during segment-based tracking instead of the previous use of road-network segments.

13.5.1 Theoretical, Constant-speed Optimal Tracking

One may distinguish between updates based on the outcomes of the associated map matching. Recall that in segment-based tracking, when the server receives an update at a position p_i, it attempts to map match the position onto the road network rn to find the most probable polyline mpl and point mp on it.

Let **MM** be the map matching function and $\mathbf{MM}(p_i, rn) = (mpl, mp)$. If **MM** fails to identify a polyline and point, tracking is done in vector mode. Assuming that the map matching is successful and $(\mathbf{MM}(p_{i-1}, rn)).mpl = (\mathbf{MM}(p_i, rn)).mpl$, where position p_{i-1} is that of the previous update, we say that the update is caused by *speed*. If the polylines differ, we say that the update is caused by a *segment change*.

The theoretical, constant-speed optimal tracking introduced here indicates how it is possible to achieve few updates with segment-based tracking in the best case, which occurs when a moving object travels on only one segment and no updates occur due to segment changes. The technique is optimal under the assumption that the speed of an object is modeled as being constant between updates.

This technique is interesting because it offers a measure of optimality. However, the technique is useful for comparison purposes only and it is not a practical technique. The technique is impractical because it assumes that the entire polyline along which a vehicle will ever move is known in advance. We are able to use this technique here because we have the entire GPS logs for each vehicle. Using these, we simply construct (very long) polylines that precisely track each vehicle "ahead of time." In practice, GPS positions are received in real time.

In Fig. 13.4, the curve for the constant-speed optimal policy gives the lower bound for the number of updates needed by the segment-based policy. The deviation of the segment-based policy using the non-modified road network from the optimal case is substantial. Using the modified road networks, the performance is significantly closer to the optimal case. For example, for a threshold of 200 m, the use of the road network modified using the direction-based approach increases the average time duration in-between consecutive updates from 32 vs. 30 s to 52 vs. 63 s (for the INFATI vs. AKTA data sets) in comparison to the use of the unmodified road network.

13.5.2 Practical Tracking Using Routes

We may assume that individuals who travel do so to reach a destination; and the same routes are often used multiple times. For example, a person going from home to work may be expected to frequently use the same route. This behavior is confirmed by the GPS logs available [10, 16].

Taking advantage of knowledge of the routes used by a moving object holds potential for reducing the number of updates caused by segment changes. Since a route is a sequence of connected, partial road segments, a route is represented simply as a polyline. Therefore, segment-based tracking that is applicable to any polylines is also directly applicable to routes.

All that is needed is to collect the routes of each user. Routes may be obtained via a navigation system. This case may occur when the user travels in unfamiliar surroundings. When the user travels in familiar surroundings, it is likely that the user travels along a route that was used previously. In this case, a system that gathers the user's routes and associates usage meta-data (e.g. times of the day and days when the routes are being used) with these can be used [2]. Such a system is able to return likely routes on request.

When using segment-based tracking with routes, we effectively assume that we know the future positions of an object. This is like in the theoretical, constant-speed optimal policy. The differences are that the polylines that represent routes are created from the road network, not from GPS logs, and that deviations from the assumed route are handled. Specifically, if an object deviates from its route, this is treated simply as a segment change. This will then most likely trigger an update.

When the route of an object is guessed successfully, the theoretical technique and the practical, segment-based technique have essentially the same performance. Slight deviations may occur because the routes used by the techniques differ: the routes used by the theoretical technique are constructed from GPS points and are more detailed, and slightly longer than those used by the segment-based technique.

Figure 13.10 illustrates the performance of segment-based tracking using routes for the INFATI and AKTA data. We conclude that exploiting knowledge of the routes used by an object can eliminate virtually all updates caused by segment changes that significantly improve the performance of the segment-based policy.

13.6 Update Reduction Using Acceleration Profiles

Even if the future trajectory of an object is known precisely and updates caused by segment changes thus are eliminated, updates still occur due to variations in speed. The reason is that segment-based tracking assumes that objects move at constant speed – it takes an update to change the speed.

In this scenario, the modeled speed of an object moving along a road is a stair function. Figure 13.5 presents the variation of a car's speed along a part of its route from home to work. The stepwise constant speed is the one used by the segment-based policy with a 90 m threshold. Each new step in the stair function represents an update. The density of the steps depends on the threshold – smaller thresholds yield more updates.

It is reasonable to expect that more accurate modeling of the speed variation of an object along its route, for example, using averages of the speeds during past traversals of the route, can help better predict the future position of the object as it moves along the route. Figure 13.6 illustrates the speed variation of one car as it traverses part of its route from home to work (the same car as in Fig. 13.5). Here, the thin lines represent the speeds for 20 traversals of the route, and the solid line represents the average speed along the route.

The figure reveals a clear pattern of how fast the car drives along different parts of the route. The geometry of the route, the driver's habits, and the traffic situation are probably the primary causes for the observed behavior. Figure 13.7 displays the geometry of the partial route. The figure contains distance measures that allow the reader to correlate the geometry with the patterns displayed in Fig. 13.6.

The first deceleration of the car happens in the preparation for negotiating a rotary. Then the car accelerates, decelerates, makes a left turn, enters a highway, and

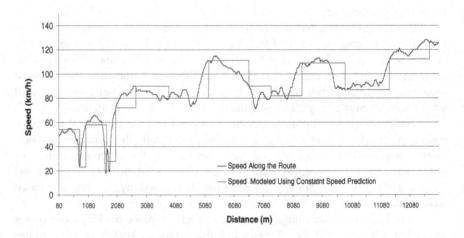

Fig. 13.5. Speed modeling using constant speed prediction

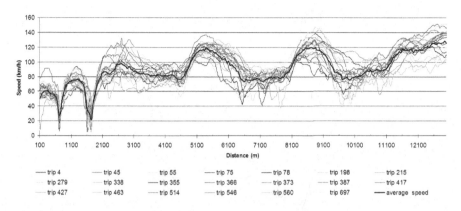

Fig. 13.6. Speed pattern for 20 traversals of a partial route

accelerates further. Note that even on the highway, the car's speed is influenced by exits from the highway and that a clear pattern can be seen. We expect this type of behavior to be typical.

The clear pattern in Fig. 13.6 indicates that tracking with better performance can be achieved by more accurate modeling of the predicted, future speed of a moving object.

Figure 13.8 illustrates another part of the same route with the same 20 traversals where no clear speed pattern exists and where constant speed prediction may work well, or at least better than prediction using variable speeds.

We consequently create an acceleration profile for capturing the average speed variation of the movement of an object along a route. While we create a profile for each combination of a route and an object using the route, it is also possible to assign profiles to the road network that are to be applied to all moving objects and for all uses of the segments of the road network. Such profiles should then be time varying. A separate software component is assumed to be present that generates frequently used routes for the moving objects being tracked [2]. Having this as a separate component is reasonable, as routes are useful for other tasks than tracking.

An acceleration profile consists of acceleration values together with the distance intervals during which these values apply. A profile is created by first dividing the average speed variation along a route into intervals where the acceleration changes sign (i.e. from positive to negative or vice versa). Then the average acceleration is calculated for each interval. An acceleration profile *apf* is then a sequence of $n + 1$ measures m_i and n accelerations a_i, $(m_0, a_0, \ldots, m_{n-1}, a_{n-1}, m_n)$. Acceleration a_i is valid at interval (m_i, m_{i+1}).

To see how an acceleration profile is used, assume that an object moves with speed v_0 and that its current location (measure) along the route is m_0 distance units after the start of the route, where m_0 belongs to the interval $[m_{begin}, m_{end})$ in which the acceleration profile has acceleration value a. Then the predicted position m_{pred}

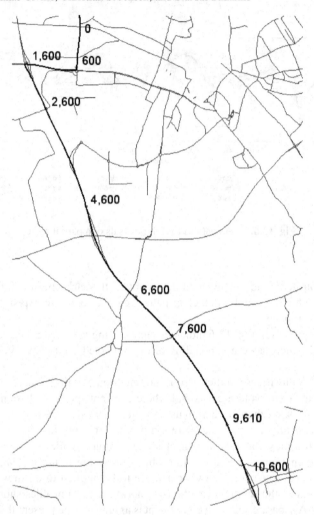

Fig. 13.7. Geometry of partial route

and the speed v_{pred} of the object within the interval $[m_{begin}, m_{end})$ at time t is given by: $m_{pred} = m_0 + v_0 t + (a/2)t^2$ and $v_{pred} = v_0 + at$.

Figure 13.9 exemplifies speed modeling when using an acceleration profile. The figure concerns the movement of one moving object along a route. We assume that segment-based tracking with a 90 m threshold is used. In this figure, the light vertical dotted lines mark updates. To provide better insight into the behavior of the policy used, we include the deviation between the real position of the moving object and its position as predicted by the policy.

The algorithm 'Predict Positions with Segments and Accelerations' (**PPSA**) takes two parameters as input, *mopa* and t, where the first parameter is a structure with five elements that are as follows: (1) A polyline, *mopa.pl*, that specifies the

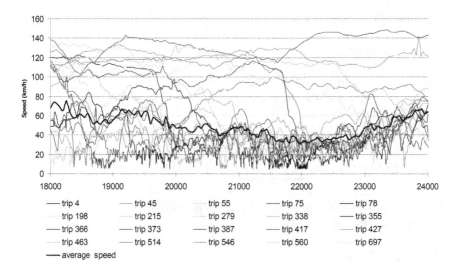

Fig. 13.8. Speed pattern for 20 traversals of a partial route

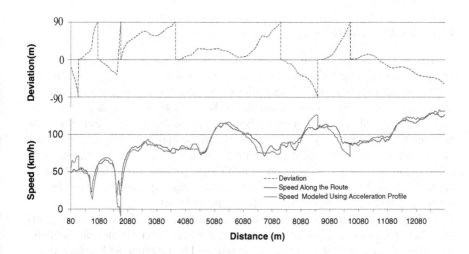

Fig. 13.9. Speed modeling using acceleration profile

geometrical representation of the moving object's route, (2) an acceleration profile, *mopa.apf*, for speed prediction along the route, (3) the location of the client, *mopa.m*, given as a measure value on the route, (4) the speed, *mopa.plspd*, of the object, and (5) the time, *mopa.t*, when the location and speed are acquired. Parameter $t > mopa.t$ is the time point for which the location of the object should be calculated. The result is the coordinates of predicted location of the object at time t.

Algorithm PPSA$(mopa, t)$

1. $m_{pred} \leftarrow mopa.m$
2. $v_{pred} \leftarrow mopa.plspd$
3. $t_{pred} \leftarrow t - mopa.t$
4. **while** $t_{pred} > 0$ **do**
5. $accel \leftarrow$ **getAcceleration** $(m_{pred}, mopa.apf)$
6. $S \leftarrow accel.end - m_{pred}$
7. $dt \leftarrow 0$
8.   **if** $v_{pred}^2 + 2 \cdot accel.a \cdot S \geq 0 \wedge accel.a \neq 0$ **then**
9. $dt_1 \leftarrow (-v_{pred} + \sqrt{v_{pred}^2 + 2 \cdot accel.a \cdot S})/accel.a$
10. $dt_2 \leftarrow (-v_{pred} - \sqrt{v_{pred}^2 + 2 \cdot accel.a \cdot S})/accel.a$
11. $dt \leftarrow \max(\{0, \min(\{dt | dt \in \{dt_1, dt_2\} \wedge dt > 0\})\})$
12.   **if** $dt = 0$ **then** $dt \leftarrow S/v_{pred}$
13. $accel.a \leftarrow 0$
14.   **if** $t_{pred} < dt$ **then** $dt \leftarrow t_{pred}$
15. $m_{pred} \leftarrow m_{pred} + v_{pred} \cdot dt + accel.a \cdot dt^2/2$
16. $v_{pred} \leftarrow v_{pred} + accel.a \cdot dt$
17. $t_{pred} \leftarrow t_{pred} - dt$
18. **if** $m_{pred} \geq \mathcal{M}mopa.pl, mopa.pl.p_{end})$ **then return** $mopa.pl.p_{end}$
19. **return** $\mathcal{M}^{-1}(mopa.pl, m_{pred})$

The algorithm first initializes temporary variables. Variables m_{pred} and v_{pred} are set to contain starting location and speed of the moving object, and variable t_{pred} initially holds the time elapsed since the time when the moving object's location was acquired. The object's movement should be predicted for this duration of time. In general, several acceleration intervals are traversed during this duration of time, meaning that different acceleration values should be applied during the prediction. The algorithm iteratively calculates the time duration required to pass through each acceleration interval and reduces prediction time t_{pred} with this duration. When the prediction time duration is exhausted (line 4), the loop stops, and the algorithm calculates and returns the coordinates of the predicted location.

In line 5, acceleration value a for the predicted location of the object m_{pred} and boundary point *end* of the acceleration interval where acceleration value a applies are retrieved and stored in *accel*; these are returned by function **getAcceleration**. In the case where m_{pred} is equal to boundary point m_i, the boundary point m_{i+1} of the next acceleration interval is returned. If there are no more acceleration intervals, an acceleration value of 0 is returned, and the boundary point is set to ∞. Note that m_{pred} is initially equal to the location of the object at the time of the update (line 1).

In line 6, the distance S to the end of the acceleration interval with acceleration $accel.a$ is calculated.

The time dt required for the object to reach the end of the acceleration interval (moving with acceleration $accel.a$) is calculated in lines 9–11. This time is calculated using the quadratic equation $accel.a \cdot dt^2/2 + v_{pred} \cdot dt - S = 0$. It has solutions only if $v_{pred}^2 + 2 \cdot accel.a \cdot S \geq 0$ (line 8), and only positive solutions are valid, as the meaning of the solution is time. If there are two positive solutions, the solution with the smaller value is the valid one (line 11). If the equation has no valid solution, the result dt is equal to 0. In this case, prediction using constant speed is performed (lines 12 and 13).

After the time required to reach the end of the acceleration interval is calculated, this time is compared to the remaining prediction time t_{pred}. If the time left for which prediction should be done, t_{pred}, is less than the time required to go a distance S, the algorithm does prediction only for time t_{pred} (line 14). Lines 15 and 16 then calculate the predicted location m_{pred} and the speed v_{pred}. The prediction time is reduced in line 17, and the loop is repeated if $t_{pred} > 0$.

Finally, the coordinates corresponding to location m_{pred} are calculated and returned. This is done in lines 18 and 19. If the predicted location m_{pred} is beyond the end of the route as described by polyline $mopa.pl$ (line 18), the end point of the polyline is returned. This is done by comparing the predicted measure on the polyline with the measure of the end point $pl.p_{end}$ of the polyline.

Experimental results for the segment-based policy using routes and acceleration profiles are presented in Fig. 13.10. These experiments are based on approximately 57,000 GPS records from the INFATI data set that correspond to the movement of five cars along different routes. In addition, approximately 36,000 GPS records from the AKTA data set, corresponding to one car moving along one route, were used. The experiments show that the use of acceleration profiles is able to improve the performance. This illustrates that when an object's movement has a clear acceleration profile and this profile is known, it is possible to more accurately predict the positions

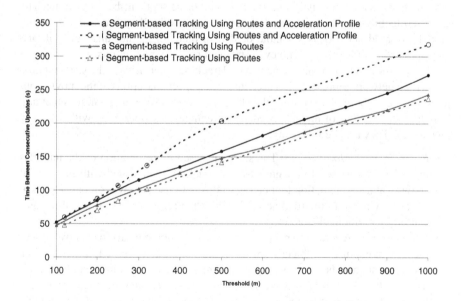

Fig. 13.10. Results using routes with and without acceleration profiles

of the object. For example, using a threshold of 500 m, the average time between up-dates is increased from 141 to 203 s with the INFATI data and from 148 to 158 s with the AKTA data. The lower benefit from using acceleration profiles for the AKTA data is likely to be due to congestion (see Fig. 13.8) as well as the majority of routes being on the highway where speed patterns are not so clear.

We note that with acceleration profiles, we outperform the previously introduced theoretical technique that is optimal only under the assumption of constant-speed prediction.

In closing, it is also worth considering a few alternatives for the speed model-ing and some implications of the alternative presented. In reality, the travel speed associated with a road segment varies during the day and different drivers may well negotiate the same segment with different speeds. By associating acceleration pro-files with routes that are specific to individual drivers, we capture the variation among drivers. And because the same route (e.g. from home to work or from work to home) is typically used during the same time of the day, the variation of speeds during the day is also taken into account fairly well. Next, if significant variations exist within the observations based on which the acceleration profile of a route is constructed, it is possible to create several speed profiles, for example, so that rush-hour and non-rush-hour profiles are available.

13.7 Conclusions

This chapter presents and empirically evaluates a range of techniques for the tracking of moving objects, including point-, vector-, and basic segment-based tracking. The proposed techniques are robust and generally applicable; they function even if no underlying road network is available or if map matching is not unsuccessful, and they apply to mobile objects with even stringent memory restrictions.

The performance of basic segment-based tracking is sensitive to the segmentation of the road network representation used and to the speed variations of the moving ob-jects. Based on these observations, the chapter describes several techniques that aim to reduce the number of updates needed for segment-based tracking with accuracy guarantees. They are the following:

- *Road Network Modification.* The segment-based representation of the underly-ing road network used in segment-based tracking is modified with the goal of arriving at a segmentation that enables objects to use as few segments as possi-ble as they move in the road network. This then reduces the number of updates caused by segment changes.
- *Use of Routes.* A route is a polyline constructed from (partial) road network seg-ments that capture an object's entire movement from a source to a destination. As segments are themselves polylines, segment-based tracking readily accom-modates the use of routes. Routes are specific to individual moving objects, and the use of routes is expected to reduce the number of updates caused by segment changes.

- *Use of Acceleration Profiles.* An acceleration profile divides a route into intervals with constant acceleration and thus enables quite accurate modeling of the speed of an object as it travels along a route. The idea underlying the use of acceleration profiles is to reduce the number of updates incurred by speed variations.

Experimental performance studies using real GPS logs and corresponding real road networks representation illustrate the following observations:

- It is possible to improve the performance of segment-based tracking by automatic resegmentation of the underlying road network representation. Experiments with three resegmentation algorithms demonstrate this and offers insight into which types of modification are most effective in reducing the number of updates.
- It is indeed attractive to use precomputed routes for the moving objects in segment-based tracking, instead of using segments from the road network representation. The GPS logs used confirm conventional wisdom that mobile users are creatures of habit (or efficiency) that frequently use the same routes through the road network to reach their destinations.
- The GPS data used also reveal distinctive speed patterns for some routes and mobile users. The experimental results show that the use of acceleration profiles is capable of increasing the performance of segment-based tracking.

13.8 Further Reading

We proceed to offer an overview of related developments in the commercial and academic arenas. We first offer an overview of 26 tracking-related products and services that we believe are representative of the current commercial state of the art. We then provide an overview of related works within the academic community that may impact future commercial offerings.

13.8.1 Commercially Available Products and Services

Table 13.1 summarizes pertinent properties of what we believe is a representative range of commercially available tracking solutions. The table captures properties of 26 solutions provided by 23 companies. The information that went into the creation of the table was obtained via the Internet during January 2006.

The first column lists company names, and the second lists product names. Starting from the third, each column concerns one product property, and a check mark in a cell indicates that the product in the row of the cell possesses the property corresponding to the column of the cell. The absence of a check mark indicates the opposite.

Columns *GPS*, *Cell*, and *WAAS* concern the means of positioning supported, with *Cell* denoting cellular network-based positioning and *WAAS* denoting the wide area augmentation system that is based on GPS, but offers higher accuracy than GPS by

Table 13.1. Properties of tracking solutions

company	product	GPS	cell	WAAS	SMS	GPRS,CDPD	satellite	phone	custom	time based	on request	sensors	spatial events
BSM	Sentinel	√	√	√						√	√	√	√
Cellfind	look4me		√		√			√			√		
Cybit	mapAmobile		√		√			√			√		
Datafactory	Compact	√						√	√	√			
	Fleetec	√								√	√	√	√
Euman	LifePilot	√					√			√	√		
Fleetella	FL1700	√					√			√	√	√	
FleetOnline	FleetOnline		√		√			√			√		
	Trimtrac	√			√					√	√	√	
Global Tracking Solutions	GTS-1000	√					√			√	√	√	√
	Sat-TDiS	√					√			√	√	√	√
GPS Fleet Solutions	Marcus	√					√			√	√	√	√
Gpsnext	Stealth tracker	√					√			√	√		
Guard Magic	VS, VG	√					√	√		√		√	
Mapbyte	Mapaphone		√		√			√			√		
Mobile knowledge	9000 MDT	√					√		√	√	√	√	√
Metro online	AVL	√					√			√	√		
Mobitrac	Mobitrac	√				√	√			√	√		√
Siemens	m.traction Senior Care Service		√		√					√		√	
Telus	Action Tracker	√					√	√		√	√		
uLocate	fleeTracker	√					√	√		√	√		
Unteh	Mobitrack	√					√	√		√	√		
Veriloaction	VL-Tracer	√					√			√	√		
Vettro	Vettro GPS	√					√	√		√	√		
Web Tech wireless	WebTech5000	√					√			√	√	√	√
2020 Fleet Management	Sentinel Live	√					√			√	√	√	

using corrections. The next three columns concern the types of communication supported, with *SMS* denoting the short messaging service, *GPRS/CDPD* denoting general packet radio service (GSM based) and cellular digital packet data, and *Satellite* denoting satellite-based communications. Then two columns follow that capture the types of terminals supported, with *Phone* denoting mobile phones and *Custom* denoting custom terminals. Finally, the last four columns characterize the type of tracking or how position updates are generated. Here, *Time-based* denotes time-based tracking, that is, updates are issued at regular time intervals; *On request* means that positions of moving objects are pulled from the clients only on request; *Sensors* means that position updates can be generated according to input from external sensors, for example, an alarm, a speedometer reading, a thermometer reading; and *Spatial events* means that position update can be generated by the moving object entering or leaving a certain region, for example, when leaving the city limits or a prespecified route or when getting into a certain range of a point of interest.

It should be noted that none of the products described in Table 13.1 provides efficient accuracy guarantees or support accuracy-based tracking. Advanced options such as *Spatial events* are usually supported by solutions involving large, custom-made terminals.

13.8.2 Related Academic Contributions

When predicting the future position of an object, the notion of a *trajectory* is typically used [12, 17, 25], where a trajectory is defined in a three-dimensional [17] or four-dimensional [21] space. The dimensions are a two-dimensional "geographical" space, a time dimension, and (possibly) an uncertainty threshold dimension. A point in this space then indicates, for a point in time, the location of an object and the uncertainty of the location. Such points may be computed using speed limits and average speeds on specific road segments belonging to a trajectory.

Wolfson et al. [25] have recently investigated how to incorporate travel-speed prediction in a database. They assume that sensors that can send up-to-date speed information are installed in the roads, and they use average real-time speeds reported every 5 minutes by such in-road sensors. This contrasts the techniques covered in this chapter that use GPS records (termed floating-car data) received from an individual object for predicting that object's movement.

Wolfson et al. [23] propose tracking techniques that offer accuracy guarantees. These assume that objects move on predefined routes already known to the objects, and route selection is done on the client side. If an object changes its route, it sends a position update with information about the new route to the server. The techniques described in this chapter go further by accommodating objects with memory restrictions, and they also work in cases where routes are not known or where map matching fails.

Lam et al. [13] present an adaptive monitoring method that takes into consideration the update, deviation, and uncertainty costs associated with tracking. The method also takes into account the cost of providing incorrect results to queries, during the process of determining when to issue updates. With this method, the moving objects that fall into a query region need close monitoring, and a small accuracy threshold is used for them. Objects not inside a query region may have big thresholds. The techniques presented in this chapter are applicable to this scenario, as they allow different objects to have different thresholds and allow thresholds to change dynamically.

A proposal for trajectory prediction by Karimi and Liu [12] assigns probabilities to the roads emanating from an intersection according to how likely it is that an object entering the intersection will proceed on them. The subroad network within a circular area around an object is extracted, and the most probable route within this network is used for prediction. When the object leaves the current subnetwork, a new subnetwork is extracted, and the procedure is repeated. In this proposal, the probabilities are global, in that the same probabilities are used for all objects; and they are history-less, in that past choices by an object during a trip are not taken into account when computing probabilities for an object.

Next, Wolfson and Yin [24] consider tracking with accuracy guarantees. Based on experiments with synthetic data, generated to resemble real movement data, they conclude that a version of the point-based tracking is outperformed by a technique that resembles basic segment-based tracking (covered in Sect. 13.3). For a small threshold of 80 m, the latter is a bit more than twice as good as the former; for larger thresholds, the difference decreases. Their dependent variable is numbers of updates per distance unit.

It should also be noted that Ding and Güting [6] have recently discussed the use of what is essentially segment-based tracking within an envisioned system based on their own proposal for a data model for the management of road network constrained moving objects.

When only low accuracy of predicted positions are needed, cellular techniques [1, 14, 15] may be used. With such techniques, the mobile network tracks the cells of the mobile objects in real time to be able to deliver messages or calls to the objects. In this approach, update is handled in the mobile network. In contrast to these techniques, this chapter assumes scenarios where higher accuracies, well beyond those given by the cells associated with the base stations in a cellular network, are needed and where positioning with respect to a road network is attractive.

References

1. Akyildiz IF, Ho JSM (1995) A mobile user location update and paging mechanism under delay constraints. ACM-Baltzer Journal of Wireless Networks 1:244–255
2. Brilingaitė A, Jensen CS, Zokaitė N (2004) Enabling routes as context in mobile services. In: Proceedings of the 12th ACM International Workshop on Geographic Information Systems, 127–136
3. Chung JD, Paek OH, Lee JW, Ryu KH (2002) Temporal pattern mining of moving objects for location-based services. In: Proceedings of the 13th International Conference on Database and Expert Systems Applications, 331–340
4. Čivilis A, Jensen CS, Nenortaite N, and Pakalnis S (2004) Efficient tracking of moving objects with precision guarantees. In: Proceedings of the International Conference on Mobile and Ubiquitous Systems: Networking and Services, 164–173. Extended version available as DB-TR-5, www.cs.aau.dk/DBTR/DBPublications/DBTR-5.pdf
5. Čivilis A, Jensen CS, Pakalnis S (2005) Techniques for efficient road-network-based tracking of moving objects. IEEE Transactions on Knowledge and Data Engineering, 17(5): 698–712
6. Ding Z, Güting RH (2004) Managing moving objects on dynamic transportation networks. In: Proceedings of the 16th International Conference on Scientific and Statistical Database Management, 287–296
7. GloVentures (2006) Glofun Games. www.glofun.com
8. Groundspeak (2006) Geocaching. www.geocaching.com
9. Güting RH, Schneider M (2005) Moving objects databases. Morgan Kaufman

10. Jensen CS, Lahrmann H, Pakalnis S, Runge J (2004) The INFATI Data. Aalborg University, TimeCenter TR-79. www.cs.aau.dk/TimeCenter
11. Jensen CS, Lee K-J, Pakalnis S, Šaltenis S (2005) Advanced tracking of vehicles. In: Proceedings of the Fifth European Congress and Exhibition on Intelligent Transport Systems June 1–3, Hannover, Germany, 12 pages
12. Karimi HA, Liu X (2003) A predictive location model for location-based services. In: Proceedings of the Eleventh ACM International Symposium on Advances in Geographic Information Systems, 126–133
13. Lam K-Y, Ulusoy O, Lee TSH, Chan E, Li G (2001) An efficient method for generating location updates for processing of location-dependent continuous queries. In: Proceedings of the Seventh International Conference on Database Systems for Advanced Applications, 218–225
14. Li G, Lam K-Y, Kuo T-W (2001) Location update generation in cellular mobile computing systems. In: Proceedings of the 15th International Parallel & Distributed Processing Symposium, 96
15. Naor Z, Levy H (1998) Minimizing the wireless cost of tracking mobile users: an adaptive threshold scheme. In: Proceedings of the Seventh Annual Joint Conference on Computer Communications, 720–727
16. Nielsen OA, Jovicic G (2003) The AKTA road pricing experiment in Copenhagen. In: Proceedings of the Tenth International Conference on Travel Behaviour Research Proceedings session 3.2, Lucerne, Switzerland
17. Trajcevski G, Wolfson O, Zhang F, Chamberlain S (2001) The geometry of uncertainty in moving objects databases. In: Proceedings of the Eighth International Conference on Extending Database Technology, 233–250
18. Voisard A, Schiller J (2004) Location-based Services. Morgan Kaufmann
19. Wikipedia (2006) Galileo positioning system. en.wikipedia.org/wiki/Galileo_positioning_system
20. Wikipedia (2006) Global positioning system. en.wikipedia.org/wiki/GPS
21. Wolfson O (2001) The opportunities and challenges of location information management. In: tech. report presented at Comp. Science and Telecommunications Board Workshop on Intersections of Geospatial Information and Information Technology. www7.nationalacademies.org/cstb/wp_geo_wolfson.pdf
22. Wolfson O, Chamberlain S, Dao S, Jiang L, Mendez G (1998) Cost and imprecision in modeling the position of moving objects. In: Proceedings of the Fourteenth International Conference on Data Engineering, 588–596
23. Wolfson O, Sistla AP, Camberlain S, Yesha Y (1999) Updating and querying databases that track mobile units. Distributed and Parallel Databases 7(3):257–287
24. Wolfson O, Yin H (2003) Accuracy and resource consumption in tracking and location prediction. In: Proceedings of the Eight International Symposium on Spatial and Temporal Databases 325–343
25. Xu B, Wolfson O (2003) Time-series prediction with applications to traffic and moving objects databases. In: Proceedings of the Third ACM International Workshop on Data Engineering for Wireless and Mobile Access, 56–60

References containing URLs are valid as of May 29th, 2006.

Index

Geography Markup Language - GML, 96